Lecture Notes on Data Engineering and Communications Technologies

Volume 67

Series Editor

Fatos Xhafa, Technical University of Catalonia, Barcelona, Spain

The aim of the book series is to present cutting edge engineering approaches to data technologies and communications. It will publish latest advances on the engineering task of building and deploying distributed, scalable and reliable data infrastructures and communication systems.

The series will have a prominent applied focus on data technologies and communications with aim to promote the bridging from fundamental research on data science and networking to data engineering and communications that lead to industry products, business knowledge and standardisation.

Indexed by SCOPUS, INSPEC, EI Compendex.

All books published in the series are submitted for consideration in Web of Science.

More information about this series at http://www.springer.com/series/15362

Paul Krause · Fatos Xhafa
Editors

IoT-based Intelligent Modelling for Environmental and Ecological Engineering

IoT Next Generation EcoAgro Systems

 Springer

Editors
Paul Krause
Department of Computer Science
Faculty of Engineering and Physical
Sciences
University of Surrey
Guildford, Surrey, UK

Fatos Xhafa 🄲
Department of Computer Science
Universitat Politècnica de Catalunya
Barcelona, Spain

ISSN 2367-4512   ISSN 2367-4520 (electronic)
Lecture Notes on Data Engineering and Communications Technologies
ISBN 978-3-030-71171-9   ISBN 978-3-030-71172-6 (eBook)
https://doi.org/10.1007/978-3-030-71172-6

This Springer imprint is published by the registered company Springer Nature Switzerland AG
The registered company address is: Gewerbestrasse 11, 6330 Cham, Switzerland

Preface

Estimates from the United Nations foresee that the world population will continue to increase to 9.7 billion by 2050. This growth poses significant challenges to agricultural production, which should, for instance, be increased by up to 60 percent during the twenty-first century. Additionally, this rise in production should be carried out cognizant of the current deteriorating environmental conditions due to aggressive traditional approaches, which have led to a devastation of soil, water, and environment. Since the start of the "Green Revolution" in the 1940s, traditional approaches have focused on the intensive, unlimited exploitation of soil and resources using technical means such as advanced machinery, pesticides, and fertilizers. With increasing awareness of the unintended consequences of such approaches, we are seeing progressively stronger advocation of approaches that support the rational and limited use of soil and resources. Our belief is that this move can be supported by an array of new technologies, most notably IoT and Cloud computing.

IoT and Cloud technologies are amongst major latest disruptive paradigms, penetrating all fields of science, engineering, businesses, industry, and everyday life activity, to name a few. IoT and Cloud systems have prompted the definition of a whole Cloud digital ecosystem referred to as Cloud-to-thing continuum computing. The key success of IoT computing together with the Cloud ecosystem is that IoT can be integrated seamlessly with the physical environment and therefore leverage innovative services unknown before. Areas such as ecological monitoring, agriculture, and biodiversity constitute a large area of the potential application of IoT and Cloud technologies.

Whilst traditional agriculture systems have aimed at increasing productivity, often with aggressive policies, new agro-ecosystems aim to increase productivity and achieve efficiency and competitiveness in modern sustainable agriculture and contribute, more broadly, to the green economy and sustainable food-chain industry. Current approaches advocate for rational and limited use of soil and resources, whereby a number of actors, from farmers to citizens, local administrations, and stakeholders aim at using ICT technologies for developing an ecosystem of services not only for increased production and benefits but most importantly for sustainable agriculture that produces both healthy food and contributes to a healthy environment.

Currently, IoT and Cloud technologies are contributing to various research and development directions in agro-ecosystems, comprising

- Fundamental approaches to modelling and simulation of environmental and eco-agricultural systems.
- IoT infrastructures for efficient and scalable monitoring and simulation of environmental and eco-agricultural units.
- Massive data processing and machine intelligence as a basis for the next generation of environmental and eco-agricultural systems.
- Data sets, benchmarking, and best practices of using IoT and Cloud computing for agro-ecosystems.
- The regulatory and policy landscape for the IoT and agriculture industry.
- The use of IoT biotechnology in developing countries to facilitate economic advances whilst restraining impact within planetary boundaries.

This book volume brings to readers 13 chapters with contributions on the benefits of using IoT and Cloud computing to agro-ecosystems from a multi-disciplinary perspective. Fundamental research as well as concrete applications from various real-life scenarios such as smart farming, precision agriculture, green agriculture, sustainable livestock and sow farming, climate threat, and societal and environmental impacts are presented. Research issues and challenges are also discussed towards envisioning efficient and scalable solutions based on IoT and Cloud technologies. The chapters of the book use various approaches based on IoT and Cloud technologies, machine learning, knowledge discovery, and decision-making systems, to achieve the goals.

The book comprises 13 chapters, summarized as follows.

- The chapter "IoT-Based Computational Modeling for Next Generation Agro-Ecosystems: Research Issues, Emerging Trends and Challenges" is an introductory chapter highlighting fundamental concepts, computational models, architectures, and definitions related to IoT-based computational modelling for next generation agro-ecosystems. The paradigms of Cloud-to-thing continuum ecosystem comprising IoT, Edge, Fog, and Cloud computing are presented. Similarly, data cycles from data gathering, processing, and analysis to knowledge generation and decision-making are introduced. Further, machine learning and stream processing, optimization, simulation frameworks, symbiotic modelling, and the digital twin as well as emerging research trends, ethics, and health and safety issues are also introduced and discussed.
- The chapter "An IoT-Based Time Constrained Spectrum Trading in Wireless Communication for Tertiary Market" addresses the study of fundamental issues in analysing, developing, and deploying IoT infrastructures, based on IoT-based spectrum trading in wireless communication in a strategic setting.
- The chapter "AgriEdge: Edge Intelligent 5G Narrow Band Internet of Drone Things for Agriculture 4.0" surveys the role of 5G NB-IoT technologies for enabling smart green agriculture 4.0 with low-latency, better connectivity speed and reliability.

- The chapter "Drones for Intelligent Agricultural Management" studies the use of drone technology for intelligent agricultural management.
- The chapter "Multi-Modal Sensor Nodes in Experimental Scalable Agricultural IoT Application Scenarios" presents multi-modal sensor nodes in experimental scalable agricultural IoT application scenarios.
- The chapter "Design Architecture of Intelligent Agri-Infrastructure Incorporating IoT and Cloud: Link Budget and Socio-Economic Impact" proposes the design and architecture of intelligent agri-infrastructure incorporating IoT and Cloud by linking budget and socio-economic impact.
- The chapter "Remote Sensing and Soil Quality" describes how Sentinel-2 satellite image data was combined with data from the UK Centre for Ecology & Hydrology Land Cover Map 2015, to train a convolutional neural network for land cover classification for the South of England as the first step in soil quality monitoring.
- The chapter "Machine Learning Modelling-Powered IoT Systems for Smart Applications" surveys machine learning modelling-powered IoT systems for smart applications with an application scenario from smart irrigation systems.
- The chapters "Enabling IoT Wireless Technologies in Sustainable Livestock Farming Toward Agriculture 4.0"–"Dynamic Shift from Cloud Computing to Industry 4.0: Eco-Friendly Choice or Climate Change Threat" contribute with real-life applications of sustainable livestock farming, decision support system for sow farm, plant leaf disease detection, and auto-medicine and climate change threats.
- The chapter "A Research Roadmap for IoT Monitoring and Computational Modelling for Next Generation Agriculture" concludes the book with an outlook on a research agenda and the road ahead to computational modelling for next generation agro-ecosystems.

In all, the chapters of the book have approached the benefits of using the paradigms of IoT and Cloud computing, data analysis, stream processing, machine learning, simulation, and modelling frameworks for a new generation of agro-ecosystems. These technologies, envisioned as game changers for agro-ecosystems, should go hand in hand with social, economic, food, ecological, and ethical sciences as well as with a new "agroculture" and community building of producers and consumers of agro-ecosystems to address the many complexities of agro-ecosystems. The achievement of the goals of such new agro-ecosystems calls for the participation and collaboration of number of actors, from farmers to citizens, local administrations, and stakeholders with the aim of using ICT technologies for developing an ecosystem of services not only for increased production and benefits but most importantly for a sustained agriculture that delivers both healthy food and a healthy environment.

Guildford, UK   Prof. Paul Krause
Barcelona, Spain   Prof. Fatos Xhafa

Acknowledgments

Professor Fatos Xhafa's work is supported by Spanish Ministry of Science, Innovation and Universities, Programme *Estancias de profesores e investigadores sénior en centros extranjeros, incluido el Programa Salvador de Madariaga 2019,* PRX19/00155. Work was partially done whilst on Leave at University of Surrey, UK.

Contents

Contributors

Konstantinos G. Arvanitis Agricultural University of Athens, Iera Odos, Athens, Greece

Mohamed Atri College of Computer Science, King Khalid University, Abha, Saudi Arabia

Olfa Ben Ahmed XLIM Institute, University of Poitiers, Chasseneuil Cedex, France

Aakashjit Bhattacharya Advanced Technology Development Centre, IIT Kharagpur, Kharagpur, West Bengal, India

Abbas Bradai XLIM Institute, University of Poitiers, Chasseneuil Cedex, France

Rachel Jane Butcher School of Electronic Engineering and Computer Science, Queen Mary University of London, London, UK

Anil B. Chowdhury Department of Master of Computer Applications, Techno India University, Kolkata, India

Sanket Dan Department of Information Technology, Kalyani Government Engineering College, Kalyani, Nadia, India

Debashis De Centre of Mobile Cloud Computing, Department of Computer Science and Engineering, Maulana Abul Kalam Azad University of Technology, Haringhata, Nadia, West Bengal, India

Meghana M. Dhananjaya Department of Computer Science and Engineering, National Institute of Technology, Durgapur, West Bengal, India

Dídac Florensa Department of Computer Science and INSPIRES, University of Lleida, Lleida, Spain

Fikreselam Gared Faculty of Electrical & Computer Engineering, Bahir Dar Institute of Technology, Bahir Dar University, Bahir Dar, Ethiopia

Lijaddis Getnet Faculty of Electrical & Computer Engineering, Bahir Dar Institute of Technology, Bahir Dar University, Bahir Dar, Ethiopia

Pritam Ghosh Department of Information Technology, Kalyani Government Engineering College, Kalyani, Nadia, India

Sukhpal Singh Gill School of Electronic Engineering and Computer Science, Queen Mary University of London, London, UK

Graham Hay Department of Computer Science, Surrey University, Guildford, UK

Rupinder Kaur Department of Chemistry, Guru Nanak Dev University, Amritsar, Punjab, India

Paul Krause Department of Computer Science, University of Surrey, Guildford, UK

Anders R. Kristensen Department of Veterinary and Animal Sciences, University of Copenhagen, Frederiksberg C, Denmark

Dimitrios Loukatos Agricultural University of Athens, Iera Odos, Athens, Greece

Mobasshir Mahbub Department of Electrical and Electronic Engineering, Ahsanullah University of Science and Technology, Dhaka, Bangladesh

Jordi Mateo Department of Computer Science and INSPIRES, University of Lleida, Lleida, Spain

Channamallikarjuna Mattihalli Faculty of Electrical & Computer Engineering, Bahir Dar Institute of Technology, Bahir Dar University, Bahir Dar, Ethiopia

Seifeddine Messaoud Laboratory of Electronics and Microelectronics, University of Monastir, Monastir, Tunisia

Kaushik Mukherjee Department of Information Technology, Kalyani Government Engineering College, Kalyani, Nadia, India

Sajal Mukhopadhyay Department of Computer Science and Engineering, National Institute of Technology, Durgapur, West Bengal, India

Subhranil Mustafi Department of Information Technology, Kalyani Government Engineering College, Kalyani, Nadia, India

Satyendra Nath Mandal Department of Information Technology, Kalyani Government Engineering College, Kalyani, Nadia, India

Adela Pagès-Bernaus Department of Business Administration, University of Lleida, Lleida, Spain;
Agrotècnio Research Center, University of Lleida, Lleida, Spain

Dimitrios Piromalis University of West Attica, Egaleo, Greece

Lluís M. Plà-Aragonès Agrotècnio Research Center, University of Lleida, Lleida, Spain;
Department of Mathematics, University of Lleida, Lleida, Spain

Kunal Roy Department of Information Technology, Kalyani Government Engineering College, Kalyani, Nadia, India

Manmeet Singh Centre for Climate Change Research, Indian Institute of Tropical Meteorology (IITM), Pune, India;
Interdisciplinary Programme (IDP) in Climate Studies, Indian Institute of Technology (IIT), Bombay, India

Vikash K. Singh School of Computer Science and Engineering, VIT-AP University, Amaravati, Andhra Pradesh, India

Francesc Solsona Department of Computer Science and INSPIRES, University of Lleida, Lleida, Spain

Eleni Symeonaki Agricultural University of Athens, Athens, Greece;
University of West Attica, Egaleo, Greece

Shreshth Tuli Department of Computer Science and Engineering, Indian Institute of Technology (IIT), Delhi, India

Fatos Xhafa Department of Computer Science, Universitat Politècnica de Catalunya, Barcelona, Spain;
University of Surrey, Guildford, UK

IoT-Based Computational Modeling for Next Generation Agro-Ecosystems: Research Issues, Emerging Trends and Challenges

Fatos Xhafa and Paul Krause

Abstract In this introductory chapter we highlight some fundamental concepts, architectures and definitions related to IoT-based Computational Modeling for Next Generation Agro-ecosystems. We distinguish and discus the paradigms of Cloud-to-thing Continuum as a large digital ecosystem comprising IoT, Edge, Fog, and Cloud Computing, data cycles from data gathering, processing and analysis to knowledge generation and decision making. Machine learning and stream processing, optimization, simulation frameworks, symbiotic modeling and the digital twin as well as emerging research trends, ethics and health & safety issues are also introduced and discussed. Challenges arising from processing and analyzing large and heterogeneous data sets are pointed out with some examples of *killer applications* from Agriculture 4.0. These concepts, models, technologies, frameworks and benchmarks are covered in the chapters of the book and exemplified with real life use cases and applications. In all, Cloud-to-thing continuum and IoT as part of it are envisaged as game changers for Computational Modeling of Next Generation Agro-ecosystems.

Acronyms

ANN	Artificial Neural Network
CI	Collective Intelligence
CM	Computational Modelling
DTwin	Digital Twin
ECA	Event-Condition-Action

Fatos Xhafa Work partially done while on Leave at University of Surrey, UK.

F. Xhafa (✉)
Universitat Politècnica de Catalunya, Barcelona, Spain
e-mail: fatos@cs.upc.edu

P. Krause
University of Surrey, Guildford, UK
e-mail: p.krause@surrey.ac.uk

© Springer Nature Switzerland AG 2021
P. Krause and F. Xhafa (eds.), *IoT-based Intelligent Modelling for Environmental and Ecological Engineering*, Lecture Notes on Data Engineering and Communications Technologies 67, https://doi.org/10.1007/978-3-030-71172-6_1

FANETs	Flying Adhoc Networks
HTM	Hierarchical Temporal Memory
IoT	Internet of Things (IoT)
IOUT	Internet of Underground Things
LR	Logistic Regression
NAB	Numenta Anomaly Benchmark
PCA	Principle Component Analysis
RDF	Resource Description Framework
RTT	Round-Trip Time
SGD	Stochastic Gradient Descent
SS	Symbiotic Simulation
UAVs	Unmanned Aerial Vehicles
VANETs	Vehicular Adhoc Networks

1 Introduction

The Cloud digital ecosystem, nowadays referred to as Cloud-to-thing continuum, has become a widespread technological platform for all fields of science, knowledge, education, business, industry, among others. Indeed, current Cloud-based technologies are penetrating to an unprecedented array of applications. Agriculture and environmental monitoring are not an exception. We are witnessing therefore how these technologies are revolutionizing agriculture and environmental monitoring, hand in hand with industrial applications, through deep digital transformations, leading to Agriculture 4.0. While traditional agriculture systems have aimed at increasing productivity, often with aggressive policies, new agro-ecosystem aim to increase productivity and achieve efficiency and competitiveness in modern sustainable agriculture and contributing, more broadly, to green economy and sustainable food-chain industry.

Traditional approaches, where the main focus has been by intensive, unlimited exploitation of soil and resources[1] (e.g. water) with more technical means such as advanced machinery, pesticides and fertilizers, current approaches advocate for rational and limited use of soil and resources, whereby a number of actors, from farmers to citizens, local administrations, and stakeholders aim at using ICT technologies for developing an ecosystem services not only for increased production and benefits but most importantly for a sustained agriculture, healthy food and environment.

Traditional approaches have led to a devastation of soil, water, and environment and it has become clear that it is based on a short/mid term vision, namely, the damage caused can be irreparable and irreversible–a kind of *"scorched earth"* model. New generation of agro-ecosystems aim just the contrary, to be long term as they use

---

[1] According to a report by World Bank, "The current food system also threatens the health of people and the planet: agriculture accounts for 70% of water use and generates unsustainable levels of pollution and waste.".

limited resources and protect the soil and environment and minimize the damages! New concepts and models of *agro-ecology, agro-ecosystems* besides smart farming, precision agriculture, etc., are being developed to embrace living and non-leaving components, diversity and complexities of the new generation of sustainable and productive agro-ecosystems. Likewise, a new culture and community building of producers and consumers of agro-ecosystems, also referred to as "agro-culture" is growing the consciences of the benefits of sustainable agriculture and agro-ecosystems.

In this introductory chapter we highlight some fundamental concepts, computational models, architectures and definitions related to IoT-based Computational Modeling for Next Generation Agro-ecosystems. The paradigms of Cloud-to-thing Continuum ecosystem comprising IoT, Edge, Fog, and Cloud Computing are presented. Similarly, data cycles from data gathering, processing and analysis to knowledge generation and decision making are pointed out. Further, machine learning and stream processing, optimization, simulation frameworks, symbiotic modeling and the digital twin as well as emerging research trends, ethics and health & safety issues are also introduced and discussed.

The remaining sections of the paper are organized as follows. In Sect. 2, we introduce basic concepts, models and paradigms that support the development of agro-ecosystems. Section 3 presents new modeling based on symbiotic computing and digital twin and anomaly detection concept as basis for agro-alert systems. Data cycle model for agro-ecosystems is outlined in Sect. 4, including a discussion on semantic data enrichment. In Sect. 5, challenges of using Machine Learning to agro-ecosystems are featured. Some further and emerging research issues are commented in Sect. 6, including data quality, provenance, adoption and cooperation with local stakeholders and policy makers and ethics issues. The chapter is ended in Sect. 7 with some conclusions.

2 Fundamental Concepts and Computational Models

2.1 Cloud-to-Thing Continuum: IoT, Edge, Fog, and Cloud Computing Technologies

A number of computing paradigms and technologies have emerged after *Internet of Things (IoT)* and Cloud computing. Indeed, the view of IoT Cloud, in which iot devices are directly connected to Cloud platforms and Data centers has shifted to a *continuum* view, where thereby various computing layers sit in between of IoT and Cloud. These layers referred to as Edge and Fog computing aim to retain the processing of data close to the source where the data is generated. Besides significantly alleviating the computing burden to the Cloud, such layers enable faster RTT processing. rtt is an important metric for the speed and reliability of network connections, whose improvements provides better support to end users. Most importantly, by processing the data close to the data source and to end-users, real time decision

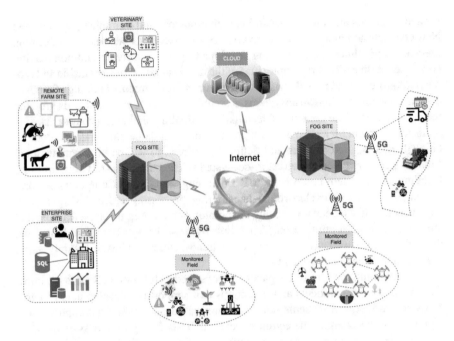

Fig. 1 Cloud-to-thing continuum for agro-ecosystems

making can be designed to support a variety of use cases. This is made possible by achieving the computing and data fabric view of Edge computing [52], where different computing granularity devices can collaboratively support both real time processing and data storage requirements.

Figure 1 depicts the Cloud-to-thing continuum concept to agro-ecosystems. In the figure, IoT layer is deployed within smart farms, smart vehicles (tractors, combines, etc.), FANETs, UAVs, VANETs, etc., which are connected to Fog servers. Processing can be done at Fog servers and reports send back to farmers, or it can be done at a more thorough level at enterprise servers / Cloud data center for decision making and end-to-end management agro-ecosystems. One particularity of Cloud-to-thing continuum for agro-ecosystems is that it has to reflect the *mobility* of the computing infrastructure. Indeed, as can be seen in Fig. 1, nodes corresponding to remote fields such as fanets, transportation fleet and vanets, *agrimotive* and farm fleet, will cover certain areas and then will be moving to other parts as seen fit for purpose. While, other nodes such as a farm site or a veterinarian center would account for in premise computing infrastructure.

It should be noted that such computing infrastructure brings the advantage of both real time processing and in-depth processing of IoT data. Real time processing enables, in particular, the detection of anomalies or abnormal events regarding farms, field, transportation fleet, etc., which would require immediate intervention by farmers and local stakeholders.

2.2   Agro-Ecosystem and Ecological Networks

Ecological networks are categorized by a large number of interactions and complex relations. As stated by Darwin[2] *"I am tempted to give one more instance showing how plants and animals, most remote in the scale of nature, are bound together by a web of complex relations".*

Since then, researcher and practitioner in the field has tried to understand and model these complex relationships and correlations among parts composing an ecological network as well as identify ecosystem functions. Many quantitative studies, however, show the difficulty in such endeavor, and how researchers quite often end up making simplifications on the ecological networks or embracing smaller scale [3].

With the ever increasing use of IoT technologies, it is believed that there will be a significant shift in the quantitative studies to advance on the complex and interactive nature of ecological networks. Indeed, IoT networks can be very useful in the data collection and monitoring ecological networks at large scale with larger number of parameters and metrics. Likewise, Cloud technologies and Machine Learning can enable simulation of a larger number of parameters in ecological networks, with higher accuracy, not possible before. It is also expected that the time and cost for such studies, often too long, and too costly, will drastically be reduced by using IoT technologies.

2.3   Connecting the Two Worlds: Under-Ground and Above-Ground

IoT, together with Cloud computing and Machine Learning, are opening up new perspectives on agro-ecosystems at large. Traditionally, a lot of efforts has been dedicated using fertilizers to increase productivity, without much knowledge of the soil, or, assuming the soil parameters are likely the same in an agro-area.

Several studies in the field have shown that taking into account the parameters of the soil/under-ground is fundamental not only to increase productivity but also to preserve the soil from aggressive fertilizers use. Such soil parameters include water, moisture, nutrients, temperature, salinity, turbidity, pH., etc. Advances in the use of IoT for agro-ecosystems has led to the new concept of IOUT [43, 48]. The particularity of iout is that it is based on autonomous devices buried under-ground and interconnected with communication and networking solutions using the soil as a medium. iout configured through different types of sensors senses and collect heterogeneous data about the soil. The resulting data stream is then transmitted to the Internet and is used for extracting knowledge about the soil, inferring soil characteristics, etc. This knowledge becomes crucial for the development of a new

---

[2]Darwin C. 1859. On the origin of species by means of natural selection, or the preservation of favored races in the struggle for life.

generation of agro-ecosystem by using resources and fertilizers based on timely decision from field information.

2.4   IoT Infrastructure and Deployment for Agro-Ecosystems

IoT technologies are crucial for the development of smart farming, precision agriculture, etc. Ensuring an IoT infrastructure is of foremost importance for agro-ecosystems. While business and industry account for high quality and robust ICT infrastructures, the agro-ecosystems lag behind in their infrastructures. On the one hand this can be explained by the need to cover vast cultivation fields, and, on the other, by the low investment on agriculture in developing countries.

Unlike other sectors such as factories, campuses, city areas, etc., where IoT infrastructures can be in place for long term, in agriculture domain, there is no one IoT infrastructure that fits all due to the agro-cycles of production. Likewise, the location of the activity in isolated farms in remote areas, makes infrastructure deployment costly and challenging. Most farmers cannot afford the cost of using such infrastructures. Yet, agriculture is a main and vital activity in such countries.

It is therefore important to find feasible solutions to support IoT infrastructures in agro-ecosystems in developing countries. A recent paper [4] (see also Chapter "*An IoT-Based Spectrum Trading in Wireless Communication for Tertiary Market*" in this book), exploits the opportunities to allocate spectrum in Wireless Communication for Tertiary Market, whereby, actors such as NGOs (Non-Governmental Organizations), institutional organizations, foundations, etc., can distribute freely their spectrum to farmers in rural areas to support smart farming, precision agriculture, etc.

The development and deployment of IoT systems for agro-ecosystems calls for a careful analysis requirements of such engineering systems. Indeed, a myriad of requirements such as mobility, vast areas of deployment, temporary allocations, agro-production cycles, low cost, optimization of IoT deployment [8], etc., are to be analyzed for such systems. As an example, the work in the Chapter of this book "*An IoT-Based Spectrum Trading in Wireless Communication for Tertiary Market*", the authors exploit the periodicity of spectrum allocations through rather short periods of times, retrieving and processing data in batches, which is deemed useful in the agro-ecosystem domain. Likewise, combination of IoT with uavs [6, 30] and IoT platforms based on edge and cloud computing [54] are envisaged as a driver for transforming traditional cultivation practices into a new precision agriculture.

Chapters of this book address various research issues, use cases and applications from IoT, Edge, Fog, and Cloud Computing Technologies in agro-ecosystems.

2.5   Life Cycle for IoT Device and Infrastructure Management

The context of agro-ecosystems, without fixed infrastructures, requires infrastructure management capabilities for adding, task(s) mapping and monitoring of new devices. Moreover, such architectures are prone to damage and failures as IoT devices are exposed to harsh climate conditions in the field. An academic perspective for education purposed has been proposed in [12], where an IoT infrastructure is presented aiming rapid building of multiple diverse IoT applications with minimum technical skills, while supporting different types of IoT sensor and actuators devices, scalability, etc.

A life cycle method need to be defined and implemented at Cloud/Fog layer to enable device integration and removal from the infrastructure.

We envisage the use of meta-data models/semantic description of devices, both existing and new ones when they join or rejoin the infrastructure. The semantic description would also be used for mapping tasks/applications to newly added devices according to device characteristics. Likewise, the semantic characterization can serve as a basis for assigning a newly added device to the appropriate semantic cluster of devices (e.g. to a geo-localization cluster, where devices are clustered according to the proximity of their geo-localization).

3   Modeling and Alert Systems for Agro-Ecosystems Based on Anomaly Detection

3.1   Symbiotic Modeling and the Digital Twin

Understanding the nature of complex systems from life, environment, engineering, agro-ecosystems, business, etc., requires the employment of a number of methods. Among such methods, CM is a de facto and *par excellence* method to achieve modeling standards. With the support of HPC, computational modeling can employ a large number of variables to accurately build a mathematical model for the phenomenon or system of interest. With the advent of iot and Cloud computing, cm has seen a shift in the ability to model even more complex systems due to:

- CI: By using a large number of IoT devices for sensing the real environment and system under study, intelligence can be built collaboratively and collectively for understanding the system behavior.
- SS: By combining ci with simulation, in ss the simulation reads data from the physical system continuously (as programmed) and update the decision variables and/or simulation parameters to improve the accuracy of the modeling.
- DTwin: is a digital replica of a physical entity or system aiming to create living digital simulation model that updates and changes as the physical counterparts

change. dt is the materialization of symbiotic simulation driven by data (also referred to as symbiotic data-driven modeling).

The use of symbiotic simulation [31] is thus emerging in many domains such as in health, industry 4.0 and manufacturing, etc. However, research into the application of symbiotic simulation in agro-ecosystems is still to be explored. As a powerful means to modeling environmental conditions, above-soil/under-soil connection and a multitude of phenomena in agro-ecosystems, it is attracting increase attention from researchers and developers in the field. In [46], the authors address the digital twins in farm management, whereby, the IoT physical objects are accompanied by Digital Twins, namely, virtual, digital equivalents to physical objects. It is argued that digital twins can play an important role in farm management systems and precise agriculture. The author in [15] addresses agri-food challenges through the integrating nature, people, and technology. Similarly, authors in [29], present a model to implement digital twins in sustainable agriculture. Through the creation of physical and digital layers of IoT-enabled structures for vertical farming, their model deploys IoT to improve productivity and other self* features related to planning, operation, monitoring, and optimization in agro-food systems. For a systematic review on digital twin, readers are referred to [18].

3.2   Anomaly Detection Algorithms

The diagnosis of faults and failures in a system has been long studied in a variety of domains such as critical infrastructures, distributed systems, hardware systems, software system and ICT infrastructures. It should be noted here that the main objective has been to study and identify potential faulty behaviors of parts of a system and therefore design robust, reliable and fault-tolerant systems [7, 34, 44].

With the fast developments in Internet technologies, the number of large scale systems and therefore of faults and failures within them has increased / is increasing significantly, namely, all such systems are prone to potential faulty behaviors. Additionally, and differently from traditional systems, in which failures could be tracked down to their root-cause(s), in modern Internet-based system it becomes very complex and challenging to identify the root-cause(s) as anomalies and failures can happen unexpectedly in hardware, software, underlying networking infrastructure, etc.

More recently, a new family of algorithms, commonly referred to as anomaly detection algorithms from data streams, has emerged to study potential anomalies not only within a system but also anomalies in external entities observed from data streams generated by means of Internet technologies, most notably, IoT and mobile technologies. Anomaly detection algorithms could be commonly classified into three groups:

1. *Detecting an anomaly consisting of a single, atomic event directly from a data stream.* In this case, the objective of a data stream processing is to spot in real-time

an anomaly from the data that could represent an abnormal or faulty behavior of an observed entity, part of a system or a system as a whole. Examples include a wide range of domains, from eHealth of patient monitoring [13] to manufacturing systems and Industry 4.0 [7]. In all such domains, there is an IoT device, such as a sensor or a group of sensors, that observer the entity and generate a single or multiple data stream, which is streamed processed for anomaly detection right from the data.

2. *Detecting an anomaly consisting of complex event from a data stream.* In this case, the anomaly is of a higher and more complex nature and cannot be identified directly from the data stream. It requires complex event processing, which is essentially building larger and more complex events entangling atomic events in the data stream. The real time requirement for anomaly detection becomes even harder in this scenario as atomic events to be correlated for complex event building could appear in distant parts of the data stream [14].

3. *Predicting an anomaly from a data stream.* Given the high complexity of detecting anomalies in real time, prediction becomes a vital approach to many systems [9, 40, 49]. In this regard, data pre-processing and semantic data enrichment play a crucial role [53].

Anomaly detection algorithms employ a variety of computational methods comprising statistical methods (Uni-variate/multi-variate anomaly detection), parametric or non-parametric probabilistic methods (Hidden Markov Models anomaly detection, Bayesian Networks anomaly detection), machine learning, ANN (ann, Unsupervised Learning –Unsupervised anomaly detection–, Ensemble-based anomaly detection), and many others.

More recently, new algorithms are emerging to better suit to the nature of IoT data streams and provide greater ability to process streaming data, among which, there is the the HTM algorithm. htm was originally offered by Numenta Platform for Intelligent Computing (NuPIC),[3] is finding numerous applications [16, 53]. Additionally, there is available the NAB [22], which is an open-source environment useful for evaluation of anomaly detection algorithms from data streams. nab benchmark contains real-world data streams with labeled anomalies and is undoubtedly an interesting source for testing HTM anomaly detection.

Anomaly detection algorithms have become essential also to IoT based monitoring of agro-ecosystems for disease detection [26], diagnosing the agriculture status [37], detection of soil moisture anomalies [11], detection of anomalous crop condition and soil variability mapping [45], etc. Unlike other domains, the anomaly detection in agro-ecosystems could have massive economical and social implications such as for preventing spreading of diseases, contaminated soil, environmental problems, etc. It should be noted however, that specific benchmarks from agro-ecosystems are needed to achieve testing and evaluations for anomaly detection algorithms in the agro-ecosystems domain.

---

[3]http://numenta.org/nupic.html.

Chapters of this book address various research issues arising from modeling, simulation and anomaly detection for intelligence in agro-ecosystems.

4 Data Cycle for Agro-Ecosystems: From IoT Data Collection to Decision Making and End-to-End Management Systems

The main premise of big data and big data streams is they would provide knowledge for the problem, phenomenon under study or the concerned business. While nowadays many technical issues with big data are well understood including the nature of big data from various domains (science, nature and environment, engineering, social life, etc.), the making use of the obtained knowledge from the data remains a critical challenge.

Indeed, unlike single studies where big data is often used as a snapshot data source used to understand an issue or a problem, in domains such as industry, agroecosystems and businesses, there should be periodicity/continuity of data gathering, processing, analysis and knowledge embedding into the management or business processes. This leads to the necessarily definition of a data cycle, which comprises various phases: data gathering, real-time data processing and off-line data processing, data feed and knowledge provision to front-end and actuator systems and knowledge/analytics integration with back-end management/business intelligence systems.

4.1 Data Collection and Acquisition

This is the process of collecting data from the agro-ecosystem. A variety of data can be obtained either automatically from sensory systems in place, which *per se* could be multi-modal data streams or data from end-users, through data collection apps and tools for any type of information (text, image, sound, video, etc.). Regarding the former, as can be seen from Fig. 1, there would be data streams coming from a remote farm, a remote monitoring field from in situ sensory devices or mobile infrastructures (fanets, vanets). Data through data collection apps and tools would be acquired from farmers, field workers, drivers, etc., who can send the data in real-time to fog nodes for immediate or historic data processing.

The data source variety, multi-modal nature of the data and data generation rate constitute most challenging issues of data processing and analysis in any modern Cloud-to-thing agro-ecosystems.

4.2   Real Time Data Pre-processing and Anomaly Detection

Data collected from Cloud-to-thing agro-ecosystems can be processed in real-time for various purposes, among which, two are salient ones: (1) for anomaly detection and (2) real time analytics. Anomalies, as can be seen in Fig. 1, can happen at various parts of an agro-ecosystem and can have various degrees of impact. A disease in a farm, when detected by the IoT data stream, should trigger a real-time alert to veterinarian site and to managers enterprise site, who would take the necessary steps to inform end-users and make real-time decisions. Likewise, events of interest from farm sites or monitored field such as fire and flooding or water leaks, plant disease are to be triggered in real time. Other anomaly events of interest such as dried areas, pollution, etc. could fall within alerting and would require further complex event data processing.

It should be noted here that real-time processing is meant for spotting anomaly events from data streams, based on either atomic event detection or sliding window-based stream processing [23] (see also Sect. 3.2). Given the nature of unlabeled IoT data streams, unsupervised real-time anomaly detection for streaming data are becoming *de facto approaches* beyond classical statistical approaches [1, 53].

Detecting more complex events corresponding to anomalies from data streams would require correlating events within the stream and therefore there is needed to give structure to data. This is envisaged here through semantic data enrichment (see Sect. 4.4).

4.3   Real Time Anomaly Prediction in Data Streams

Unsupervised learning methods for anomaly detection, such as HTM algorithm, have been shown to be useful also for anomaly prediction by computing the prediction error and anomaly likelihood in the data stream. Anomaly prediction is extremely useful as it warns about anomaly occurrence within a time period. However, high accuracy anomaly prediction from data streams using unsupervised / supervised learning require training data, which is typically not available for data streams.

Therefore there are typically four tasks to be completed here for anomaly prediction through a proper training data set:

- Detecting anomalies in data stream through unsupervised method(s).
- Labeling a data set with detected anomalies for the prediction algorithm. Anomaly detection has to be of high accuracy for the prediction algorithms to perform accurately.
- Training prediction algorithm(s) either by training historical data or continuous training.
- Using pre-trained algorithms for anomaly detection in data stream.

A variety of machine learning methods have been studied in the literature for real time anomaly prediction such as anomaly prediction based on PCA, LR, SGD, among others. The selection of the right method (pca,lr,sgd ...) would depend on several factors and parameters and would require a proper study based on the objectives and characteristics of data stream (e.g. continuous *vs.* categorical data, data points streams, single *vs.* multiple stream, etc.) attributes and of real-time requirements of applications on top of data stream.

4.4 Data Structuring and Semantic Data Enrichment

Data stream pre-processing is a must for any domain, the main reason being that raw data could contain errors, missing values, duplication and futile information. Likewise, in many case it is desirable to perform a data normalization process due to data scale variety in multi-streams aiming to create a structured *data lake*. This data pre-processing task can be seen as an independent task from real-time anomaly detection, by taking advantage of the collaborative computing efforts at edge computing devices and fog servers [53].

As part of this pre-processing pipeline, semantic data enrichment would play an important role to give structure to data and enrich it further. Semantic data enrichment is the process of adding metadata information to content so that data becomes machine readable and data can be logically linked to each other. Adding contextual information, such as time, location, data source, etc., is particularly of interest to enable data interpretation in context. Altogether, by adding a semantic data layer to original data it is possible to build semantic reasoning, complex event processing and rule based systems, such as ECA. eca is commonplace for rule-based systems.

The foundation of semantic data enrichment come from the semantic web, where it has been successfully used so that data in web pages can be read and interpreted by computers. With the fast developments in IoT technologies, semantic technologies are gaining attention in the research community to explore the advantages that such technologies can bring to IoT data streams. Among such advantages, beyond giving meaning to the data, there are the efficient reasoning and mining, linking contents from different IoT devices and enabling complex event processing from IoT data streams.

In the case of semantic web, RDF can be used for purpose. rdf frameworks cannot be used in a straight forward-way to IoT data, however. Indeed, data at websites is rather static, well identified and located, while IoT data is dynamic, more volatile and generated by a large number of devices. A number of data annotation techniques such as JSON for Linked Data (JSON-LD), Entity Notation (EN) and Header-Dictionary-Triples (HDT), Sensor Markup Language (SML), TripleWave and others are useful for the semantic data enrichment process for IoT data streams from agro-ecosystems.

Another issue of interest with the IoT data stream processing and semantic data enrichment is to address the question of what data should be persistently stored in Cloud data centers for more in depth analysis and future studies. Given the nature of the IoT data, which is mainly for reading, data aggregation and analyzing, the noSQL databases seem the right choice for most scenarios in IoT data streams from agro-ecosystems. Indeed, noSQL databases would accommodate the main requirements on this type of data, namely, data heterogeneity, semantic interoperability, scalability, high data rate performance and real time processing.

4.5 Knowledge and Analytics Integration with Back-End Management and Business Intelligence Systems

The final stage of the data cycle is the integration process of Knowledge and Analytics with Back-End Management and Business Intelligence Systems. The final users, comprising, engineers, managers and stakeholders will use analytics and knowledge for informed decision-making for operational and strategic purposes, optimizing the resources and increased efficiency of production processes (historical, current, and predictive perspectives).

The integration is a complex system, especially if legacy and/or external information and management systems are to be used. The integration process could target separately either business intelligence or knowledge management or jointly both of them. Indeed, analytics and business intelligence relies on modern data centers and intelligent processing, most prominently machine learning, of structured and enriched data. While, knowledge management requires advanced content management systems to make the knowledge available for using and sharing in a scalable and reliable way to end users for achieving organizational objectives (performance, competitiveness and innovation).

In the agro-ecosystem data analytics [33] together with knowledge management [38] are technology drivers and enablers for digital transformation in the field. Examples in the literature evidence the increasing efforts by the community of researcher and developers to advance the state of the art in the field, yet, a myriad of challenges ranging from technological to policy making [36, 47].

Chapters of this book address approaches for agro-ecosystems based on IoT data collection, analysis, mining and machine learning for decision making and agro-management systems.

5 Machine Learning: Killer Applications Are from Agriculture

Machine Learning has become a cornerstone for data-centric systems in all fields, where vast amounts of data, either in batches or in streams, are processed to distill valuable knowledge for decision making. Agro-ecosystems are not an exception. In fact, according to professionals from John Deere company –the 200-year-old tractor manufacturer– the world's hardest machine learning problem it's agriculture. For instance, classification problems are certainly challenging in agriculture due to different conditions under which plants are grown. Even if an accurate classifier would work for an area, it may fail to give accurate results for a nearby area. Probes for measuring the parameters in an area, might not reveal key observations, leading therefore to patterns that might not hold. Indeed, this stresses the fact that soil is not homogeneous even in adjacent areas, but vary dramatically. In the same vein, it becomes very complex to accurately detect residues left on the soil after a harvest, the amount of residues left depends on "*where*" and "*when*", namely it's not the same amount everywhere in the field and it changes overtime.

The complexity of processing data with Machine Learning in agriculture comes as well due to the huge amount of data that can be collected just for a handful of acres! Therefore the use of Machine Learning for agriculture come with needs for in-house modern and scalable Cloud infrastructures or to use off-premise Cloud services. In all, Machine Learning is a game-changing technology for agro-ecosystems!

Collaboration with farmers and other involved actors is deemed crucial aggregate data from different sources and thus to provide support at scale, rather than to limited areas and users.

6 Emerging Research Trends and Further Issues

As a new field of research and development, IoT for agro-ecosystems is a dynamic and fast growing research. A number of issues, beyond infrastructure, Big Data and Machine Learning are emerging and require efforts to understand and address them. They are by no means new research trends as actually they are found in other productivity sectors but their resolution needs to take into account the specifics of agro-ecosystems.

6.1 Data Quality and Benchmarks for Agro-Ecosystems

As in other fields, data quality is paramount for extracting meaningful and useful information and knowledge for agro-ecosystems. Unlike other fields such as Industry 4.0, eHealth, etc., where data quality models and standards are found with abundance, in agro-ecosystem there are no developed data quality models and benchmarks although there are various studies and initiatives such as to benchmark agricultural irrigation practices [20]. Again, the challenge is that benchmarks from some areas or countries may not be extrapolated to other areas or countries [28, 41, 51]. Procedures about data processing and cleansing from missing values and errors should be defined in the context of agro-ecosystems. Other issues include data provenance, traceability, curation as well as Open Data, Interoperability, and Standardization.

Finally, cost-effective solutions for collection, processing, anomaly detection and curation of data will be central to design and development of benchmarks for agro-ecosystems.

6.2 Digital Agro-Ecosystems Adoption, Cooperation with Local Stakeholders and Policy-Makers

Collaboration among various actors of agro-ecosystems, including farmers, local administrations, policy makers and multi-stakeholders, is a key driver to a successful adoption of IoT technologies [19, 24, 39, 50]. New issues not known before in agro-ecosystems such as innovation, trust, business value and risks are to be addressed.

On the other hand, the digitalization process of agro-ecosystems calls for development of digital sharing platforms and digital networks among the various actors. Open platforms would be of special value here to involve also citizens and other non-governmental actors. Such platforms will enable development of strategies collectively among actors. Finally, digital platforms will encourage the integration with services from Digital Marketplace, FinTech sectors and other external sectors [2, 36, 47].

6.3 Ethics, Health & Safety Issues in Agro-Ecosystems

The new generation of agro-ecosystems is expected to have various impacts such as health and safety, socio-economical, food safety, food chain production systems and environmental impacts [21].

Some of these potential impacts have to do with the pervasive use of IoT technologies in the field, for instance, related to use of 5G IoT technology [10, 32].

Significant impacts are expected on the social-economical side, where a number of actors (farmers, consumers, stakeholders, local administration, businesses of food chain industry, supply chain, etc.) will have to envisage benefits and risks of the technology impact [17, 35].

Data privacy, data sharing, ethics and safety are essential issues in agro-ecosystems as such systems have direct impact on food, people and sustainability. Therefore, such issues will require proper addressing in the new generation of agro-ecosystems hand in hand with their design, development and deployment [5, 25, 27, 42].

7 Conclusion

In this introductory chapter, we have highlighted some fundamental concepts, architectures and definitions related to IoT-based Computational Modeling for Next Generation Agro-ecosystems. We have analyzed the benefits of using the paradigms of Cloud-to-thing Continuum as a large digital ecosystem comprising IoT, Edge, Fog, and Cloud Computing as well as data analysis based on a formal data cycle definition. Then, Machine Learning and stream processing, simulation and modeling frameworks are introduced and discussed. We finalize the chapter by describing various emerging research trends, ethics and health & safety issues. The concepts, models and technologies for agro-ecosystems introduced in this chapter are covered in the chapters of the book and exemplified with real life use cases and applications. Cloud-to-thing continuum and IoT as game changers for the development of Next Generation agro-ecosystems should go hand in hand with social, economic, food, ecological and ethical sciences.

Glossary

Artificial Neural Network (ANN) is a computing system inspired by biological neural networks used for modeling complex relationships between inputs and outputs or to find patterns in data.

Collective Intelligence (CI) is shared or group intelligence that emerges from the collaboration, collective efforts, and competition of many individuals and appears in consensus decision making.

Computational Modelling (CM) is a mathematical model in computational science that requires extensive computational resources to study the behavior of a complex system by computer simulation.

Digital Twin (DTwin) is a digital replica of a physical entity or system aiming to create living digital simulation model that updates and changes as the physical counterparts change.

Event-Condition-Action (ECA) represents the structure of active rules in event driven architecture and active database systems.

Flying Adhoc Networks (FANETs) is an Ad Hoc Network connecting Unmanned Air Vehicle.

High Performance Computing (HPC) is the use of super computers and parallel processing techniques for solving complex computational problems, which cannot otherwise be solved within reasonable time by single computers.

Hierarchical Temporal Memory (HTM)   is a biologically inspired algorithm for prediction, anomaly detection and classification.

Internet of Things (IoT)   is a system of interrelated computing devices, mechanical and digital machines, objects, animals or people that are provided with unique identifiers and the ability to transfer data over a network without requiring human-to-human or human-to-computer interaction.

Internet of Underground Things (IoUT)   is a system of interrelated sensor devices buried under-ground and interconnected with communication and networking solutions using the soil as a medium.

Logistic Regression (LR)   is a statistical model for estimating the parameters of a logistic model (a form of binary regression).

Numenta Anomaly Benchmark (NAB)   is a benchmark of data streams with labeled anomalies for evaluation of anomaly detection algorithms.

Principle Component Analysis (PCA)   is a statistical procedure that uses an orthogonal transformation to convert a set of observations of possibly correlated variables into a set of values of linearly uncorrelated variables called principal components.

symbiotic data-driven modelling   is a methodology in which symbiotic simulation is driven by data.

Resource Description Framework (RDF)   is a family of World Wide Web Consortium specifications originally designed as a metadata data model.

Round-Trip Time (RTT)   is the time duration that takes to complete a request from a user to an application and back again to the user.

Stochastic Gradient Descent (SGD)   is an iterative method for optimizing an objective function with suitable smoothness properties.

Symbiotic Simulation (SS)   is a methodology in which there is a close relationship between a physical system and the simulation system that represents it.

Unmanned Aerial Vehicles (UAVs)   is an aircraft without a human pilot on board.

Vehicular Adhoc Networks (VANETs)   is an Ad Hoc Network connecting vehicles and vehicles with road side units.

Acknowledgements Fatos Xhafa's work has been supported by Spanish Ministry of Science, Innovation and Universities, Programme "*Estancias de Profesores e investigadores senior en centros extranjeros, incluido el programa Salvador de Madariaga 2019*", PRX19/00155. Work partially done during the Leave at University of Surrey, UK.

References

1. Ahmad A, Lavin A, Purdy S, Agha Z (November 2017) Unsupervised real-time anomaly detection for streaming data. Neurocomputing 262(1):134–147
2. Anshari A, Almunawar MN, Masri M, Hamdan M (2019) Digital marketplace and fintech to support agriculture sustainability. Energy Procedia, Vol 156, 2019, pp 234–238, https://doi.org/10.1016/j.egypro.2018.11.134
3. Bender SF, Wagg C, van der Heijden MGA (2016) Review an underground revolution: biodiversity and soil ecological engineering for agricultural sustainability. Trends Ecol Evol 31(6):440–452

4. Bikash Chowdhury A, Xhafa F, Rongpipi R, Mukhopadhyay S, Kumar Singh V (2019) Spectrum trading in wireless communication for tertiary market. INCoS 2018: 134-145. Advances in intelligent networking and collaborative systems. In: The 10th international conference on intelligent networking and collaborative systems (INCoS-2018), Bratislava, Slovakia, September 5–7, 2018. Lecture Notes on Data Engineering and Communications Technologies 23, Springer 2019

5. Burkhardt J (2008) Chapter 3 - The ethics of agri-food biotechnology: how can an agricultural technology be so important?. In: David K, Thompson PB (eds) Food science and technology, What Can Nanotechnology Learn From Biotechnology?, Academic Press, 2008, pp 55–79, https://doi.org/10.1016/B978-012373990-2.00003-0

6. Boursianis AD, Papadopoulou MS, Diamantoulakis P, Liopa-Tsakalidi A, Barouchas P, Salahas G, Karagiannidis G, Wan Sh, Goudos SK. Internet of things (IoT) and agricultural unmanned aerial vehicles (UAVs) in smart farming: a comprehensive review. In Press Internet of Things Journal, Elsevier. https://doi.org/10.1016/j.iot.2020.100187

7. Carletti M, Masiero Ch, Beghi A, Susto GA (2019) A deep learning approach for anomaly detection with industrial time series data: a refrigerators manufacturing case study. Procedia Manufact 38:233–240

8. Chehri A, Chaibi H, Saadane R, Hakem N, Wahbi M (2020) A framework of optimizing the deployment of IoT for precision agriculture industry. Procedia Comput Sci 176:2414–2422. https://doi.org/10.1016/j.procs.2020.09.312

9. Corizzo R, Ceci M, Japkowicz N (2019) Anomaly detection and repair for accurate predictions in geo-distributed big data. Big Data Res 16:18–35

10. Gonzalez-de Santos P, Ribeiro A, Fernandez-Quintanilla C, Lopez-Granados F, Brandstoetter M, Tomic S, Pedrazzi S, Peruzzi A, Pajares G, Kaplanis G et al (2017) Fleetsof robots for environmentally-safe pest control in agriculture. Precis Agric 18(4):574–614

11. Greifeneder F, Khamala E, Sendabo D, Wagner W, Zebisch M, Farah H, Notarnicola C (2019) Detection of soil moisture anomalies based on Sentinel-1. Phys Chem Earth, Parts A/B/C 112:75–82

12. Gunasekera K, Borrero AN, Vasuian F, Bryceson KP (2018) Experiences in building an IoT infrastructure for agriculture education. Procedia Comput Sci 135:155–162, ISSN 1877-0509, https://doi.org/10.1016/j.procs.2018.08.161

13. Hawley-Hague H, Boulton E, Hall A, Pfeiffer K, Todd Ch (2014) Older adults' perceptions of technologies aimed at falls prevention, detection or monitoring: A systematic review. Int J Med Inform 83(6):416–426

14. Helmer S, Poulovassilis A, Xhafa F (2011) Reasoning in event-based distributed systems. Springer Series Studies in Computational Intelligence, Vol 347. Springer

15. Hofmann T (2017) Integrating nature, people, and technology to tackle the global agri-food challenge. J Agricult Food Chem 65(20):4007–4008. https://doi.org/10.1021/acs.jafc.7b01780

16. Hole KJ (2016) Anomaly Detection with HTM. In: Anti-fragile ICT systems. Simula Springer-Briefs on Computing, Vol 1. Springer, Cham

17. Jakku E, Taylor B, Fleming A, Mason C, Fielke S, Sounness Ch, Thorburn (2019) If they don't tell us what they do with it, why would we trust them? Trust, transparency and benefit-sharing in Smart Farming. NJAS - Wageningen J Life Sci 90–91. https://doi.org/10.1016/j.njas.2018.11.002

18. Jones D, Snider Ch, Nassehi A, Yon J, Hicks B (2020) Characterising the digital twin: a systematic literature review. CIRP J Manufact Sci Technol, vol 29(Part A):36–52. https://doi.org/10.1016/j.cirpj.2020.02.002

19. Khatri-Chhetri A, Pant A, Aggarwal PK, Vasireddy VV, Yadav A (2019) Stakeholders prioritization of climate-smart agriculture interventions: evaluation of a framework. Agricultural Systems, vol 174. https://doi.org/10.1016/j.agsy.2019.03.002

20. Kitta E, Bartzanas T, Katsoulas N, Kittas C (2015) Benchmark irrigated under cover agriculture crops. Agricult Agricult Sci Procedia 4:348–355

21. Klerkx L, Rose D (2020) Dealing with the game-changing technologies of Agriculture 4.0: How do we manage diversity and responsibility in food system transition pathways? Global Food Security, Vol 24. https://doi.org/10.1016/j.gfs.2019.100347

22. Lavin A, Ahmad S (2015) Evaluating real-time anomaly detection algorithms - the Numenta anomaly benchmark. In: Proceedings of the 14th international conference on machine learning application, Miami, Florida, IEEE 2015
23. Li G, Wang J, Liang J, Yue C (2018) Application of sliding nest window control chart in data stream anomaly detection. Symmetry 10(4):113. https://doi.org/10.3390/sym10040113
24. Hannachi M, Fares M, Coleno F, Assens Ch (2020) The "new agricultural collectivism": How cooperatives horizontal coordination drive multi-stakeholders self-organization. J Co-operat Organiz Manag8(2). https://doi.org/10.1016/j.jcom.2020.100111
25. Mark R (2019) Ethics of using AI and big data in agriculture: the case of a large agriculture multinational. The ORBIT J 2(2):1–27. https://doi.org/10.29297/orbit.v2i2.109
26. Mattihalli Ch, Gedefaye E, Endalamaw F, Necho A (2018) Plant leaf diseases detection and auto-medicine. Int Things Elsevier Vols 1–2:67–73
27. Mepham B (2012) Agricultural Ethics. In: Chadwick R (ed) Encyclopedia of applied ethics, (Second Edition). Academic Press, pp 86–96. https://doi.org/10.1016/B978-0-12-373932-2.00347-1
28. Mekonnen MM, Hoekstra AY, Neale ChMU, Ray Ch, Yang HS (2020) Water productivity benchmarks: The case of maize and soybean in Nebraska. Agricult Water Manag 234. https://doi.org/10.1016/j.agwat.2020.106122
29. Monteiro J, Barata J, Veloso M, Veloso L, Nunes J (2018) Towards sustainable digital twins for vertical farming. In: International conference on digital information management (ICDIM), 2018, pp 234–239. https://doi.org/10.1109/ICDIM.2018.8847169
30. Mukherjee A, Misra S, Sukrutha A, Narendra M, Raghuwanshi NS (2020) Distributed aerial processing for IoT-based edge UAV swarms in smart farming. Comput Netw 167. https://doi.org/10.1016/j.comnet.2019.107038
31. Onggo BS, Mustafee N, Smart A, Juan AA, Molloy O (2018) Symbiotic simulation system: hybrid systems model meets big data analytics. In: Proceedings of the 2018 winter simulation conference (WSC '18). IEEE Press, pp 1358–1369
32. Park A, Jabagi N, Kietzmann J (2020) The truth about 5G: It's not (only) about downloading movies faster!. Business Horizons. https://doi.org/10.1016/j.bushor.2020.09.009
33. Pham X, Stack M. (2018) How data analytics is transforming agriculture. Bus Horiz 61(1):125–133. https://doi.org/10.1016/j.bushor.2017.09.011
34. Puig V, Escobet T, Sarrate R, Quevedo J (2015) Fault diagnosis and fault tolerant control in critical infrastructure systems. In: Intelligent monitoring, control, and security of critical infrastructure systems 2015: 263–299, Studies in Computational Intelligence, 565, pp 263–299, Springer
35. Regan A (2019) 'Smart farming' in Ireland: A risk perception study with key governance actors, NJAS - Wageningen J Life Sci 90–91. https://doi.org/10.1016/j.njas.2019.02.003
36. Rose DC, Sutherland WJ, Barnes AP, Borthwick F, Ffoulkes Ch, Clare Hall C, Moorby JM, Nicholas-Davies Ph, Twining S, Dicks LV (2019) Integrated farm management for sustainable agriculture: lessons for knowledge exchange and policy. Land Use Policy, Vol, 81, 2019, pp 834–842, ISSN 0264-8377, https://doi.org/10.1016/j.landusepol.2018.11.001
37. Singh S, Chana I, Buyya R (2020) Agri-Info: cloud based autonomic system for delivering agriculture as a service. Internet of things, Vol, 9, Elsevier
38. Skobelev PO, Simonova EV, Smirnov SV, Budaev DS, Voshchuk GYu, Morokov AL (2019) Development of a Knowledge Base in the "Smart Farming" System for Agricultural Enterprise Management. Procedia Comput Sci 150:154–161, ISSN 1877-0509. https://doi.org/10.1016/j.procs.2019.02.029
39. Sparrow AD, Traoré A (2019) Limits to the applicability of the innovation platform approach for agricultural development in West Africa: Socio-economic factors constrain stakeholder engagement and confidence. Agricult Syst 165:335–343, https://doi.org/10.1016/j.agsy.2017.05.014
40. Stock CA, Pegion K, Vecchi GA, Alexander MA, Tommasi D, Bond NA, Fratantoni PS, Gudgel RG, Kristiansen T, O'Brien TD, Xue Y, Yang X (2015) Seasonal sea surface temperature anomaly prediction for coastal ecosystems. Progress Oceanogr 137(Part A):219–236

41. Thomas IA, Buckley C, Kelly E, Dillon E, Lynch J, Moran B, Hennessy T, Murphy PNC (2020) Establishing nationally representative benchmarks of farm-gate nitrogen and phosphorus balances and use efficiencies on Irish farms to encourage improvements. Sci Total Environ 720. https://doi.org/10.1016/j.scitotenv.2020.137245

42. Thompson PB, Noll S (2014) Agricultural ethics and social justice. In: Neal K, Van Alfen (eds) Encyclopedia of agriculture and food systems. Academic Press, pp 81–91. https://doi.org/10.1016/B978-0-444-52512-3.00128-5

43. Trang H, Dung L, Hwang S (2018) Connectivity analysis of underground sensors in wireless underground sensor networks. Ad Hoc Netw 71:104–116

44. Vavilis S, Egner A, Petković M, Zannone N (2015) An anomaly analysis framework for database systems. Comput Secur 53:156–173

45. Venteris ER, Tagestad JD, Downs JL, Murray CJ (2015) Detection of anomalous crop condition and soil variability mapping using a 26 year Landsat record and the Palmer crop moisture index. Int J Appl Earth Obs Geoinf 39:160–170

46. Verdouw C, Kruize W (2017) Digital twins in farm management: illustrations from the FIWARE accelerators Smart AgriFood and Fractals. In: Conference: 7th Asian - Australasian conference on precision agriculture, 2017. https://doi.org/10.5281/zenodo.893662

47. Vik J (2020) The agricultural policy trilemma: On the wicked nature of agricultural policy making. Land Use Policy 99. ISSN 0264-8377, https://doi.org/10.1016/j.landusepol.2020.105059

48. Vuran MC, Salam A, Wong R, Irmak S (2018) Internet of underground things in precision agriculture: Architecture and technology aspects. Ad Hoc Netw 81:160–173

49. Wang X, Ahn SH (2020) Real-time prediction and anomaly detection of electrical load in a residential community. Appl Energy 259:Article 114145

50. Wilhelm JA, Smith RG, Jolejole-Foreman MC, Hurley S (2020) Resident and stakeholder perceptions of ecosystem services associated with agricultural landscapes in New Hampshire. Ecosyst Serv 45. https://doi.org/10.1016/j.ecoser.2020.101153

51. Williams R, Walcott J (1998) Environmental benchmarks for agriculture? Clarifying the framework in a federal system - Australia. Land Use Policy 15(2):149–163

52. Xhafa F. The Vision of Edges of Internet as a Compute Fabric. Chapter 1, In: Advances in Edge Computing: Massive Parallel Processing and Applications. Book Series: Advances in Parallel Computing Series. IOS Press

53. Xhafa F, Kilic B, Krause P (2020) Evaluation of IoT stream processing at edge computing layer for semantic data enrichment. Future Generat Comput Syst 105:730–736. https://doi.org/10.1016/j.future.2019.12.031

54. Zamora-Izquierdo MA, Santa J, Martinez JA, Martinez V, Skarmeta AF (2019) Smart farming IoT platform based on edge and cloud computing. Biosyst Eng 117:4–17. https://doi.org/10.1016/j.biosystemseng.2018.10.014

Fatos Xhafa is Professor of Computer Science at the Technical University of Catalonia (UPC), Barcelona, Spain. He received his PhD in Computer Science in 1998 from the Department of Computer Science of BarcelonaTech. He was a Visiting Professor at University of Surrey (2019/2020), Visiting Professor at Birkbeck College, University of London, UK (2009/2010) and a Research Associate at Drexel University, Philadelphia, USA (2004/2005). Prof. Xhafa has widely published in peer reviewed international journals, conferences/workshops, book chapters and edited books and proceedings in the field (Google h-index 52 / i10-index 270, Scopus h-index 41, ISI-WoS h-index 33, as of December 2020). He is awarded teaching and research merits by Spanish Ministry of Science and Education, by IEEE conference and best paper awards. Prof. Xhafa has an extensive editorial and reviewing service for international journals and books from major publishers as a member of Editorial Boards and Guest Editors of Special Issues. He is Editor in Chief of the Elsevier Book Series "Intelligent Data-Centric Systems" and of the Springer Book Series "Lecture Notes in Data Engineering and Communication Technologies". He is a member of IEEE Communications Society, IEEE Systems, Man & Cybernetics Society and Emerging Technical Subcommittee of Internet of Things. His research interests include parallel and distributed algorithms, massive data processing and collective intelligence, IoT and networking, P2P and Cloud-to-thing continuum computing, optimization, security and trustworthy computing, machine learning and data mining, among others. He can be reached at http://www.mailto:fatos@cs.upc.edu and more information can be found at WEB: http://www.cs.upc.edu/ fatos/ Scopus Orcid: http://orcid.org/0000-0001-6569-5497

Paul Krause is Professor in Complex Systems at SURREY. He has over forty years' research experience in the study of complex systems in a wide variety of domains, in both industrial and academic research laboratories. Currently his research work focuses on distributed systems for the Digital Ecosystem and Future Internet domains. He has over 120 publications and is author of a textbook on reasoning under uncertainty. He is leader of the recently founded Digital Ecosystems research group at Surrey which, although based in the Department of Computer Science, collaborates strongly with other disciplines throughout the University. He has been working in and leading strong interdisciplinary teams since 2006 in the EU funded DBE and OPAALS projects, more recently in the RCUK funded projects ERIE (for Evolution and Resilience of Industrial Ecosystems) and MILES projects. These last two were funded under the Complexity in the Real World, and Bridging the Gaps, programmes respectively. He is currently active in the above mentioned TASCC, SPEAR and HBP projects. He also has forty years' experience as a volunteer in practical nature conservation projects. He is a Fellow of the Institute of Mathematics and its Applications, and a Chartered Mathematician.

An IoT-Based Time Constrained Spectrum Trading in Wireless Communication for Tertiary Market

Anil B. Chowdhury, Vikash K. Singh, Fatos Xhafa, Paul Krause, Sajal Mukhopadhyay, and Meghana M. Dhananjaya

Abstract In this chapter, we study some research issues from IoT-based spectrum trading in Wireless Communication in a *strategic* setting. We consider the scenario in which there are multiple secondary users (such as non-governmental organizations (NGOs), institutional organizations, foundations, etc.) having available un-utilized spectrum and multiple tertiary users (such as small farms, agricultural enterprises or people residing in different localities). Tertiary users provide preferences over the subset of all the available secondary users (NGOs, hereafter). Based on their preference ordering, the tertiary users are allocated the best possible NGOs among the available ones and under the restrictions that each user is assigned to at most one NGO. However, it is to be noted that, in this model, the allocated spectrum might not be available through out a long period of time but rather for a short duration of time within a time period. Therefore, tertiary users have to be able to work off-line and access cached data at the Edges of Internet even if the Internet access is not

A. B. Chowdhury (✉)
Department of Master of Computer Applications, Techno India University, Kolkata, India
e-mail: abchaudhuri007@gmail.com

V. K. Singh
School of Computer Science and Engineering, VIT-AP University, Amaravati, Andhra Pradesh, India
e-mail: vikash.singh@vitap.ac.in

F. Xhafa
Department of Computer Science, Universitat Politècnica de Catalunya, Barcelona, Spain
e-mail: fatos@cs.upc.edu

University of Surrey, Guildford, UK

P. Krause
Department of Computer Science, University of Surrey, Guildford, UK
e-mail: p.krause@surrey.ac.uk

S. Mukhopadhyay · M. M. Dhananjaya
Department of Computer Science and Engineering, National Institute of Technology, Durgapur, West Bengal, India
e-mail: sajmure@gmail.com

M. M. Dhananjaya
e-mail: meghna0394@gmail.com

© Springer Nature Switzerland AG 2021
P. Krause and F. Xhafa (eds.), *IoT-based Intelligent Modelling for Environmental and Ecological Engineering*, Lecture Notes on Data Engineering and Communications Technologies 67, https://doi.org/10.1007/978-3-030-71172-6_2

available. For the purpose of storing and retrieving the cached data several algorithms are designed and their computational complexity is analyzed. In order to empirically measure the efficacy of the proposed mechanisms the simulations are carried out and are compared with the benchmark mechanism. The proposed allocation mechanisms are envisaged as especially useful tools for emerging scenarios of smart farming and precision agriculture, where in situ infrastructures are not available.

List of Acronyms with Explanation

STOM-OSM	Spectrum Trading withOut Money in One Sided Market
STOM-TSM	Spectrum Trading withOut Money in Two Sided Market
NGO	Non-Governmental Organisation
URL	Uniform Resource Locator
LSR	Linear Search based Retrieval
HR	Hashing based Retrieval
CHR	Consistent Hashing based Retrieval
IC	Incentive Compatible
IoT	Internet of Things
TOM-SA	Truthful Optimal Mechanism for Spectrum Allocation
TOM-SA-TR	Truthful Optimal Mechanism for Spectrum Allocation with Total Random
TOM-SA-HR	Truthful Optimal Mechanism for Spectrum Allocation with Half Random
TOM-SA-TTR	Truthful Optimal Mechanism for Spectrum Allocation with Two-Third Random

1 Introduction

With the emergence of Edge computing technologies, most notably 5G technology, the demand of spectrum has significantly increased. It has been found that spectrum utilization is highly dynamic in nature [1]. To meet an unprecedented demand of the mobile network, cellular network, etc., the bands are overloaded in most of the parts of the world. However, it has been observed by the regulatory bodies such as *Federal Communications Commission* (FCC) that the other frequency bands (such as of *military*, *institutions*, etc.) are only used occasionally and hence insufficiently utilized. In the past, several research works reveal this under utilization of spectrum [2–4]. Therefore, one of the challenges in the field of spectrum trading is: *how to efficiently and fully utilize spectrum*? One solution could be to redistribute the un-utilized spectrum to users that can make use of it, especially new emerging users such as farmers, small businesses in agriculture or supporting Internet connectivity of users in remote areas, etc. The re-distribution of the un-utilized or under-utilized spectrum to the secondary users (such as an organization, NGOs, etc.) held by the *primary users* (such as service providers *Vodafone, Airtel*, etc.) has been an active area of research for decades in the field of wireless communications. By seeing the rapid increase in the wireless traffic, the primary users of the system placed their under-utilized spectrum (also known as *white space*) in the spectrum market for use, in order to meet the increasing demand of the wireless spectrum.

More formally, one can say that the primary users leased out their under-utilized spectrum to the secondary users on demand (hereafter, both primary users and secondary users will be referred to as agents). The research question that we address in this chapter is *how to achieve an efficient distribution of spectrum among the strategic agents?* By *strategic* we mean that agents will try to use their game strategies in the system in order to gain. Existing literature has studied the problem of spectrum trading in a *strategic setting*. For instance. spectrum trading in cognitive radio has been studied by utilizing the concept of game theory [5], where various game theoretical models have been adopted such as *evolutionary game, non-coperative game* [6], and *Stackelberg game* [7]. Likewise, a solution concept that has been utilized in the past for similar types of scenarios is auction *mechanism* [8–12].

From the above discussion, it is clear that to date existing works mainly consider the allocation of spectrum held by the primary users (such as the service providers) to the secondary users, but not beyond that. However, it may be the case that after receiving the spectrum from the primary users, the secondary users can consider two scenarios: (1) either utilize the complete spectrum for their own purpose, or (2) re-distribute the spectrum that is under-utilized by them. In the literature not much emphasis has been given to the second scenario. So, re-distributing the un-utilized spectrum by the secondary users to the community (farmers, people working in agriculture, small businesses, etc.) that really need it gives rise to another market for spectrum allocation, which we refer to as the *tertiary market* , while the users present in the tertiary market are called *tertiary users*.

To the best of our knowledge, in our previous work [13], for the first time, the discussion moved beyond the primary and secondary users to the tertiary users, in a *strategic setting*. The aim is to provide the Internet facility to farmers, small businesses in agriculture, and other end-users in remote areas, free of cost. As a prerequisite, we assume here that tertiary users provide their preferences over all the NGOs or on a subset of NGOs carrying spectrum. We assume that users are rational in their behaviour. Also, the preferences of each user over the available NGOs is his/her *private* information, namely, the preference ordering is only known to the user and not known to others. Once the preference list of the users are available, the goal is to allocate the NGOs to the tertiary users, under the restrictions that each of the tertiary users is allocated to a single NGO for a short period of time. In the model discussed in [13], it is assumed that both the NGOs and the tertiary users are available throughout a short time period, such as day or less.

As it can be seen, in our model money is not involved in any sense. Therefore this is a favourable setting that can be investigated with the concept of mechanism design without money [14–19]. For this purpose, *truthful mechanism* the *truthful* mechanisms are proposed that also results in *optimal* and *The core* allocation (see definitions in Sect. 2.3).

Our discussion covers the *evolution of spectrum allocation market* and the problem scenario in an IoT-based spectrum allocation, which is useful for various emerging scenarios at the Edges of Internet.

Our contribution in this chapter can be summarized as follows:

- The IoT-based *spectrum allocation problem* is studied as a two fold process and is modelled by utilizing the concept of mechanism design without money.
- Various algorithms, namely, *Truthful Optimal Mechanism for Spectrum Allocation* (TOM-SA) motivated by [16, 20], *Linear Search-based Retrieval* (LSR), *Hashing based Retrieval* (HR), and *Consistent Hashing based Retrieval* (CHR), are proposed, analyzed and empirically evaluated.

The rest of the chapter is organized as follows. In Sect. 2, we overview the works carried out in the recent past in the direction of spectrum allocation and discuss the main scenario. The extension of the model for the spectrum allocation problem is presented in Sect. 3. The experiments and results are outlined in Sect. 4. We end the chapter in Sect. 5 with some concluding remarks and directions for future works.

2   Spectrum Trading in Tertiary Market

In this section, we start by introducing the concept of evolution of spectrum allocation market in Sect. 2.1, next, we state the research problem and main scenario in Sect. 2.2. Further, we discuss the work carried out in [13] in the direction of spectrum allocation in tertiary market. In Sect. 2.3 some important definitions are given that will be used throughout the rest of the chapter. The proposed algorithms for the considered scenario are presented in Sect. 2.4. Further, in Sect. 2.5, the results that were obtained in [13] are discussed.

2.1   *Evolution of Spectrum Allocation Market*

The general framework of the evolution of spectrum allocation is depicted in Fig. 1. A governmental body is assumed to be the owner of the spectrum and distributes it to several service providers such as *Vodafone, Airtel*, etc., so that they can provide their services to end-users. In the literature, and therefore in this chapter, the service providers are referred to as *primary users*.

As discussed earlier, it is not the case that all of the spectrum that is allocated to and owned by the primary users is fully utilized and eventually, a portion of the spectrum will be under-utilized. If this is the case, then it may give rise to a next level of spectrum trading, whereby the under-utilized spectrum of the primary users will be re-distributed to the secondary users (such as *NGOs, an organization*, etc.) in order

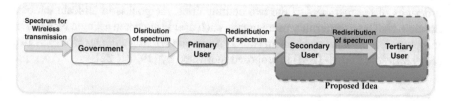

Fig. 1 System Evolution

to meet the increasing demand of wireless spectrum. Similar to the primary users, the secondary users have the flexibility to re-distribute their un-utilized spectrum to tertiary users, leading to another market for the spectrum trading, called the tertiary market and its users are the tertiary users. This chain of distribution of spectrum from first to tertiary users is graphically shown in Fig. 1. The motivation behind such evolution of the spectrum is supported by emerging scenarios at the Edges of Internet such as eHealth, smart farming, eLearning and others for people living in remote areas where there is no terrestrial Internet infrastructure in place. The scenarios from agro-ecosystems, in particular, can benefit from this setting given that for certain applications an Internet service is only temporarily needed, such as for profiling a land, monitoring a field, imageries, etc.

2.2 Problem Statement and Scenario

We consider that there are multiple secondary users (such as non-governmental organizations (NGOs), an organization etc.) having un-utilized spectrum and multiple tertiary users (such as such as farmers, small businesses in agriculture, etc., located in different remote areas). Each of the tertiary users provide preferences over all of a subset of the available NGOs. Based on the preference ordering of the users, each of the users is allocated the best possible NGOs among the available ones (each user is assigned to at most one NGO). However, it is to be noted that, in our model, the allocated spectrum to the users will not be available throughout long periods but rather, for a short duration of time. Also, reliability could be an issue here, namely, it is not fully ensured that users will be having access to the Internet through this freely distributed spectrum at all times. If this is the case, the users can still have access to the previously cached information utilizing their IoT devices.

Given the above discussed scenario the goal is to design an algorithm for storing and retrieving cached data / web pages from the IoT devices in an efficient manner.

For this set-up, a *truthful* mechanism namely *Spectrum Trading withOut Money in One Sided Market* (STOM-OSM) is proposed for allocating the NGOs to the users (each user is allocated a single NGO) motivated by [15, 21–23]. Whereas, in the second case, both the tertiary users and the NGOs are providing the preference ordering over all the members of the opposite community. Given the preference orderings of the members of the two communities, the goal is to allocate the best possible NGOs to the tertiary users (again, each user is allocated a single NGO). For this purpose, a *truthful* mechanism namely *Spectrum Trading withOut Money in Two Sided Market* (STOM-TSM) is proposed motivated by [19, 21, 22].

2.3 Definitions and Terminology

Here, the definitions and terminology used throughout the chapter are given.

Definition 1 *(Truthful)* A mechanism is truthful, if the agents cannot gain by misreporting their true preference ordering.

Definition 2 *(Pareto optimal)* An allocation is Pareto optimal, if in that allocation we cannot make anyone better off without making someone else worse off.

Definition 3 *(Blocking Coalition)* If a subset of participating agents form a coalition and re-allocates among themselves through some internal re-allocation, and the participating agents in the coalition makes themselves better off then that coalition is called a blocking coalition.

Definition 4 *(The Core)* This property ensures that the resulting allocation will be free from blocking coalitions.

Definition 5 *(Perfect Matching (from graph theory))* A perfect matching of a graph is a matching in which every vertex of the graph is incident to exactly one edge of the matching.

2.4 Proposed Mechanisms: STOM-OSM and STOM-TSM

In this section, we analyze the mechanisms proposed for the above discussed problem scenarios. The proposed mechanisms, namely *Spectrum Trading withOut Money in One Sided Market* (STOM-OSM) is motivated by [15, 21–23] and *Spectrum Trading without Money in Two Sided Market* (STOM-TSM) is motivated by [19, 21, 22].

In Sect. 2.4.1, the central idea of STOM-OSM is presented. Further the outline of STOM-TSM is discussed in Sect. 2.4.2.

2.4.1 Outline of STOM-OSM Algorithm

In order to allocate the spectrum held by the NGOs to the tertiary users, the central idea is as follows in the listing below:

STOM-OSM

1. First of all randomly allocate a distinct user to each of the NGOs.
2. next each of the users will point to his/her (henceforth his) most preferred NGO from his preference ordering, out of the available NGOs.
3. Step 1 and 2 together results in a directed graph. Once the directed graph is formed, next goal is to identify the directed cycle(s) in a graph.
4. Following the directed cycle(s) in a graph, allocate the NGOs to the users.
5. Users along with the assigned NGOs will be removed from the spectrum market.
6. Steps 2–5 are repeated until each of the users are not allocated to the NGO.

2.4.2 Outline of STOM-TSM Algorithm

In order to allocate the spectrum held by the NGOs to the tertiary users, the central idea of STOM-TSM is as follows in the listing below:

STOM-TSM

1. Each user requests for the most preferred NGO from his/her preference list.
2. NGOs hold only one offer and rejects all others.
3. Each rejected user removes the already requested NGO from his preference list.
4. If there are no new rejections, finish. Otherwise, iterate until allocation is completed.

2.5 Analysis of the Algorithms

In this section, we describe the results presented in [13]. First, it is shown that STOM-OSM is truthful. It means that the participating users cannot gain by mis-reporting their privately held preference lists over the NGOs. Further, it has been shown that the allocation resulted by the STOM-OSM is a unique core allocation. Finally, it is shown that the STOM-OSM is Pareto optimal (see definitions in Sect. 2.3).

Proposition 1 *STOM-OSM is truthful.*

Proposition 2 *STOM-OSM results in a unique core allocation.*

Proposition 3 *STOM-OSM is Pareto optimal.*

On the other hand, as regards STOM-TSM, first it is shown that the matching resulted by STOM-TSM is a perfect match. Next, it is shown that STOM-TSM is truthful for the requesting party, that is, the members of the requesting party cannot gain by mis-reporting their preference ordering.

Proposition 4 *STOM-TSM results in a perfect matching.*

Proposition 5 *STOM-TSM is truthful for the requesting party.*

3 IoT-Based Spectrum Trading in Tertiary Market

In this section, we present the formal statement of our problem for the IoT context.

- We assume a number m of NGOs having under-utilized spectrum, a number n of IoT devices, and a number k of tertiary users. It might be the case that either $m \neq k$ or $m = k$.
- The set of NGOs is represented as $\mathcal{N} = \{\mathcal{N}_1, \mathcal{N}_2, \ldots, \mathcal{N}_m\}$, the set of IoT devices is denoted as $\mathcal{I} = \{\mathcal{I}_1, \mathcal{I}_2, \ldots, \mathcal{I}_n\}$ and the set of tertiary users is given as $\mathcal{U} = \{\mathcal{U}_1, \mathcal{U}_2, \ldots, \mathcal{U}_k\}$.
- In this set-up, each user gets at most one NGO and each NGO can provide its service to one user.

Each user $\mathcal{U}_i$ has a preference ordering over the subset of NGOs given as $S \subseteq \mathcal{N}$. The strict preference ordering of the ith user is denoted by $\succ_i$ over the set S, where $\mathcal{N}_j \succ_i \mathcal{N}_l$ means that user $\mathcal{U}_i$ prefers $\mathcal{N}_j$ over $\mathcal{N}_l$. It should be noted that it may be the case that the preference ordering could have ties. The ties in the preference ordering of the user $\mathcal{U}_i$ is denoted by $=_i$, where $\mathcal{N}_j =_i \mathcal{N}_l$ means that user $\mathcal{U}_i$ prefers equally to $\mathcal{N}_j$ and $\mathcal{N}_l$. The preference orders of the users is *private* information.

The following notations are used in terms of preference ordering:

1. The preference ordering of the user having both strict preference and ties is represented by the symbol $\succeq_i$.
2. The preference ordering of the user having only strict preference is represented as $\succ_i$.
3. The preference ordering of the user having only ties is denoted as $=_i$.

Given the strict preference ordering of the users, the goal is to allocate the NGOs to the users. We assume that users are strategic in nature, that is, users may misreport their privately held preference ordering in order to gain. Let us denote the allocation vector as $a = \{a_1, a_2, \ldots, a_f\}$, where $f = \min\{m, k\}$.

In this model, it is considered that the allocated spectrum of any NGO will be available for a rather short span of time, say within a day. Although this might seem a limitation, the aim of this restriction is to enable giving Internet services to a larger number of tertiary users and optimizing the allocation of the spectrum. For the case of emerging applications at the Edges of Internet, this spectrum allocation will be still worthwhile as in many scenarios, some tasks can be accomplished within a limited amount of time (e.g. imagery, sensing, communication and data feed, etc.).

Let us denote, for any user $\mathcal{U}_i$, the start time of the spectrum allocation as s_i and corresponding finish time as f_i. More formally, we can say the time span required by any user $\mathcal{U}_i$ is given as $\tau_i = [s_i, f_i)$. The time span vector of all the users is given as $\tau = \{\tau_1, \tau_2, \ldots, \tau_k\}$.

By considering a limited time interval for spectrum allocation, with the aim of coverage and allocation optimization, the question of access to data outside such interval arises. We envisage that users can still access cached data distributed among multiple Edge devices (e.g. a Mobile Access Server), formally denoted as $\mathcal{I} = \{\mathcal{I}_1, \mathcal{I}_2, \ldots, \mathcal{I}_n\}$. Obviously, efficient storing and retrieving of cached data (an example of cached data would be reports sent to a farmer) at Edge devices is a matter of importance in this setting to ensure continuity of users activity and avoid business disruption.

For the above discussed problem scenario a *truthful* mechanism is proposed that also satisfies the other economic properties such as *the core* and *optimality* (For definitions see Sect. 2.3).

3.1 Proposed Mechanisms

As our proposed model is a two fold, we have proposed mechanisms for the two folds separately. For the first fold, we have proposed a *truthful* mechanism namely *Truthful Optimal Mechanism for Spectrum Allocation* (TOM-SA) motivated by [16, 21]. The TOM-SA is *truthful*, along with this it satisfies the other economic properties such as *Pareto optimality* and *The Core*.

For the second fold, we have considered two different scenarios. In the first scenario, the number of IoT devices i.e. n is fixed. For this first case, two algorithms are designed namely *Linear Search-based Retrieval* (LSR) and *Hashing based Retrieval* (HR) motivated by [24] to retrieve the information (such as cached data) stored in the IoT/Edge devices. The first algorithm will be used as a motivation for a faster algorithm for retrieving our desired information. Futher, we have considered that n varies over time corresponding to a more dynamic assumption, i.e. the IoT/Edge devices can join and leave the system on regular basis. For this an algorithm is proposed namely *Consistent Hashing based Retrieval* (CHR) motivated by [25].

These two scenarios are summarized in the listing below.

1. When n is fixed:

   - We can access the caches through linear search (see Sect. 3.1.2 for details), when a user requests for cached data off-line.

     - **Problem**: Searching is not efficient.
     - **Solution**: We can access the caches efficiently using the concept of hashing motivated by [24] (see Sect. 3.1.2 for details).

2. When n is variable:

   - The concept of consistent hashing is used to efficiently access the caches.

It should be noted that we are assuming here that IoT/Edge devices used here are either newly deployed or are already deployed for the other purposes. In the upcoming sections first the mechanism proposed for the first fold is discussed and presented. Next the mechanisms for the second fold are presented.

3.1.1 Proposed Mechanism for First Fold: TOM-SA

In this section, the proposed mechanism i.e. TOM-SA is discussed and presented.

Outline of TOM-SA
First, a random number is assigned to every user. Next, based on the random number assigned, a user is selected and is assigned the most preferred NGO from the preference list. The process iterates until the list becomes empty.

TOM-SA

1. Firstly, n distinct random numbers are generated and assigned to the users.
2. Based on the assigned random number, users are sorted in increasing order.
3. After that, in each iteration a user is selected from the sorted ordering and it is checked that whether the selected users' preference list is empty or not.

- If the selected users' preference list is not empty, then the most preferred NGO among the available ones from the selected users' preference list is allocated to him. Once the allocation is done, the selected user along with the assigned NGO is removed from the spectrum market.
- Otherwise, an unallocated user is removed from the spectrum market.

4. Step 3 is repeated until the users list becomes empty.

Detailed description of TOM-SA
In line 1, variables and data structures are initialized to 1 and ϕ respectively.

In lines 2–4, the generated numbers from 1 to k are stored in $\mathcal{G}$. Next, the list $\mathcal{G}$ is randomized using lines 5–7 of Algorithm 1. Using lines 8–11 each of the users is assigned the distinct random numbers between 1 to k that is stored in $\mathcal{G}$ sequentially. In line 12, a list of users given as $\mathcal{U}$ is sorted based on the random numbers assigned to the users. Line 13 depicts that TOM-SA ends once all the users are processed (or the users list become empty). In line 14, a user is selected sequentially using $pick()$ function based on the assigned number. In line 15 the user's preference list is checked. In line 16, a best NGO is selected among the available NGOs from the preference list of a user under consideration by using $Select\_best()$ function. In line 17, $\mathcal{F}$ maintains the selected user-NGO pairs and removes the selected users and NGOs from their respective preference lists. In line 18, the allocated NGO is removed from the remaining users list and sets $\hat{f}$ and $\hat{r}$ to ϕ. Next, if the condition in line 15 fails then lines 19–21 of Algorithm 1 gets executed. The algorithm ends, at line 23, where the final user-NGO pairs are returned.

Algorithm 1 TOM-SA $(\mathcal{U}, \mathcal{N})$

1: $d \leftarrow 1, \hat{f} \leftarrow \phi, \hat{r} \leftarrow \phi, \mathcal{G} \leftarrow \phi$
2: **for** $i = 1$ to k **do**
3:     $\mathcal{G} \leftarrow \mathcal{G} \cup \{i\}$
4: **end for**
5: **for** $i = 1$ to k **do**
6:     swap $\mathcal{G}[i]$ with $\mathcal{G}[Random(i, k)]$
7: **end for**
8: **for** each $\mathcal{U}_i \in \mathcal{U}$ **do**
9:     Assign $(\mathcal{U}, \mathcal{G}[d])$
10:    $d = d + 1$
11: **end for**
12: Sort $(\mathcal{U}, \mathcal{G})$ ▷ Sort $\mathcal{U}$ based on random number assigned.
13: **while** $\mathcal{U} \neq \phi$ **do**
14:    $\hat{f} \leftarrow pick(\mathcal{U})$ ▷ Sequentially picks the user from $\mathcal{U}$.
15:    **if** $\succ_i \neq \phi$ **then**
16:       $\hat{r} \leftarrow Select\_best(\succ_i)$
17:       $\mathcal{F} \leftarrow \mathcal{F} \cup (\hat{f}, \hat{r}); \mathcal{U} \leftarrow \mathcal{U} \setminus \hat{f}; \mathcal{N} \leftarrow \mathcal{N} \setminus \hat{r}$
18:       $\succ_i \leftarrow \succ_i \setminus \hat{r}, \forall \mathcal{U}_i \in \mathcal{U}; \hat{f} \leftarrow \phi; \hat{r} \leftarrow \phi$
19:    **else**
20:       $\mathcal{U} \leftarrow \mathcal{U} \setminus \hat{f}; \hat{f} \leftarrow \phi$
21:    **end if**
22: **end while**
23: return $\mathcal{F}$

Running time analysis of TOM-SA

The *random* number generator in lines 2–11 of Algorithm 1 is motivated by [24] and it takes time equal to $O(k)$. The statements in the *while* loop in line $14 - 21$ will execute for k times, as there are k users. In line 12, the sorting takes time equal to $O(k \log k)$. The *while* loop in lines 13–22 is bounded above by $O(k^3)$. So, the overall running time of TOM-SA is $O(k) + O(k \log k) + O(k^3) = O(k^3)$.

Example 1 The detailed functioning of TOM-SA is illustrated with an example in Fig. 2. The number of users that are participating in the system is 4 i.e. $n = 4$ and the number of available NGOs is 3 i.e. $k = 3$. The preference lists of the users present in a set $\mathcal{U}$ is shown in Fig. 2a. Using lines 2–11 of Algorithm 1, the numbers generated randomly for the users and are assigned to them as shown in Fig. 2a. Now, based on the assigned random number, first the user $\mathcal{U}_3$ is selected and assigned NGO $\mathcal{N}_1$ from his reported preference ordering. Similarly, the rest of the users $\mathcal{U}_1, \mathcal{U}_4$, and $\mathcal{U}_2$ are selected in the given order and are assigned the NGOs $\mathcal{N}_2, \mathcal{N}_3$, and none respectively. The final allocation of user-NGO pairs is shown in Fig. 2b.

3.1.2 Proposed Mechanisms for Second Fold: LSR, HR, and CHR

In the second fold the focus will be on how to efficiently store and retrieve cached data to/from the IoT/Edge devices. For this purpose we have proposed the algorithms,

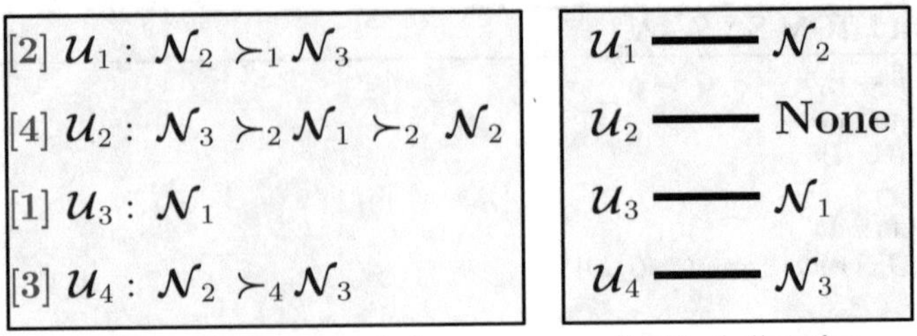

[2] $\mathcal{U}_1 : \mathcal{N}_2 \succ_1 \mathcal{N}_3$	$\mathcal{U}_1 \underline{\quad\quad} \mathcal{N}_2$
[4] $\mathcal{U}_2 : \mathcal{N}_3 \succ_2 \mathcal{N}_1 \succ_2 \mathcal{N}_2$	$\mathcal{U}_2 \underline{\quad\quad}$ None
[1] $\mathcal{U}_3 : \mathcal{N}_1$	$\mathcal{U}_3 \underline{\quad\quad} \mathcal{N}_1$
[3] $\mathcal{U}_4 : \mathcal{N}_2 \succ_4 \mathcal{N}_3$	$\mathcal{U}_4 \underline{\quad\quad} \mathcal{N}_3$
(a) Strict Preference ordering	(b) Final allocation

Fig. 2 Detailed illustration of TOM-SA

namely, *Linear Search based Retrieval* (LSR), *Hashing based Retrieval* (HR), and *Consistent Hashing based Retrieval* (CHR). In the upcoming subsections the algorithms are discussed in detailed manner.

Linear Search based Retrieval (LSR)

In LSR, the probe will be based on a *linear search*, that is, cached data is searched sequentially through each of the IoT devices until the requested information is found. In that case we have to search for all the n IoT devices and for each IoT device $\mathbb{I}_i$ another n_i slots. If data is found, it is made available to the user otherwise not. We depict the main steps in the listing below:

LSR

- User request information (a *file*, a *document*, a *web page*, URL, etc.).
- Sequentially search through the n IoT devices.
- If the requested information is found on any of the available IoT devices then return it to the user.
- Otherwise, display a message that the requested information is not found.

As can be seen from the listing in Algorithm 2, the input to the LSR is the set of IoT devices and the set of users. In line 1, the flag variable is set to 0. In line 2, the requested information is held in r. Using lines 3–11 of Algorithm 2 the requested information is searched.

In lines 3–11, the requested information held in r is searched in ith IoT device and, if found, the information is returned. If the requested information is not found in any of the IoT devices then the sentence at line 13 gets executed and the algorithm ends.

Running time of LSR

In this the running time of LSR is determined. In line 1 of Algorithm 2 the initialization is done in constant time. Likewise, the requested information is stored in r data structure in constant time. Lines 3–11 perform the actual searching algorithm, which take time of $O(n^2)$. Computations at lines 12–14 is bounded above by $O(1)$. So, the overall running time of LSR is $O(n^2)$.

Algorithm 2 LSR $(\mathcal{I}, \mathcal{U})$

1: boolean flag = 0
2: $r \leftarrow$ requested information.
3: **for** $i = 1$ to n **do**
4:     $l = length\ (\mathcal{I}_i)$ $\triangleright \mathcal{I}_i \in \mathcal{I}$ is the ith IoT device.
5:     **for** $j = 1$ to l **do**
6:         **if** $r == \mathcal{I}_i[j]$ **then**
7:             flag=1;
8:         **end if**
9:         **return** $\mathcal{I}_i[j]$
10:     **end for**
11: **end for**
12: **if** flag $== 0$ **then**
13:     requested information not found.
14: **end if**

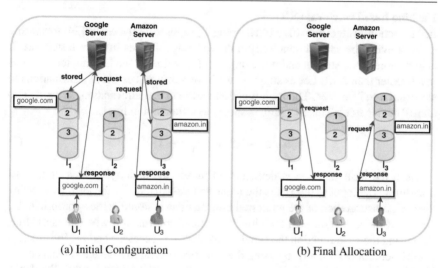

(a) Initial Configuration   (b) Final Allocation

Fig. 3 Detailed Illustration of LSR

Example 2 Let us go thorugh the LSR algorithm with the help of an example. We assume for sake of illustration that the requested information is cached at a Google edge/fog server or Amazon edge/fog server. Following the set-up discussed in Example 1, say users $\mathcal{U}_1$ and $\mathcal{U}_3$ are assigned the spectrum for the time span $\tau_1 = [8, 10)$.

It can be seen in Fig. 3 that during this course of time users $\mathcal{U}_1$ and $\mathcal{U}_3$ are requesting for the web pages google.com and amazon.in respectively. Based on the requests made by the servers, the web pages are fetched from the respective servers and made available to the users. Along with providing the requested web pages to the users, one copy of the web pages is cached to the local cache (or IoT device) as shown in Fig. 3.

Now, assume that the spectrum will not be made available to these users during the rest of the day but they want to access the requested web pages such as google.com and amazon.in at a particular time on that day. If the Internet is not available the requested web pages will be fetched from the local cache (or IoT/Edge device). As it is not known in which particular IoT device the web page is stored, so for determining this the LSR will be used in the running example. Using lines 3–11 of Algorithm 1, the searching will be done in the three IoT devices. In our example first the requested web page google.com is searched in IoT device $\mathcal{I}_1$, it can be seen that the web page google.com is available in the first slot of the $\mathcal{I}_1$ IoT device, so it is returned to the user $\mathcal{U}_1$.

Similarly, for the requested web page of user $\mathcal{U}_3$ the search is made starting from the IoT device $\mathcal{I}_1$. The web page amazon.in is present in the second slot of the $\mathcal{I}_3$ IoT device and is returned.

Hashing based retrieval (HR)

In this section an algorithm based on hashing is proposed motivated by [24] for storing and retrieving the cached data to/from the IoT/Edge devices in more faster way. It is to be noted that, storing and retrieving of information by the hashing technique is much faster than LSR. Let us suppose, we have X key values i.e. X informations to store and n IoT devices. The hash function will map one information, *say* $x \in X$ to an IoT device $h(x)$. This mapping is shown as a function in Eq. 1.

$$h : X \rightarrow \{0, 1, \ldots, n - 1\} \tag{1}$$

Once the hash function is defined,[1] let us see the underlying idea of the HR algorithm. The input to the HR is the requested information by the user. Now, using the hash function each of the requested informations is stored at the appropriate IoT device. If the user requested the already accessed information when he is off-line, then using the hash function the suitable IoT device is accessed and the information is retrieved. More formally, by using the hash function an IoT device is reached and if an information is found it is made available to the user otherwise not. We depict the underlying idea of HR algorithm in the listing below:

HR

- User request an information.
- By using the hash function we go to the IoT device $\mathcal{I}_i$ and if the requested web page is found then return it to the user.
- Otherwise, display a message "an information is not found".

The detailed illustration of HR is presented in Algorithm 3. The hashing based approach is very fast when we can accommodate all the keys (e.g. URLs) to some n IoT devices and n is fixed a priori. Using line 1 of Algorithm 3, the hash value for

---

[1] We can implement the hash function by a classical division or multiplication method. For simplicity, we will use the division method. An implementation could be $h(x) = x(\mod)n$.

an information is calculated. Once the particular IoT device is known based on the hash value, then the capacity of that IoT device is determined. Using lines 4–10, the linear search is done in the selected IoT device and once the information is found, it is returned otherwise not.

Algorithm 3 HR $(\mathcal{U}, \mathcal{I})$

1: $h(x) = x(\mod)n$
2: $l = length(\mathcal{I}_{h(x)})$
3: $h(x) = h(x)(\mod)l$
4: **for** $i = 1$ to n **do**
5:     **for** $j = 1$ to l **do**
6:         **if** $x = \mathcal{I}_{h(x)}[j]$ **then**
7:             **return** $\mathcal{I}_{h(x)}[j]$
8:         **end if**
9:     **end for**
10: **end for**

Example 3 Let us see the HR algorithm with the help of an example. The initial set-up is given in Fig. 4. In this set-up we have 4 users $\mathcal{U}_1, \mathcal{U}_2, \mathcal{U}_3$, and $\mathcal{U}_4$ and 4 IoT devices denoted as $\mathcal{I}_0, \mathcal{I}_1, \mathcal{I}_2$, and $\mathcal{I}_3$. Each of the users requests a web page when they are accessing to the Internet. In the running example, users $\mathcal{U}_1, \mathcal{U}_2, \mathcal{U}_3$, and $\mathcal{U}_4$ requested for the youtube.com, yahoo.in, amazon.in, and google.com respectively. The requested web pages are made available to the respective users and one copy is cached in the IoT devices as shown in Fig. 4.

The web page amazon.in having id as 12 is cached to slot 0 as $h(x) = 12(\mod)4 = 0$, where we are using $x(\mod)n$ as a simple hash function for purposes

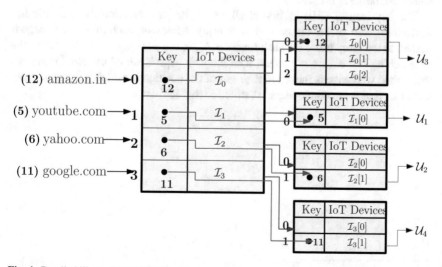

Fig. 4 Detailed illustration of HR algorithm

of illustration.[2] Similarly, the web pages youtube.com, yahoo.in, and google.com having IDs 5, 6, and 11 respectively are stored in slot 1, slot 2, and slot 3 respectively. As the capacity of the IoT device $\mathcal{I}_0$ is 3, so the web page will be hashed to decide at which particular slot in $\mathcal{I}_0$ the web page amazon.in will be stored. So, we get $h(x) = 12(\mod)3 = 0$ and is stored in 0th slot of $\mathcal{I}_0$. Similarly, we can calculate for the other IoT devices. Now, when say $\mathcal{U}_1$ is requesting the web page youtube.com in the offline mode then the requested web page is fetched from $h(x) = 5(\mod)4 = 1$ and then $h(x) = 1(\mod)1 = 0$. More formally, the web page retrieved from the first slot of IoT device $\mathcal{I}_1$. The final configuration is depicted in Fig. 4.

Consistent Hashing based Retrieval (CHR)

Until now, we have discussed the scenario where the number of IoT devices i.e. n is fixed. However, in some applications, a scenario could be, some new IoT devices are dynamically added to the system (on a regular or spontaneous basis). Suppose after the addition of some k number of IoT devices, the number of caches increased to $(n + k)$ instead of n. In such situation, for most of the keys $x \in X$ we have $h(x)(\mod)n \neq h(x)(\mod)(n + k)$. It means that, many cached data may be relocated to other IoT devices when the number of IoT devices are changed (increased/decreased) dynamically. This motivates us to propose the solution using the idea of consistent hashing in our scenario.

In our scenario, there are multiple tertiary users that are requesting the web pages (such as google.com, youtube.com etc.). The requested web pages are fetched from the respective servers and are made available to the users. Along with providing the web pages to the users, the web pages are stored to the IoT/Edge devices for their future use. However, the question that is arising here is, *how the web pages are to be stored and retrieved to/from the IoT/Edge devices?* For this purpose the idea of consistent hashing is used.

The key idea of CHR is, first of all, using the hash function the available IoT devices (or web cache) are hashed to an array. After that, each time a web page is hashed to the array using the hash function. Now, if a web page is hashed to the slot to which an IoT device is present, then the web page is stored therein. Otherwise, the hashed web page is moved to the right of the position to which it is hashed until an IoT device is encountered and is stored to that IoT device.

---

[2]It is to be noted that the IDs of the web pages are generated randomly for the example purpose.

CHR

1 Find the hash value of the IoT device (or web caches) and place it in an array according to the hash value.
2 Find the hash value of the keys (or web pages) and place it in an array according to the hash value.
3 If the position of the IoT device and key is the same, assign the key to that IoT device.
4 Else, assign the key to the IoT device which is closest to the hashed key on its right side.

CHR in detail

In this section the CHR algorithm is discussed in detail. In line 2 of Algorithm 4, the array $\mathcal{A}$ is initialized to empty set ϕ.

Algorithm 4 CHR $(\mathcal{U}, \mathcal{I})$

1: **begin**
2: $\mathcal{A} \leftarrow \phi$
3: **for** each $\mathcal{I}_i \in \mathcal{I}$ **do**
4:     $h(x_i) = x_i (mod) n$ ▷ Calculating the hash value of the IoT device $\mathcal{I}_i \in \mathcal{I}$ with id x_i.
5:     $\mathcal{A}[h(x_i)] = \mathcal{I}_i$
6: **end for**
7: **for** each $w_j \in \mathcal{W}$ **do**
8:     $h(w_j) = w_j (mod) n$ ▷ Calculating the hash value of the web page $w_j \in \mathcal{W}$.
9:     $\mathcal{A}[h(w_j)] = w_j$
10:     **for** each $\mathcal{I}_i \in \mathcal{I}$ **do**
11:         **if** $h(w_j) == h(x_i)$ **then**
12:             Assign the web page hashed to the array index $h(x_i)$ to the IoT device present at $h(x_i)$.
13:         **else**
14:             Assign the web page w_j to the IoT device y_l which is closest to the right side of hashed web page under consideration.
15:         **end if**
16:     **end for**
17: **end for**
18: **end**

Using lines 3–6 of Algorithm 4, the hash values for the IoT devices are calculated and are placed at the suitable index in the array. Using lines 7–17 the hash values of the web pages are determined and are placed at the particular indexes in an array. Further, in line 11 a check is made whether the hash value calculated for the web page w_i is equal to the hash value of any of the IoT devices $\mathcal{I}_i \in \mathcal{I}$. If yes, then assign the web page w_i to the IoT device $\mathcal{I}_i$. Otherwise, use line 14 of Algorithm 4.

Example 4 Let us understand the CHR algorithm with the help of an example. In this set-up we have 4 users $\mathcal{U}_1$, $\mathcal{U}_2$, $\mathcal{U}_3$, and $\mathcal{U}_4$ and 4 IoT devices $\mathcal{I}_0$, $\mathcal{I}_1$, $\mathcal{I}_2$, and $\mathcal{I}_3$. Each of the IoT devices has an ID, say $\mathcal{I}_0$ has id as 46, $\mathcal{I}_1$ has id as 38, $\mathcal{I}_2$ has id as 30, and $\mathcal{I}_3$ has id as 44. Given the ids of the IoT devices, each of the

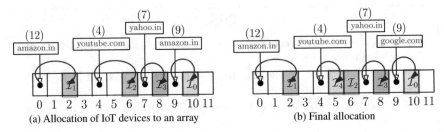

Fig. 5 Detailed illustration of CHR

IoT devices is allocated an index in an array using the hash function $x(\mod)12$. Following this, the IoT devices $\mathcal{I}_0$, $\mathcal{I}_1$, $\mathcal{I}_2$, and $\mathcal{I}_3$ are allocated the index 10, 2, 6, and 8 respectively as shown in Fig. 5a. In the running example, users $\mathcal{U}_1, \mathcal{U}_2, \mathcal{U}_3$, and $\mathcal{U}_4$ requested youtube.com, yahoo.in, amazon.in, and google.com respectively. The web pages amazon.in, youtube.com, yahoo.in, and google.com have ids as 12, 4, 7, and 9 respectively.

Following the hash function $x(\mod)12$, the web page amazon.in is assigned to the index 0 in the array $\mathcal{A}$, as no IoT device is assigned to the 0th index of $\mathcal{A}$, so the web page amazon.in will be shifted to the right of the index currently allocated to it. In such case, the web page amazon.in is assigned to the IoT device $\mathcal{I}_1$ present at index 2. Similarly, the allocation for the web pages youtube.com, yahoo.in, and google.com are done as shown in Fig. 5a.

Now, in the running example, consider that the IoT device $\mathcal{I}_4$ is added into the system with id 41. So applying the hash function $41(\mod)12$, we get that the IoT device $\mathcal{I}_4$ is assigned to the index 5 in an array $\mathcal{A}$. If this is the case, then the allocation made to the web page youtube.com will be changed to $\mathcal{I}_4$ from $\mathcal{I}_2$. This is due to the reason that on moving to the right of the index assigned to the web page youtube.com we get $\mathcal{I}_4$ as the first cache. The final allocation is shown in Fig. 5b.

3.2 Realistic Implementation of Proposed Framework

In our country, the schools in urban areas are complemented by great infrastructure, facilities, and teachers, however, the schools in rural areas lack in terms of primary facilities such as toilets and proper teaching staffs. Due to this, there is a lack of motivation among students and in many cases is a significant cause of poor quality of education in rural schools.

Currently, in our country, some non-profit organizations like *eVidyaloka*,[3] etc. and several government schemes such as *Study Webs of Active Learning for Young*

---

[3]https://www.evidyaloka.org.

Aspiring Minds (Swayam),[4] *Swayam Prabha*,[5] and many more are trying to provide the quality learning to the students living in rural areas by bringing online education to rural classrooms. In *eVidyaloka*, the "each one, teach one" ideology encourages educated professionals (or experts) to take time out of their busy schedules to deliver online lectures to students from rural areas. In this, the participating experts does not charge any fees (mainly *free of cost*) in exchange of their service. The general practice in such existing systems is that, based on their availability, the associated experts (say *volunteers*) delivers the lectures online to the students sitting in the rural areas. However, one of the drawbacks that persists in such system is that the students have access to these online lectures only when they are having access to the Internet. But, what if the students want to revisit the already delivered lectures in near future when they are offline (or the Internet facility is unavailable)? In such case, our proposed framework could be deployed into the existing system of online education. More formally, at the time of availability of Internet the streaming of lectures will be done and also the lectures will be cached to the available *IoT/Edge* devices for future references. If the students want to revisit a particular lecture in future in an offline mode, then the requested lecture will be searched in the available *IoT/Edge* devices (using CHR) and once found will be made available to them. So, the existing online education system could be strengthened if our proposed framework is deployed in such scenarios and also help in increasing the satisfaction level of students in terms of learning.

4 Experiments and Computational Results

In this section, we compare the proposed mechanism with the benchmark mechanism (*random* mechanism) through simulations. The experiments are carried out to provide a simulation based on the preference orderings of the users, that are generated randomly. For the purpose of random generation of preference orderings of users the *Random* library of *Python* is used. In the *random* mechanism, for each user the NGO is randomly picked up from his preference ordering and is allocated to him. The process terminates once all the users are processed.

4.1 Simulation Set-Up

For simulation purposes, we have considered some fixed number of NGOs and fixed number of tertiary users. In this chapter, for the simulation we have considered the scenario that there are equal number of NGOs and tertiary users.

The performance of TOM-SA is measured using the following metrics:

---

[4]https://swayam.gov.in/.

[5]https://www.swayamprabha.gov.in/.

Fig. 6 Comparison of
TOM-SA, TOM-SA-TR,
TOM-SA-HR, and
TOM-SA-TTR

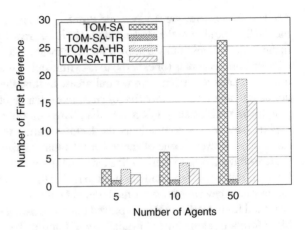

- **Number of first preference**: the total number of tertiary users getting their first preference.
- **Number of the kth preference**: the number of tertiary users getting their kth preference where $k = 1, \ldots, n$.

4.2   Analysis of the Results

In Fig. 6, the number of agents is represented on $x - axis$ represents and the $y - axis$ represents the number of agents getting their first preference. From Fig. 6, we can see that the number of agents getting their first preference in case of TOM-SA is: (a) more than the number of agents getting first preference in case of one-half random (Random-half); (b) more than the number of agents getting first preference in case of two-third random (Random-two-third) and (c) more than the number of agents getting first preference in case of total random (Random-total). With the increase in the number of agents, when using the randomly allocated NGOs, the number of agents getting first preference in the system is decreasing. As this is natural from the construction of the random mechanism.

In Figs. 7a–d, we have compared the TOM-SA, *Total Random Mechanism*, *Half Random Mechanism*, and *Two-third Random Mechanism*. By *Total Random Mechanism* it is meant that the NGO will be allocated to all of the participating users randomly from their preference list. By *Half Random Mechanism* it is meant that first half of the users will be allocated using TOM-SA and the other half is allocated using random mechanism. By *Two-Third Random Mechanism* it is meant that first one-third of the users will be allocated using TOM-SA and the rest two-third of the users is allocated randomly.

From Fig. 7a–d, it is clear that the number of users getting first preference in case of TOM-SA is higher than that in TOM-SA-HR, which in turn is higher than

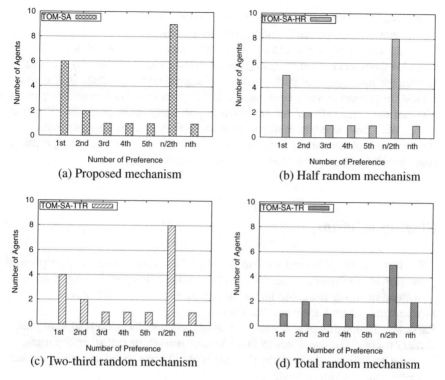

Fig. 7 Comparison of number of users getting the kth preference

TOM-SA-TTR as well as higher than TOM-SA-TR. This is due to the fact that the percentage of agents allocated using TOM-SA is decreasing from TOM-SA to TOM-SA-TR. Similar behaviour can be seen for the remaining number of preferences. So, we can conclude that the TOM-SA results in the best allocation among the available NGOs to users.

5 Conclusion

In this chapter, first we have discussed some recent developments in spectrum allocation in the tertiary market. This problem emerges as a result of the evolution of spectrum allocation from first to tertiary market. The importance of spectrum allocation in the tertiary market stems from the usefulness of under-utilized spectrum in the secondary market. Various emerging scenarios supported by IoT-based spectrum trading are studied in a *strategic* setting of tertiary users (such as small farms, agricultural enterprises or people residing in different localities). The designed algorithms, based on tertiary users' preferences ordering, achieve allocations satisfying truthfulness, Pareto optimality and the core properties.

In our future work we would like to identify, analyze and design some relevant scenarios from smart farming, precision agriculture and more broadly from agro-ecosystems and study the performance of the proposed algorithms in a realistic context. In particular, some core engineering requirements about mobility, periodicity and temporary allocations are interesting to study in the proposed allocation model. We believe that spectrum trading in the tertiary markets opens up interesting opportunities for agro-ecosystems especially in countries or remote areas within a country with limited or non-existent Internet services.

Glossary of Terms

IoT   It is a system of interrelated computing devices, object, animals, or people that are provided with unique identifiers and the ability to transfer data over a network without requiring human-to-human or human-to-computer interaction.

Mechanisms   Also known as Algorithms.

Rational   Rationality of a user means that user chooses the actions so as to gain.

Tertiary users   The users present in the tertiary market are called tertiary users.

Private information   The information that is only known to the user under consideration and not known to others.

Spectrum   It refers to the invisible radio frequencies that wireless signals travel over.

References

1. Moscibroda T, Murty R, Chandra R, Bahl P (2011) SenseLess: a database-driven white spaces network
2. Yi C, Cai J, Zhang G (2015) Online spectrum auction in cognitive radio networks with uncertain activities of primary users. IEEE international conference on communications (ICC), London, UK, 2015
3. Zhang Y, Song L, Pan M, Dawy Z, Han Z (2017) Non-cash auction for spectrum trading in cognitive radio networks: Contract theoretical model with joint adverse selection and moral hazard. IEEE J Sel Areas Commun 35(3):643–653
4. Akyildiz IF, Lee WY, Vuran MC, Mohanty S (2006) Next generation/dynamic spectrum access/cognitive radio wireless networks: A Survey 50(13):2127–2159
5. Han Z, Niyato D, Saad W, Ba sar T, Hjorungnes A (2011) Game theory in wireless and communication networks theory, models, and applications. Cambridge Univ. Press, Cambridge, U.K
6. Niyato D, Hossain E, Han Z (2009) Dynamics of multiple-seller and multiple-buyer spectrum trading in cognitive radio networks: A game-theoretic modeling approach. IEEE Trans Mobile Comput 8(8):1009–1022

7. Xie R, Yu FR, Ji H (2012) Spectrum sharing and resource allocation for energy-efficient hetero-geneous cognitive radio networks with femtocells. In: Proceedings of the IEEE international conference communication (ICC), Ottawa, Canada, June 2012, pp 1661–1665

8. Feng X, Chen Y, Zhang J, Zhang Q, Li B (2012) TAHES: A truthful double auction mechanism for heterogeneous spectrum. IEEE Trans Wireless Commun 11(11):4038–4047

9. Hyder CS, Jeitschko TD, Xiao L (2016) Bid and time strategyproof online spectrum auctions with dynamic user arrival and dynamic spectrum supply. In: 2016 25th international conference on computer communication and networks (ICCCN), Waikoloa, HI, pp1–9, Aug 2016. https://doi.org/10.1109/ICCCN.2016.7568496

10. Yang D, Xue G, Zhang X (2016) Group buying spectrum auctions in cognitive radio networks. IEEE Trans Veh Technol 66(1):810–817

11. Zhou X, Zheng H (2009) Trust: A general framework for truthful double spectrum auctions. In: 28th IEEE international conference on computer communications (INFOCOM), Rio de Jenerio, Brazil, pp 999–1007, April 2009

12. Zhang Q, Jia J, Zhang Q, Liu M (May 2009) Revenue generation for truthful spectrum auction in dynamic spectrum access, In: Proceedings of 10th ACM international symposium on mobile ad hoc networking and computing (MobiHoc). New Orleans, Louisiana, USA, pp 3–12

13. Chowdhury AB, Xhafa F, Rongpipi R, Mukhopadhyay S, Singh VK (2019) Spectrum trading in wireless communication for tertiary market. In: Xhafa F, Barolli L, Gregus M (eds) Advances in intelligent networking and collaborative systems. Lecture Notes on Data Engineering and Communications Technologies, vol 23. Springer, Cham

14. Bandyopadhyay A, Mukhopadhyay S, Ganguly U (2016) Allocating resources in cloud com-puting when users have strict preferences. In: 2016 International conference on advances in computing, communications and informatics (ICACCI), Jaipur, pp 2324–2328, Sept 2016. https://doi.org/10.1109/ICACCI.2016.7732401

15. Roughgarden T (2013) Lecture #9: Beyond Quasi-Linearity, Lectures Notes on Algorithmic Game Theory. October 2013

16. Roughgarden T (2016) Lecture #1: The Draw and College Admissions, Lecture Notes on Incentives in Computer Science. September 26 2016

17. Schummer J, Vohra RK (2017) Mechanism design without money. Algorithmic game theory. Cambridge University Press, New York, pp 243–267

18. Singh VK, Mukhopadhyay S, Das R (2018) Hiring doctors in e-healthcare with zero budget. The 12th International Conference on P2P. Parallel, Grid, Cloud and Internet Computing (3PGCIC). Barcelona, Spain, Springer International Publishing, Cham, Palau Macaya, pp 379–390

19. Singh VK, Mukhopadhyay S, Xhafa F, Sharma A, Roy A (2018) Hiring expert consultants in e-healthcare: an analytics-based two sided matching approach. In: Thanh Nguyen N, Kowalczyk R (eds) Transactions on computational collective intelligence XXX. Lecture Notes in Computer Science, vol 11120. Springer, Cham, pp 178–199

20. Nisan N, Roughgarden T, Tardos E, Vazirani VV (2007) Algorithmic game theory. Cambridge University Press

21. Gale D, Shapley LS (1962) College admissions and the stability of marriage. Am Math Monthly 69(1):9–15

22. Shapley LS, Scarf H (1974) Cores and indivisibility. J Math Econ 1:23–37

23. Singh VK, Mukhopadhyay S, Xhafa F (2019) A mechanism design framework for hiring experts in E-Healthcare. Enterprise Information Systems, Taylor & Francis 00:1–51

24. Cormen TH, Leiserson CE, Rivest RL, Stein C (2009) Introduction to algorithms, 3rd edition. MIT press

25. Roughgarden, T, Valiant G (2017) Lecture #1: Introduction and Consistent Hashing, Lecture Notes on The Modern Algorithmic Toolbox, April 3, 2017

Anil Bikash Chowdhury is an Associate Professor in the Department of Master of Computer Applications at Techno India University, Kolkata, West Bengal, India. He is pursuing his Ph.D. in Engineering in the Department of Computer Science and Engineering of National Institute of Technology (NIT) Durgapur, India. His research interests include Algorithmic Game Theory and its Applications. He can be reached at abchaudhuri007@gmail.com.

Vikash Kumar Singh is an Assistant Professor in the Department of Computer Science and Engineering at Siksha 'O' Anusandhan (Deemed to be University), Bhubaneswar, Odisha, India. He received his Ph.D. in Engineering in 2019 from the Department of Computer Science and Engineering of National Institute of Technology (NIT) Durgapur, India. He has published papers in many reputed conferences and Journals such as IEEE HealthCom, IEEE AINA, LNCS Transactions on Computational Collective Intelligence (Springer), Journal of Ambient Intelligence and Humanized Computing (Springer), and Enterprise Information Systems (Taylor & Francis). His research interests include Mechanism Design (A sub-field of Game Theory), Healthcare, Crowdsourcing, Participatory Sensing, Cloud and), Fog Computing, Massive Data Processing, Intelligent Data), Analysis, Decision Making, Multi-armed Bandits, and Optimization. He can be reached at vikashsingh@soa.ac.in and more information can be found at Scopus Orcid: https://orcid.org/0000-0002-8747-1627.

Fatos Xhafa is Professor of Computer Science at the Technical University of Catalonia (UPC), Barcelona, Spain. He received his PhD in Computer Science in 1998 from the Department of Computer Science of BarcelonaTech. He was a Visiting Professor at University of Surrey (2019/2020), Visiting Professor at Birkbeck College, University of London, UK (2009/2010) and a Research Associate at Drexel University, Philadelphia, USA (2004/2005). Prof. Xhafa has widely published in peer reviewed international journals, conferences/workshops, book chapters and edited books and proceedings in the field (Google h-index 52 / i10-index 270, Scopus h-index 41, ISI-WoS h-index 33). He is awarded teaching and research merits by Spanish Ministry of Science and Education, by IEEE conference and best paper awards. Prof. Xhafa has an extensive editorial and reviewing service for international journals and books from major publishers as a member of Editorial Boards and Guest Editors of Special Issues. He is Editor in Chief of the Elsevier Book Series "Intelligent Data-Centric Systems" and of the Springer Book Series "Lecture Notes in Data Engineering and Communication Technologies". He is a member of IEEE Communications Society, IEEE Systems, Man & Cybernetics Society and Emerging Technical Subcommittee of Internet of Things. His research interests include parallel and distributed algorithms, massive data processing and collective intelligence, IoT and networking, P2P and Cloud-to-thing continuum computing, optimization, security and trustworthy computing, machine learning and data mining, among others. He can be reached at fatos@cs.upc.edu and more information can be found at WEB: http://www.cs.upc.edu/ fatos/ Scopus Orcid: http://orcid.org/0000-0001-6569-5497.

Paul Krause is Professor in Complex Systems at SURREY. He has over forty years' research experience in the study of complex systems in a wide variety of domains, in both industrial and academic research laboratories. Currently his research work focuses on distributed systems for the Digital Ecosystem and Future Internet domains. He has over 120 publications and is author of a textbook on reasoning under uncertainty. He is leader of the recently founded Digital Ecosystems research group at Surrey which, although based in the Department of Computer Science, collaborates strongly with other disciplines throughout the University. He has been working in and leading strong interdisciplinary teams since 2006 in the EU funded DBE and OPAALS projects, more recently in the RCUK funded projects ERIE (for Evolution and Resilience of Industrial Ecosystems) and MILES projects. These last two were funded under the Complexity in the Real World, and Bridging the Gaps, programmes respectively. He is currently active in the above mentioned TASCC, SPEAR and HBP projects. He also has forty years' experience as a volunteer in practical nature conservation projects. He is a Fellow of the Institute of Mathematics and its Applications, and a Chartered Mathematician.

Sajal Mukhopadhyay is an Associate Professor in the Department of Computer Science and Engineering at NIT Durgapur, India. He received his Ph.D. in Engineering in 2013 from the Department of Computer Science and Engineering of NIT Durgapur, India. He has published papers in many reputed conferences and Journals such as IEEE HealthCom, IEEE AINA, LNCS Transactions on Computational Collective Intelligence (Springer), Journal of Ambient Intelligence and Humanized Computing (Springer), and Enterprise Information Systems (Taylor & Francis). He has received the Young Faculty Research Fellowship under Visvesvaraya Ph.D. Scheme and the best paper awards. His research interests include Algorithmic Game Theory and its Applications to Healthcare, Crowdsourcing, Participatory Sensing, Cloud Computing, Spectrum Trading, Multi-Armed Bandits, IoT, and Bipartite Matching. He can be reached at sajal@cse.nitdgp.ac.in and more information can be found on the WEB: https://www.nitdgp.ac.in/faculty/0db34405-8fab-4320-9f1d-f084b362f6e5.

Meghana M Dhananjaya is a M.Tech student in the Department of Computer Science and Engineering (CSE) at NIT Durgapur, India since 2019. She has work experience of two years as senior system engineer in Infosys and was part of the projects which were under the domains Networking Media and Block-chain. She is broadly interested in algorithmic mechanism design and its applications (game theory with incomplete information).

AgriEdge: Edge Intelligent 5G Narrow Band Internet of Drone Things for Agriculture 4.0

Aakashjit Bhattacharya and Debashis De

Abstract The previous three industrializations had a major impact on the agricultural sector. There was a transformation from indigenous farming to mechanized farming. With the advent of Industry 4.0, the fourth generation of the Agriculture revolution has taken place with the introduction of precision farming, and the backbone of this revolution is mainly the Internet of Things. Edge Intelligence in combination with Agriculture Big-Data Analysis and modern evolving communication technologies like 5G Narrow Band (NB) IoT is going to play a significant role in the progress of Agriculture 4.0 through real-time applications as a simpler waveform like NB-IoT consumes less power, have high reliability, have wider deployment capability and have lower latency. Energy-efficient and Real-time application of technologies in the agricultural sector is the need of the hour for a smart green industrial revolution through Agriculture 4.0. The induction of Multilingual Voice User Interface control enabled drones in combination with dynamic sensors in 5G era, will prove to be a game-changer in the agricultural sector. Edge Intelligence enabled drones will introduce the concept of ubiquitous computing in the agricultural domain, will be quite beneficial for the farmers as it will provide real-time alerting, monitoring systems in addition to real-time precision farming.

Keywords 5G · NB-IoDT · Agriculture 4.0 · Precision Agriculture · Contactless Multilingual Voice User Interface · Edge Intelligent Drone

A. Bhattacharya
Advanced Technology Development Centre, IIT Kharagpur, Kharagpur 721302, West Bengal, India
e-mail: aakashjit.bhattacharya021@kgpian.iitkgp.ac.in

D. De (✉)
Centre of Mobile Cloud Computing, Department of Computer Science and Engineering, Maulana Abul Kalam Azad University of Technology, Haringhata, Nadia 741249, West Bengal, India
e-mail: dr.debashis.de@gmail.com

© Springer Nature Switzerland AG 2021
P. Krause and F. Xhafa (eds.), *IoT-based Intelligent Modelling for Environmental and Ecological Engineering*, Lecture Notes on Data Engineering and Communications Technologies 67, https://doi.org/10.1007/978-3-030-71172-6_3

Acronyms

3GPP	3Rd Generation Partnership Project
4G	4Th Generation mobile network
5G	5Th Generation mobile network
BSNL	Bharat Sanchar Nigam Limited
CoAP	Constrained Application Protocol
CS	Compressive sensing
DOS	Denial of Service
DTLS	Datagram Transport Layer Security
EU	European Union
GPS	Global Positioning System
GSM	Global System for Mobile Communications
IoT	Internet of Things
IoDT	Internet of Drone Things
LAN	Local Area Network
LDR	Light Dependent Resistor
LED	Light Emitting Diode
LTE	Long-Term Evolution
M2M	Machine-To-Machine
MQTT	Message Queue Telemetry Transport
NB	Narrow Band
NB-IoT	Narrow Band IoT
Node MCU	Node Micro Controller Unit
PAN	Personal Area Network
RAT	Radio Access Technology
TCP	Transmission Control Protocol
UAC	User Access Control
UAV	Unmanned Aerial Vehicle
UDP	User Datagram Protocol
URL	Uniform Resource Locator
WAN	Wide Area Network
WEP	Wired Equivalent Privacy
Wi-Fi	Wireless Fidelity
WPA	Wi-Fi Protected Access
WPA 2	2nd Generation WPA

1   Introduction

Agriculture is the source of livelihood for many countries and it plays a significant role in the economy of many countries. Other than growing crops, it is also associated with animal breeding, land cultivation. Agriculture is not only associated with food but is also associated with the fiber and medicine sector. Agriculture is a vibrant area for the nourishment of mankind. It is of significance to innovate techniques that help in the subsistence of agriculture. Blockchain based IoT can ameliorate the food chain.

Advancement in the field of agriculture in some countries took after an agriculture revolution took place in those countries. Like after the Green Revolution that took place in India in the earlier 1960s, led to an increase in higher-yielding varieties of seeds due to improved technologies in the agriculture sector. Similarly, after Golden Fiber Revolution that took place in India, jute was started to be used as a raw material in the fabric industry, and even today jute is used as a raw material in making strong threads and jute products. 5G is an evolving technology and is going to play a vital role in the field of Internet of Things (IoT) [1], Industrial IoT, Industry 4.0 [2], Precision farming [3], and Agriculture 4.0 as shown in Fig. 1 to provide real-time updates. Agriculture as a whole has been influenced by centuries of continuous cultivation of exotic plants in various regions across the world due to various historical events and human interventions [4].

Skylo's partnership with Bharat Sanchar Nigam Limited (or BSNL) to launch World's First Satellite-Based NB-IoT in India, is going to play a very vital role in the advancement of the implementation of the Digital India program in India. Through this partnership, program machines will be turned into always-connected smart objects. Through this NB-IoT project, millions of sensors and machines will be connected from space. This will help the business owners to make smart timely decisions as they will be able to understand, manage, and predict what is going to happen. To date, Skylo is integrating the system with trucks, tractors, trains, and buses. After the system is commercialized and available to the local public, this system can be integrated with Drones for smart agriculture, smart surveillance system, etc. In Skylo

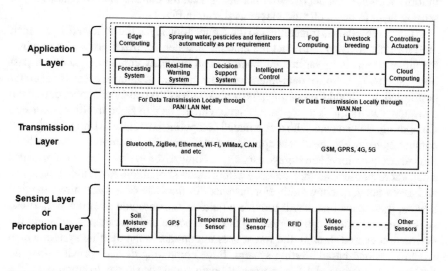

Fig. 1 Layers of internet of drone things in the application of smart agriculture

there are sensors fitted that transmit data to Skylo Satellite Network, and from there it sends data to the destined group of people. The Skylo platform provides cross-device support to visualize the collected data and based on the data received the end-user is given the ability to take a quick appropriate on-spot decision. From an economic point of view, Skylo provides the world's cheapest satellite-based solutions to connect sensors with machines [5].

The architecture of the Internet of Things can be divided into 3 layers mainly the Sensing layer, the Transport layer, and the Application layer. The Sensing Layer is responsible to collect real-time data from the surrounding through sensors, and with the help of the transport layer, the processed or raw data is forwarded to the Application layer. Compressive sensing (CS) based IoT has also a 3-layered architecture, but there is a slight difference in the second layer. Layer 1 is called the CS-Based Acquisition layer, Layer 2 is called the CS-Based Reconstruction Layer. In this layer, after the data is collected, the raw data gets filtered out and only the processed and important data are transmitted to reduce the bandwidth consumption. The next layer is Layer 3, and just like any other IoT architecture, this layer also deals with the application of the collected data through actuators and devices. Any project related to the Internet of Things domain means there is somewhere a hardware-software interaction within the system and with every passing day, the wireless sensor networks are getting famous for their simple form factor and lower cost of design. By the application of Agricultural Big Data, on Agriculture, there will be optimization in the agricultural economy, that will continuously promote to achieve sustainable optimization of industrial development. With the advent of advanced communication technologies like 5G, the combination of the internet of things, big data, artificial intelligence, mobile internet, and cloud computing is getting closer and closer. 5G will help in the rapid growth of smart agriculture. Various means of communications in form of PAN/LAN net or WAN net being used for transmission of data from the sensing layer to the application layer as shown in Fig. 1.

After the live data is transmitted to the Application layer, the collected data is then used for data analysis to predict the amount of water, fertilizer, pesticides required for different plants in various temperature and humidity conditions based on past observations. This type of data-intensive and computation-intensive task is generally performed in a Cloud Computing platform. Whereas the real-time application of the collected data for quicker response is done with the help of Edge Intelligence or Fog Computing. Edge means with a distributed computing paradigm to bring storage and computation close to the area of application. Although Edge computing devices have limited computational and storage capacity in comparison to cloud servers, it provides a quicker response to address an issue within a very short duration of time or instantly and saves network bandwidth. For instance, in the case of a forest fire detection system, these edge nodes will provide a quicker response and alert the authorities, so that the authorities can provide a quicker response and thus help in reducing the destruction caused through the forest fires in contrast to the alerting system which would be Cloud-based alerting system. Edge computing devices provide ubiquity thus helps to provide real-time computation on any device at any location.

With the evolution of 5G communication technology, drone-based smart precision agriculture based on the Internet of Drone Things (IoDT) [6] will get enhanced due to lower network latency and can provide Edge Intelligence enabled solution to Agriculture 4.0. With the help of Edge Intelligence and a ubiquitous network of drones, it will smartly, ubiquitously, and autonomously provide the right amount of water, fertilizer, and pesticides through image processing and other dynamic sensors as shown in Fig. 2 based on forecasting concept and agriculture big data analysis. Some of the parameters which will be taken into considerations are atmospheric humidity, temperature, crop condition soil type moisture level in the soil. It will also provide smart real-time farm monitoring especially during the night time to provide

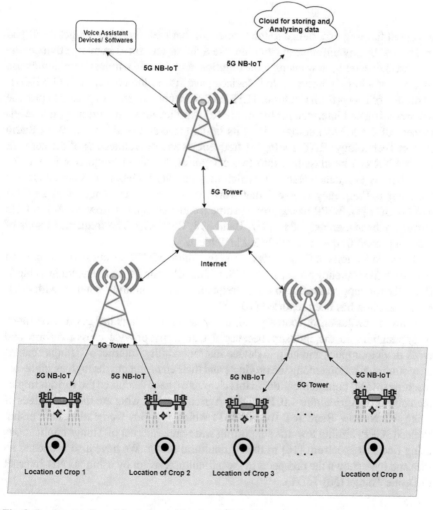

Fig. 2 Implementation of Agriculture 4.0 using 5G NB-IoDT

Table 1 5G NR Frequency bands and their Frequency Ranges [9]

Frequency Bands	Frequency Range in MHz
Band n77	3300–4200
Band n78	3300–3800
Band n79	4400–5000
LTE Band 46	5150–5925

Table 2 Countries and their Frequency Bands [9]

Countries	Frequency Range in MHz
US	3550–4200
China, EU, Japan	3600–4200 and 4400–4900

a secured farming experience to the farmers so that their crops do not get destroyed or affected by any unforeseen situation like a fire or unwanted human activities.

Ultra-low latency, low energy consumption, and reliable wireless communication support for a huge number of IoT devices gave rise to the concept of 5G NB-IoT. Evolution of Long-Term Evolution (LTE) and 5G technologies is expected to provide a newer advanced interface for the future IoT Applications, but currently, the development of the 5G technologies is in its initial stage and is aiming at using Radio Access Technology (RAT) for higher frequencies and re-architecting of the network [7]. 5G NR can be classified into two categories, based on Frequency Range. At first, there is Frequency Range 1, which includes sub-6 GHz frequency bands and the other is Frequency Range 2 that works in the frequency range of 24.25 GHz to 52.6 GHz [8]. 3GPP has approved some frequency bands below the Sub-6 GHz frequency bands, namely, Bandn77, Band n78, Band n79. The frequency bands of these approved frequencies are listed in Table 1.

These collections of frequencies range the future 5G NR spectrum that is catered from certain frequency ranges in the US, China, EU, and Japan as shown in Table 2. Recently for supporting Sub-6 GHz Frequency bands, an array antenna with dual-band operation has been reported [10–15].

With the evolution of Industry 4.0, the Internet of Things has got a wide range of applications in the industry because it can easily provide cross-platform and cross-device support, and help in device interoperability. Internet of Things enables easy communication among the products and their environment through machine-to-machine (M2M) communication. In this chapter we have discussed the various implementations of Agriculture 4.0, Precision Agriculture and what are the advantages of using 5G Narrow Band IoT (or NB-IoT) which is a 3rd Generation Partnership Project(3GPP) aiming towards supporting wide-area Internet of Things applications using licensed spectrum [16] in the Agricultural sector. We have also proposed an idea and have shown the prospects in the agricultural sector by using 5G NB-Internet of Drone Things (NB-IoDT).

NB-IoT can be categorized into several bands based on the frequency of its uplink and downlink. The main advantage of using NB-IoT is that it provides better connectivity to a massive number of devices along with longer battery life, low complexity, and low costing. Actuators or devices can be easily operated using various Bot APIs available like Telegram Bot, Line Bot, WhatsApp Bot, or Messaging Apps or websites in a real-time environment. Some of these bots are paid and some are available free of cost.

In the IoT domain, the two most commonly used fundamental protocols are CoAP and MQTT. These protocols are best suited for low energy consumption devices. These protocols have little overhead and huge extensibility. As both the CoAP and MQTT protocols are Internet protocols they both have transport layer security [17]. There is always a hardware-software interaction in the Internet of Things domain. MQTT protocol works in the Publisher-Subscriber model. In simple terms, it means that the publisher publishes content at a regular or a given interval of time. And the subscribers who are subscribed to that publisher will be able to receive the updates through a broker. CoAP (or Constrained Application Protocol) design helps simple constrained devices to join IoT through a constrained network that is having low bandwidth and low availability. It supports asynchronous message exchange with low overhead. Very simple syntactic analysis can be done using the CoAP protocol. It has proxy and caching capabilities also. CoAP protocols work where TCP based protocols like MQTT fails to perform. It supports networks having billions of nodes. The DTLS parameters used for security purpose in CoAP is equivalent to 128-bit RSA keys. CoAP is by default bounded to UDP (or User Datagram Protocol) and optionally to DTLS.

In recent years drones have got a wide range of applications across multiple domains and fields [18–24]. Sometimes it is claimed that the history of the first drone dates back to 1849 when the Australians launched around 200 pilot-less balloons against Venice [25, 26]. Again in [26, 27] it has been told that Unmanned Aerial Systems can be traced back to 1896 when Samuel Langley's unnamed drone flew over the Potomac River. It has got several applications in the field of military missions [25, 28], precision farming [29, 30], crop monitoring [31], disaster management [32], etc. The application of drones in these fields is comparatively quite cheaper in comparison to manned aerial vehicles [33]. With the introduction of drones in the agricultural domain, farmers need not have to get exposed to harmful pesticides, as pesticides and fertilizers get sprayed over the crops with the help of these drones. Again, these drones can be used for aerial surveying of the crops and detect un-towards incidents occurring in the field or farm, like forest fire detection. Thus, the overall introduction of drones in the field of agriculture has revolutionized the advancement of Agriculture 4.0.

2 Literature Review

Diverse types of sensors like temperature, humidity sensor, soil moisture sensor, water level sensor, LDR sensor, and various other actuators like LED, submersible water pump, along with an ESP8266 Wi-Fi module to transmit data to the ThingSpeak cloud platform has been used [34]. They have designed an automated system to maintain the moisture level in the soil. There are many disadvantages to the system, like when the main motive is just to maintain the level of moisture of the soil, they have unnecessarily used the DHT11 sensor. Readings taken from DHT 11 sensor do not have any contribution to the system, other than just collecting atmospheric humidity data unlike the LDR sensor which is used to sense the intensity of light in the surrounding and turn on and off the LED accordingly to provide an ample amount of light to the plants. In addition to this, their system is working to read available moisture in the soil from the soil moisture sensor and water the plants accordingly. They have used Way2SMS for notification purposes, which provides only 2 free SMS per day [35], which hinders the path of free 2-way communication between the user and the system. Overall, this system does not seem to be bandwidth-efficient, because they are at first sending data to the cloud and again fetching that data back to a local Web Server. So, their system is consuming a lot of network bandwidth. This 2-way journey of the data from the device to the Cloud and again back to the local Web Server is making it more and more time consuming and as a result of that, this system is going to send an alert notification to the end-user very late. Although they have said that by storing the data locally, they use the data for data analytics purposes, but they can export the dataset from the ThingSpeak Cloud platform for further research purposes and there is no other added requirement for this local web server. Using an additional web server is making the system less power efficient. If they wanted to store data locally, they could have parallelly transmitted the data to both Cloud and the local web server thus reducing the utilization of the network bandwidth. They have used Wi-Fi as a means of communication with the internet, but Wi-Fi connectivity will fail to provide a higher latency unlike 5G NB-IoT enabled devices.

Arduino Micro-controller along with ESP8266 Wi-Fi module for controlling the actuators and for communication purpose respectively [36]. They have a water flow sensor, soil moisture sensor, and temperature (ds18b20) sensor. Temperature (ds18b20) sensor can sense temperature from $-55\,°C$ to $+125\,°C$ [37]. This system can automatically toggle the state of their water pump and sprinklers from Power on to Power off state and vice-versa based on the level of moisture in the soil and temperature and can also be operated in manual mode through a website. The main disadvantage of this system is that multiple crop types are not supported in a single system. If the farmer plans to sow seeds of a different crop type, then the system will have to be reprogrammed with the new moisture level of the new crop variant. This system requires an active Wi-Fi connection for communication purposes. So, in case the Wi-Fi router gets compromised, hackers or cybercriminals can cause malfunctioning of the devices. They have also not implemented a simple authentication before sending any command to the device, so any unauthorized user can easily operate the system if they get the website URL or IP Address. Proper UAC could have been implemented to increase the security of the system and prevent unauthorized access to the system.

Smart Farm system in [38] has been divided into three components—hardware, web application, mobile application. The use of the hardware components comprising of DHT22, 1 channel relay module, Solenoid, Node MCU is used to control the actuators and collect the live data of the crops. The second component is the web application which has been used to visualize the collected live data from the IoT devices from various villages. This interface is used by an admin level user to manipulate the conditions of water needed for each of the crops. The real-time data that is collected will be used to monitor the level of moisture in the soil required for a particular crop and in the future, these collected data will be used to predict the amount of water that will be required for a particular variety of crops. The third component, that is the mobile application will be used by the farmers to operate the actuators in both manual and automatic mode of operation. In this project, they have used Node MCU for the transmission of the data to the server. They have used the Apriori algorithm available in Weka [39] to extract the association rules from the collected live data and have used linear regression to model the relationship between several inputs and outcome variables. The input variables include temperature, humidity, and soil moisture. The model is shown in Eq. (1):

$$y = \beta_0 + \beta_1 x_1 + \beta_2 x_2 + \beta_3 x_3 + \ldots + \beta_p x_p + \in, \tag{1}$$

where y is the outcome of the variable. $B_n x_n$ represents the product of the change in y with a unit change in x_n for n = 1 to p-1.

In this system, they have used the Line Application for the notification purpose. But the main problem with this application is that the Line API is not available to free of cost, and there is a subscription charge, so this could be replaced with Telegram Bot as it's API is available free of cost and can address 30 requests per second. Nothing has been mentioned about the proper UAC of the system, which can make illegal users access the system.

UAV based mapping system for precision agriculture has been implemented using fog computing [40]. Both static and dynamic sensor deployment has been done to collect data. The data collected by UAVs are transmitted to a broker which sends the data to the other brokers to execute scheduled policies based on the analysis. Fog computing has been implemented for delay-sensitive applications, which has helped them to get early warnings and provided them data-driven monitoring system. This proposed work is based on INET5 and OMNet++4 which is the extension of FogNetSim++ [41]. The fog devices are tracked by a broker with the help of data collected from the fog nodes. This proposed framework provides a complete farming ecosystem inclusive of fog locations, UAVs, and sensors.

A new architecture of cloud-based autonomic information system named AgriInfo for agriculture domain [42]. This system has been designed to collect information from pre-configured IoT/Edge devices at the user level and process it based on certain fuzzy rules and store the processed data in the cloud for future reference. These fuzzy rules are kept on updating based on future research and developments. This system collects mainly nine types of data, which are—weather, crop, soil, irrigation, fertilizer, productivity, equipment, and cattle. This system can automatically allocate resources at the infrastructure level after identifying the QoS requirement request of the user. In other words, AgriInfo has been designed to provide Agriculture-as-a-Service (AaaS) through web and mobile-based applications. AgriInfo has been broadly divided into two sub-systems:—User Subsystem and Cloud Subsystem. The User Subsystem provides an interface for the end-user to interact with the system. The users are classified into three categories- agriculture expert, agriculture officer, and farmer. Agriculture experts answer user queries and update the fuzzy rules. The agriculture officers are the government officials who provide the latest information regarding the new schemes, policies, and rules that are passed by the government. In this system, farmers are important entities, they take the advantage of the system by asking queries and getting automatic replies after analysis. At first, the users will have to do the registration to the system, and after that when they login into the system, the homepage of the user will be displayed. The users can monitor any data related to their domain and get benefitted from them without visiting the agriculture help center. All the queries asked by the users are updated on the database and after analysis, an automated reply is given to the user via their pre-configured devices. The main advantage of this system is that no technical expertise is required to use this system. This application has been developed on CloudSim [43] to validate the proposed system through real-time mobile and web applications. Table 3 gives a brief comparison of the surveyed papers.

Table 3 Comparison Table of Agriculture state of the art models

	Proposed model	[34]	[36]	[38]	[40]	[42]
GUI Based	✓	X	✓	✓	N.A	✓
Voice Interface	✓	X	X	X	X	X
Drones Used	✓	X	X	X	✓	X
Static Sensor	X	✓	✓	✓	✓	✓
Dynamic Sensor	✓	X	X	X	✓	X
Communication medium	5G NB-IoT	Wi-Fi	Wi-Fi	Wi-Fi	Wi-Fi, GSM Module, and short-range communication	Through JADE Agent
Contactless interaction with the device	✓	X	X	X	X	X

3 Advantage of Dynamic Sensors Over Static Sensors in Agriculture 4.0

In some of the smart irrigation and precision farming systems, static soil moisture sensors along with the water controllers are kept scattered in the field. This estimates the amount of water that is to be sprinkled on the field quite inaccurate.

As shown in Fig. 3, we have pictorially represented farmland in form of a square and divided the square into sub-squares representing multiple zones based on the level of moisture in the soil to demonstrate some real-world challenging scenarios in the case of smart agriculture. The yellow-colored square boxes represent the areas that are having less moisture content and the blue-colored square boxes represent areas that are having excess moisture content and the green-colored box represents an area with an optimal amount of soil moisture level. We have highlighted some scenarios to demonstrate the disadvantages of the existing systems.

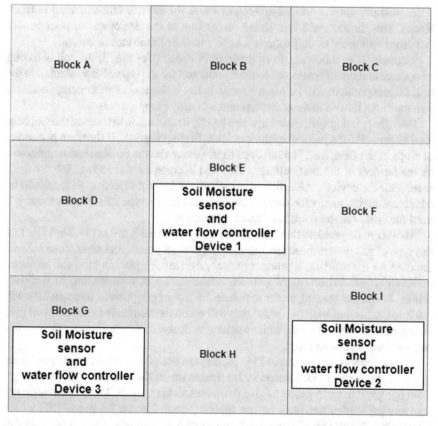

Fig. 3 Demonstrates the various scenarios that can affect the crops from existing smart irrigation systems

Scenario 1: When Device 1 in Block E, locally detects the soil moisture level in the soil, it finds that there is very less moisture content in the soil, so it starts pumping water in Block E. Now, Block B, Block C, Block D, Block H, Block I are already having a huge amount of moisture in the soil but they are adjacent to Block E. So, when the water pump in Block E starts pumping water, from the general property of water, it will flow to the adjacent areas as water does not get soaked in some types of soil quickly, which means that the adjacent block of lands which are already having high moisture level will get this excess amount of water. Now, this can adversely affect the crops, because with an excess amount of water in the soil, crops like wheat which does not require much water will get destroyed, and even if they do not get destroyed, we will get poor quality wheat, from which the farmers will not be able to earn much profit.

Scenario 2: Here we will assume that we are planting rice crops that require a huge amount of water. Suppose the device is installed in Block B or Block C or Block D or Block H or Block I, where there is an excess amount of water. Then this device will not pump in water. Now the blocks A, E, F which are having a low amount of water, will not get the desired amount of water for the rice crops planted in those blocks. This, in turn, will lead to the destruction of the rice crops, or poor-quality rice crops will grow as they require a huge amount of moisture in the soil.

Scenario 3: In this scenario as depicted in Block G of Fig. 3, the soil is having an optimal amount of moisture required in the soil for the type of crop planted in the soil. So, the problem will be when a scenario like a Scenario 1 will occur, excessive water will get flowed to these regions and will affect crop growth.

Thus, the existing systems are giving a locally best solution in terms of the moisture level of the soil for a particular block of land. In simple words, if there are N number of crops in the field, then N number of these devices has to be used and maintained, by the farmers or the horticulturist. This will become a quite tedious job for them to monitor the devices. Also, this solution will not be cost-effective, as there will be additional maintenance for these N devices. So, in this type of scenario, there is a need for drones in the domain of smart agriculture.

Another main problem that occurs with the systems designed in [34, 36, 38] is that this type of soil moisture sensors has metallic contact points, and when these sensors are used for a considerable amount of time, the metallic plates of the soil moisture sensors gets eroded and they give inaccurate readings, thus leading to incorrect results. People, in general, prefer to reduce the cost of peripherals, so when they will use low-quality soil moisture sensors, it will erode the terminals easily and will give wrong results. Several times while working with soil moisture sensors, these types of problems have been faced.

All the models mentioned in [34, 36, 38] are not giving fertilizers or pesticides to the crops, instead, the farmers or horticulturists or the gardener will have to go from one plant to other plants to give pesticides and fertilizers. Many farmers do not take any preventive measures before applying pesticides on the plants [57] which affects their health [58–62]. So, applications of these models of smart farming can prove harmful to the farmers. Another problem with the systems mentioned in [34, 36, 38] is that the sensors are fixed at a given location, so the sensors might fail when there is no power supply. In [40] they have used both static and dynamic sensors, and that makes maintenance of sensors much easier and less hectic in comparison to farmlands where all the deployed sensors are static. In [42], the proposed architecture AgriInfo is only able to process English and do not have multilingual support. So it will be difficult for the farmers who talk in local languages to communicate with the devices. Although handheld devices ease the access to a particular system, AgriInfo does not have the option for handheld devices.

4 Security Issues in Agriculture 4.0

All the models mentioned in [34, 36, 38] are using Wi-Fi for communication purposes. For encryption purposes, Wi-Fi uses popular protocols like WEP, WPA, and most recently it uses WPA2 for encryption purposes. Wi-Fi, which is a wireless connection possesses various cybersecurity threats [44–53]. Several tools are available in the market to hack into Wi-Fi networks. The most common and advanced tool in Kali Linux [54–56]. So, if anyone hacks into the Wi-Fi router, then the hacker can easily compromise all the systems that are connected to the Wi-Fi. If the hacker is having some ill intentions, he/she can easily make the device malfunction without the knowledge of the farmer and in turn will destroy the crops. These systems have not implemented proper User Access Controls, so any unauthorized user can access the system with ease in case they get access to the Mobile App or the website URL.

5 Advantage of Using 5G NB-IoDT Over Wi-Fi and Wi-Fi 6 in Agriculture 4.0

5.1 Comparison of Download Speed Among Wi-Fi, 4G, and 5G Network

In the latest analysis by OpenSignal it has been found that in the USA, UK, Spain, Australia, Kuwait, Switzerland, South Korea, Saudi Arabia, 5G download speed is better than Wi-Fi connection. It has been forecasted that cellular technology will improve faster than fixed networks and Wi-Fi connections [63]. Data collected between January 22- April 21, 2020, is shown in Table 4.

From Fig. 4, we can graphically see that the average download speed of 5G is more than that of 4G and Wi-Fi connectivity. So, the initiatives to implement smart agriculture systems using 5G Narrow Band Internet of Drone Things is of utmost importance to provide a smart, ubiquitous, latency-free communication between the end-user and smart agriculture devices.

Table 4 Average download speed in Mbps across Wi-Fi, 4G, and 5G connections [63]

Countries	Wi-Fi Download Speed	4G Download Speed	5G Download Speed
USA	59.8	27.7	52.3
UK	34.1	24.9	138.1
Spain	47.0	28.8	146.8
Australia	25.6	44.1	163.9
Kuwait	26.7	16.7	185.1
Switzerland	73.9	45.0	201.9
South Korea	74.5	53.7	224.0
Saudi Arabia	21.4	24.4	291.2

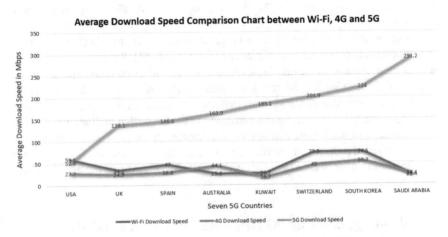

Fig. 4 Graphical representation of Average download speed across seven 5G countries

5.2 Advantage of 5G NB-IoDT Over Wi-Fi and Wi-Fi 6 from Cyber Security Perspective

Although we know that Wi-Fi 6 [64] is also one of the evolving technologies with 5G networks with a significantly higher speed of operation in comparison to the current Wi-Fi connectivity, it can have certain aspects from the cybersecurity point of view. Wi-Fi 6 is the future generation implementation of Wi-Fi connectivity that follows IEEE 802.11ax [65–71] standards and works in the frequency range of 5925–7125 MHz. But the main disadvantage of using Wi-Fi 6 is that if the router gets compromised then it will not take much time for the hacker to hack into all smart irrigation systems that are connected to that particular Wi-Fi 6 router and make the devices malfunction. Popular cyber-attacks like flooding DOS [72, 73] (or Denial-of-Service) attacks on these connected smart irrigation systems can be done through flooding requests through the compromised Wi-Fi or Wi-Fi 6 router as shown in Fig. 5.

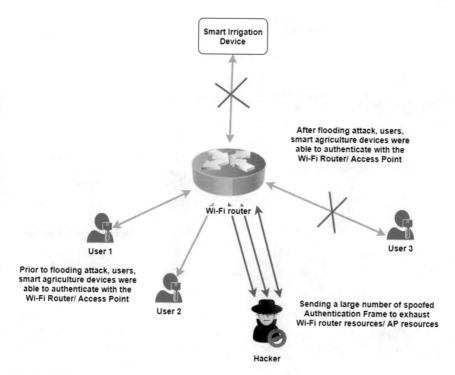

Fig. 5 DOS attack on the Wi-Fi router using flooding attack

This will not only make the smart irrigation devices inaccessible by the farmers but can also lead to quick drainage of the batteries that are fitted on those smart irrigation devices through the flooding of illegal requests sent by the hackers. Remote Code Execution attack can be performed on a Wi-Fi router [74–76] to which multiple smart irrigation devices are connected. Once the attack is successful, through the Wi-Fi router, the attacker can perform Man-In-The-Middle Attack [77] and can either make the smart irrigation devices malfunction by modifying the requests sent by the user to the smart devices or modify the live data that is being sent to the end-users from the smart irrigation device as shown in Fig. 6.

This is possible because when an IoT device gets connected to a Wi-Fi router it's IP Address, and MAC Address is stored in the Wi-Fi router, and by just hacking into a single router [77–79], a hacker can easily get details of all the connected IoT devices and can easily plant those malicious codes in them without the idea of the farm owner or horticulturist. By malfunctioning of these IoT devices, the attacker can hinder the growth of good quality crops. So, these attacks can be minimized by individually connecting each smart irrigation device to the internet through independent GSM modules connected to each smart irrigation device. So even if the hacker hacks into a single device, the chances of hacking into another adjacent device are negligible. So, although Wi-Fi 6 is having better connectivity in comparison to the current Wi-Fi

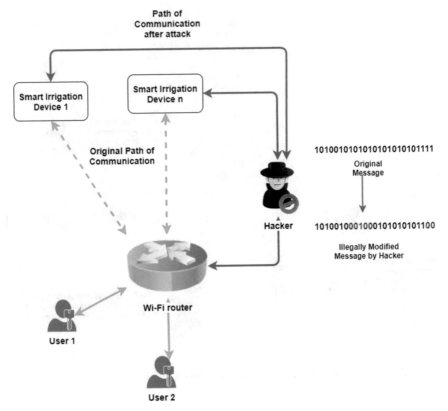

Fig. 6 Man-In-The-Middle Attack on Smart Agriculture System after compromising Wi-Fi router

connections, using it can be quite challenging from the security point of view. So, there is a need for 5G NB IoT connectivity of individual devices.

5.3 Contactless Voice User Interface for 5G NB-IoDT

There are many popular voice assistants available in the market like Apple Siri, Google Assistant, Microsoft Cortana, Amazon Alexa [98]. The help of various other smart home devices like Google Nest, Google Nest Mini, Amazon Echo to name a few. We can combine the power of both 5G NB-IoDT along with that of the voice assistants to make a much faster and cheaper approach in controlling drones for precision agriculture, which means giving the right amount of fertilizer, water, pesticides, etc. at the right time and at the right plant or crop in right quantity based on several parameters like temperature, soil moisture, humidity, etc. through several sensors using various computing models [80–92] as shown in Fig. 7. Precision Agriculture provides better quality of crops [93–97].

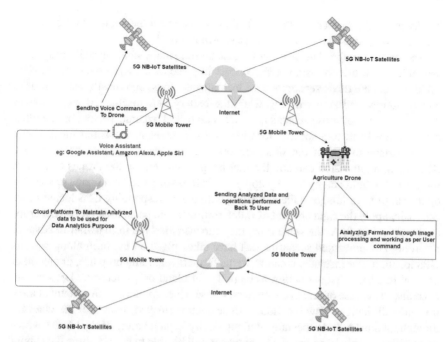

Fig. 7 Block diagram of contactless Voice User Interface enabled Agriculture 4.0 through 5G NB-IoDT

APIs and SDKs of these popular assistants like Apple's Siri, Amazon's Alexa, Microsoft's Cortana, Google's Assistant [98] are available, which can ease the process of controlling devices with voice commands. Using these popular services offered by Amazon, Google, and Microsoft, or any other organization, we can provide a Voice User Interface to a Smart Precision Agriculture system. Some of these Voice Assistant provides multilingual support, so these voice-assistants can be used to provide multilingual support to the proposed model. As the voice-based user interface has become a very popular User Interface, using this system will help the end-users not to rely on common user interfaces like Graphical User Interface or Command Line Interface which requires physical contact. During this global outbreak of CovID-19 pandemic where the e-commerce companies and food delivery apps are switching towards contact-less delivery just to reduce the spread of CovID-19 through physical contact [99–101], this approach will be quite helpful to reduce the spread of similar types of harmful diseases from contact, and this play a very vital in reducing the spread of contaminating diseases like CovID-19 through the food supply chain or product supply chain since the farmers or the horticulturists may be infected with CovID-19 and from that, it might lead to the spreading of this virus. If required any farming organization can either design their Voice Assistants or make use of the readily available voice assistants to control the drones. Another major advantage of using these Voice Assistants is that they provide multilingual support, so it will be easier for the farmers or the horticulturist to control the drones in their native language using these voice assistants as shown in Fig. 7.

Another advantage of using drones is that there are many smart agriculture systems available, but those are using static sensors in place of the dynamic sensors. Now static sensors are those sensors whose locations are fixed and mostly work on fixed battery supply and stop working when the battery gets discharged which makes the maintenance of those sensors quite a hectic task. Dynamic sensors on the other hand are not fixed to a specific location, so their maintenance is not much hectic in comparison to that of the static sensors. We can easily fit various sensors like IR camera, night vision camera, thermal imaging camera on the drones to capture images of the field and by then by applying Edge Intelligence and image processing on the set of farm images collected along with the corresponding GPS location, we can easily apply the right amount of water, fertilizer, and pesticides, etc. on the crops as shown in Fig. 8. At the same time, they can also alert the farm owner in advance regarding any unwanted scenarios that have taken place on the farm along with its GPS location. For instance, it can warn the farmers regarding crop fire, or any other unusual incidents that are taking place in the farmland or horticulture. Let's cite an example, in the case of grapevine owners, whose wines spread a huge amount of area, it is quite difficult for few individuals to be continuously vigilant over the vineyard throughout the day and especially at night. So, by implementing 5G NB-IoDT with a night vision enabled camera, the farm owner will be able to get real-time data about his or her vineyard and will be alerted as soon as any unusual or unwanted scenario takes place like fire in the grape vineyard.

After the collection of the images and sensor data, we can process the data in the drones, using the concept of Edge computing [102–110] as shown in Fig. 8 as Edge

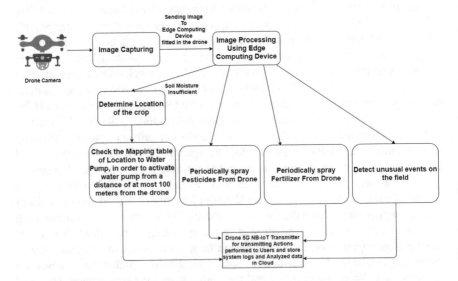

Fig. 8 Edge Intelligence for processing the Images and Sensor data using

computing plays a very crucial role in handling real-time challenges [111]. These drone-based systems can even detect events like forest fires or fires in the wines or farmland and alert the owners in advance to reduce further damage. If there are fixed sensors that will detect forest fires, there is a high chance that these systems might fail if there is a huge fire and the batteries fixed in these systems can fail and the farm owner or the grape wine owner may not be even informed about the fire in the initial stage and that can cause a huge loss to the owners.

5.4 Satellite-Based 5G NB-IoDT Implementation

With the introduction of Satellite-based 5G NB-IoDT, the monitoring of crops in the areas where there is no proper network coverage has also been made possible as shown in Fig. 9 through satellite communication. In the remotest of the islands, where there is no network coverage, Satellite-based 5G NB-IoT enabled drones can be used for smart alerting and monitoring of the crops. The drones can be designed in such a way that they will try to communicate with the nearest 5G mobile tower, but in case there is no nearby 5G mobile tower, it will automatically switch to another mode of operation where it will use satellite-based 5G NB-IoT. This will add flexibility to the ubiquitousness of the drone sensors by providing network connectivity in the remotest of places where there is no proper network connectivity.

Satellite-based 5G NB-IoT network will provide a very fast, reliable communication with a very low latency, which will make it an ideal choice over 4G and Wi-Fi to

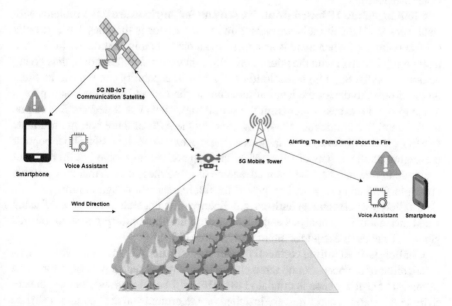

Fig. 9 Forest Fire Detection System Using Smart Farming Drone

be used in real-time applications like this, especially in areas where there is no mobile network available. By applying Machine Learning, our system will understand that what action is required to be performed by the drone and the analyzed result will be sent back to the drone and the drone will perform the required task and forward both the analyzed data and the action performed both the user and store the results in the Cloud platform for future reference. For notification purposes, we can use Telegram Bot as it's APIs are all available free of cost. The Telegram Messenger API supports at most 30 messages per second [112]. Its APIs are available free of cost to use unlike Line and WhatsApp APIs.

6   Future Open Research Challenges

Challenge 1: Smart water pump. The main disadvantage of the agriculture system is that, if there is excess water in the farmland, there is no backup system to pump out the excess water. So, water pumps can be installed in the farms that can pump out excess water from the farmland. The point of contact of the input of these water pumps should be placed at a certain height above the soil so that only the excess water gets pumped out of the soil. The output of the water pumps should be done on a specific channel to conserve the excess water that is getting pumped out and use the water later. This will prove to be quite beneficial, especially in the desert areas, where there is not much moisture content in the soil and also have a scarcity of water. IoDT based smart watering sprinkling systems needs to develop for proper water distribution.

Challenge 2: BOT based Adaptive sensing for Agriculture 4.0. Problems with static sensors like soil moisture sensor and water sensor is that they have metallic contact points and when there is an excessive amount of moisture content in the soil, their point of contact with the micro-controllers may start to malfunction, thus giving incorrect results. So, a high-resolution and zoom-in capability camera can be fitted on the drones, to detect the level of moisture in the soil and start the water pumps to pump out the excess water from the soil. Popular bots like Telegram Bot can be used to send live notifications to the farmers. As the APIs of these bots are available free of cost, it will not cost the farmers a single penny to use its APIs. Other popular messaging platform-bots of Line and WhatsApp are also available but they are not available to free of cost. The main advantage of using these bots is that they provide cross-platform and cross-device support for future generation Agriculture.

Challenge 3: Green Agriculture 4.0. Energy-efficient techniques of IoDT using zonal thermal pattern analysis and adaptive crop health monitoring system towards green IoT needs to design for smart agriculture [113–117].

Challenge 4: Realtime connectivity for Agriculture 4.0. for Narrowband characterization of the near-ground radio channel for wireless sensor networks for smart farming at 5G-IoT bands is crucial [118]. This will help to provide better connectivity to the static sensors that are installed on the ground and are sending real-time data to the Cloud or the End-Users. And with the inclusion of 5G NB-IoT, unlike 4G connectivity, there will be no network congestion and network interference issues. Besides 5G NB-IoT will provide support for satellite communication, which means the smart devices can work in sync with the end-users and cloud in places where there is poor network connectivity through satellite communication. 6G technology is under development for ultra-low latent network.

Challenge 5: 6G Smart Agriculture Appliances for IoDT. IoDT-enabled smart appliances development under industry 4.0. References [119, 120] will be developed for real-time machine Learning and ubiquitous computing-based solution for agriculture. 6G network research is on to provide better connectivity and better bandwidth efficiency.

Challenge 6: Sensor Cloud for Agriculture 4.0. Deployment of AgriEdge based on sensor mobile Edge computing/ sensor cloud platform. Sensors can increase agricultural productivity. Hence, a collaborative approach of the sensor cloud can be most effective in agriculture. Agri sensors increase the agricultural yield, so a collaborative method of sensor edge cloud can be most effective in agriculture [116, 121–123].

Challenge 7: Geospatial Mobile Edge Computing(GMEC). GMEC is used for developing a ubiquitous sensor network platform using IoT in precision agriculture [124, 125]. The geographic information is collected from various sources and with this, IoT establishes communication to the entire world through the Internet. The information will be helpful in the maintenance of the farmland by applying the required amount of fertilizer at the right time in the right place. The main goal in combining the Geospatial technology with IoT for precision is to monitor and predict the critical parameters such as water quality, soil condition, ambient temperature and moisture, irrigation, and fertilizer for improving crop production. It can be expected that with the help of Geospatial and IoT in smart farming, the prediction of the amount of fertilizer, weeds, and irrigation will be accurate and it helps the farmers in making decisions related to all the requirements in terms of control and supply [126]. The application of groundwater for domestic and agricultural uses has increased day by day. However, geology and anthropogenic activities can impact groundwater quality. But information on quality evaluation and suitability classification of groundwater based on water quality index (WQI) is limited. Thus, study evaluated the spatial variability of groundwater quality and its suitability for potable and irrigation purposes. Water samples will be collected from wells across the study area and analyzed for physicochemical properties. Suitability for domestic and irrigation uses will be determined. Regular quality monitoring of the groundwater is recommended to avert likely deterioration indicated by some quality parameters [127].

Challenge 8: Edge Intelligence for Agriculture 4.0(EI Agri 4.0). For precision agriculture, intelligence at the edge level provides a real-time solution for agriculture problems [128].

Challenge 9: Qos for Agriculture 4.0. QoS aware Internet of Agricultural Things needs to implement an autonomous system [6, 129–131]. QoS level is implemented in the MQTT protocol and it functions with different parameters provide a grade of IoT system to set the cost parameters QoS_0, QoS_1, and QoS_2 [132].

QoS_0—at most once: The minimal QoS also known as "fire and forget" level is zero. This service level guarantees a best-effort delivery for agriculture data with no guarantee of delivery as the recipient does not acknowledge receipt of the message and the message is not stored and re-transmitted by the sender. It provides the guarantee as to the underlying TCP protocol.

QoS_1—at least once: QoS level 1 guarantees that a message is delivered at least one time to the receiver of agriculture devices. The sender stores and retransmits the message multiple times until it gets a PUBACK (Publisher Acknowledgement) packet from the receiver.

QoS_2—exactly once: QoS_2 is the highest, safest but slowest level of service in MQTT for agriculture 4.0, which guarantees that each message is received only once by the intended recipients by at least 2 requests or response flows a secured four-way handshake between the sender and the receiver. Harmonization between the sender and receiver is performed by packet identifier of the original PUBLISH message to deliver the message of agriculture data. Ensuring proper QoS level of IoDT ensures proper quality of Experience (QoE) of farmers.

Challenge 10: Blockchain for Agriculture 4.0 (BCAgri 4.0). BCAgri 4.0 for 5G-enabled IoT for industrial automation for food supply chain management is essential [133–138]. Blockchain-based e-agricultural systems with distributed ledger systems for storage organization, agricultural conservation data integrity is protected in information administration [136–138]. IoDT and Blockchain technology as two rapidly emerging fields can ameliorate the state of the food supply chain [139–143]. Blockchain ensures security of IoDT based Agriculture 4.0. 6G enabled futuristic Quantum blockchain will provide higher security for IoDT in near future.

Challenge 11: AgriDew. Dew computing based smart IoDT will be used for future Agriculture even if the internet connectivity is not stable [144–146]. Lightweight dew computing paradigm needs to design to manage heterogeneous wireless sensor networks with various types of UAVs [147].

Challenge 12: Drone Swarm optimization. Particle swarm optimization (PSO) and Glowswarm optimization (GSO) based flight path optimization is important area of research to provide edge intelligence to autonomous IoDT [148–156].

7   Conclusion

With the increase in World population, there is a higher demand for operational efficiency in almost all spheres of life, especially in the agricultural sector. Thus, this is compelling us to shift towards a smarter approach to agriculture with the evolution of the Internet of Drone Things in agriculture domain. 5G NR can be classified into two categories, based on Frequency Range. At first, there is Frequency Range 1, which includes sub-6 GHz frequency bands and the other is Frequency Range 2 that works in the frequency range of 24.25–52.6 GHz. This frequency band varies from country to country. So, there is a need to shift to these latest network architectures as they have lower latency and helps in developing real-time applications. We have proposed an implementation of a smart agriculture model using 5G NB-IoDT. This comprises of implementation idea, basic system architecture, etc. We address certain challenges of the surveyed models in the area of smart agriculture.

During this CovID-19, when there is a global pandemic, we have suggested a contact-less idea of smart irrigation using 5G NB-IoDT, voice assistants, edge computing, and Machine Learning techniques. The recent introduction of satellite-based 5G NB-IoT by Skylo and BSNL will play a significant role in the field of smart agriculture by providing 5G NB-IoT connectivity even to the remotest of the places where there is network connectivity or the network connectivity is very weak. In the remotest islands where there is no proper internet connectivity, the introduction of this satellite-based 5G NB-IoT will prove to be a boon in the arena of smart agriculture. We hope that crops will grow with the help of cyber-physical systems, will reduce human interaction between the food supply chain.

Glossary

Agriculture 4.0 4th Generation Agricultural Revolution using cyber-physical systems, internet of things, cloud computing, machine learning, artificial intelligence

DHT Sensor to calculate atmospheric temperature and humidity

Drone/UAV An unmanned flying vehicle that is either autonomous or is manually controlled from distance by an operator

Fuzzy Logic It is a form of many-valued logic in which the truth values of variables may be any real number between 0 and 1 both inclusive that is where the truth value may range between completely true and completely false

Industry 4.0 4th Generation Industrial Revolution that includes automation and data exchange via Cyber-Physical systems, Internet of Things, Cloud Computing

LDR Sensor to detect light intensity

Multilingual Knowing multiple languages

Real-time When an operation is to be performed in a time-bounded way

WEP Security algorithm for IEEE 802.11 wireless networks

Wi-Fi Used for wireless internet connectivity

Wi-Fi 6 6th generation Wi-Fi connectivity

WPA Security standard for users of computing devices equipped with wireless internet connections

Ubiquitousness Capability to physically move computing services and devices with us

References

1. Manimegalai R (2020) An IoT Based smart water quality monitoring system using cloud. In: 2020 International conference on emerging trends in information technology and engineering (ic-ETITE) (pp. 1–7). IEEE
2. I-Scoop, 5G and IoT in 2018 and beyond: the mobile broadband future of IoT. Available on line 14 Jan 2018. https://www.i-scoop.eu/internetof-things-guide/5g-iot/
3. Abozariba R, Broadbent M, Mason K, Argyriou V, Remagnino P (2019) An integrated precision farming application based on 5G, UAV and deep learning technologies. In: Computer analysis of images and patterns: CAIP 2019 international workshops, ViMaBi and DL-UAV, Salerno, Italy, September 6, 2019, Proceedings (Vol 1089, p 109). Springer Nature
4. Fagan B (2019) The Little Ice Age: how climate made history 1300–1850. Hachette UK
5. https://www.businesswire.com/news/home/20201210005965/en/Skylo-Partners-with-BSNL-to-Launch-World%E2%80%99s-First-Satellite-Based-IoT-Network-in-India. Accessed 14 Dec 2020
6. Mukherjee A, Dey N, De D (2020) EdgeDrone: QoS aware MQTT middleware for mobile edge computing in opportunistic Internet of Drone Things. Comput Commun 152:93–108
7. Akpakwu GA, Silva BJ, Hancke GP, Abu-Mahfouz AM (2017) A survey on 5G networks for the Internet of Things: Communication technologies and challenges. IEEE Access 6:3619–3647
8. https://en.wikipedia.org/wiki/5G_NR_frequency_bands. Accessed 14 Dec 2020
9. Liu HY, Huang CJ (2019) Wideband MIMO antenna array design for future mobile devices operating in the 5G NR frequency bands n77/n78/n79 and LTE band 46. IEEE Antennas Wirel Propag Lett 19(1):74–78
10. Shi X, Zhang M, Xu S, Liu D, Wen H, Wang J (2017) Dual-band 8-element MIMO antenna with short neutral line for 5G mobile handset. In: 2017 11th European conference on antennas and propagation (EUCAP) (pp 3140–3142). IEEE
11. Li Y, Luo Y, Yang G (2017) 12-port 5G massive MIMO antenna array in sub-6GHz mobile handset for LTE bands 42/43/46 applications. IEEE Access 6:344–354
12. Wong KL, Lin BW, Li BWY (2017) Dual-band dual inverted-F/loop antennas as a compact decoupled building block for forming eight 3.5/5.8-GHz MIMO antennas in the future smartphone. Microw Optic Technol Lett 59(11):2715–2721
13. Li Y, Luo Y, Yang G (2018) Multiband 10-antenna array for sub-6 GHz MIMO applications in 5-G smartphones. IEEE Access 6:28041–28053
14. Guo J, Cui L, Li C, Sun B (2018) Side-edge frame printed eight-port dual-band antenna array for 5G smartphone applications. IEEE Trans Anten Propag 66(12):7412–7417
15. Li Y, Yang G (2019) Dual-mode and triple-band 10-antenna handset array and its multiple-input multiple-output performance evaluation in 5G. Int J RF Microw Comput Aided Eng 29(2):e21538
16. Chen M, Miao Y, Hao Y, Hwang K (2017) Narrow band internet of things. IEEE Access 5:20557–20577
17. Zamfir S, Balan T, Iliescu I, Sandu F (2016) A security analysis on standard IoT protocols. In: 2016 international conference on applied and theoretical electricity (ICATE) (pp 1–6). IEEE
18. Kumar V, Vijay Kumar Lab. https://www.kumarrobotics.org/. Accessed 14 June 2020
19. Javaid AY, Sun W, Alam M (2015) Single and Multiple UAV Cyber-Attack Simulation and Performance Evaluation. EAI Endorsed Trans Scalable Inf Syst 2(4):e4
20. Boursianis AD, Papadopoulou MS, Diamantoulakis P, Liopa-Tsakalidi A, Barouchas P, Salahas G, Karagiannidis G, Wan S, Goudos SK (2020) Internet of Things (IoT) and Agricultural Unmanned Aerial Vehicles (UAVs) in Smart Farming: A Comprehensive Review. Internet of Things, p 100187
21. Schiavullo R (2018) Ehang 184 world first self driving taxi car to flight autonomously at low altitude. Genesis 11(04):591

22. Blackmore S (2014) Farming with robots 2050. In Presentation delivered at Oxford Food Security Conference (Vol. 592).
23. Cohn P, Green A, Langstaff M, Roller M (2017) Commercial drones are here: the future of unmanned aerial systems. McKinsey & Company
24. Luppicini R, So A (2016) A technoethical review of commercial drone use in the context of governance, ethics, and privacy. Technol Soc 46:109–119
25. Mairaj A, Baba AI, Javaid AY (2019) Application specific drone simulators: Recent advances and challenges. Simul Model Pract Theory 94:100–117
26. Goodman JM, Kim J, Gadsden SA, Wilkerson SA (2015) System and mathematical modeling of quadrotor dynamics. In: Unmanned Systems Technology XVII (Vol 9468, p 94680R). International Society for Optics and Photonics
27. Gundlach J (2012) Designing unmanned aircraft systems: a comprehensive approach. American Institute of Aeronautics and Astronautics
28. Smith S (2018) Military and civilian unmanned aerial vehicles (drones), https://tinyurl.com/y87et7ck. Accessed 14 Aug 2018
29. Long S (2019) Drones and Precision Agriculture: The Future of Farming
30. A Meola (2017) Exploring agricultural drones: The future of farming is precision agriculture, mapping & spraying - Business Insider, 2017. https://tinyurl.com/ya6cjswm. Accessed 29 Nov 2020
31. Anderson C (2014) Agricultural Drones-MIT Technology Review
32. Reich L (2016) How Drones are being used in disaster management?" Geo awesomeness, [Online]. Available. https://geoawesomeness.com/dronesfly-rescue/. Accessed 25 Nov 2020
33. Vogeltanz T (2016) A survey of free software for the design, analysis, modelling, and simulation of an unmanned aerial vehicle. Arch Comput Methods Eng 23(3):449–514
34. Guchhait P, Sehgal P, Aski VJ (2020) Sensoponics: IoT-enabled automated smart irrigation and soil composition monitoring system. In Information and communication technology for sustainable development (pp 93–101). Springer, Singapore
35. https://www.way2sms.com/pricing. Accessed 12 Dec 2020
36. Singh P, Saikia S (2016) December. Arduino-based smart irrigation using water flow sensor, soil moisture sensor, temperature sensor and ESP8266 WiFi module. In 2016 IEEE Region 10 Humanitarian Technology Conference (R10-HTC) (pp 1–4). IEEE
37. Bamodu O, Xia L, Tang L (2017) An indoor environment monitoring system using low-cost sensor network. Energy Procedia 141:660–666
38. Muangprathub J, Boonnam N, Kajornkasirat S, Lekbangpong N, Wanichsombat A, Nillaor P (2019) IoT and agriculture data analysis for smart farm. Comput Electron Agric 156:467–474
39. https://www.cs.waikato.ac.nz/ml/weka/. Accessed 29 Nov 2020
40. Rani SS, Janet J, Ramya KC, Sitharthan R, Kesavan T, Shrivastava S (2020) UAV based mapping system for precision agriculture. In: IOP Conference series: materials science and engineering (Vol 937, No. 1, p 012035). IOP Publishing
41. Jerin ARA, Kaliannan P, Subramaniam U (2017) Improved fault ride through capability of DFIG based wind turbines using synchronous reference frame control based dynamic voltage restorer. ISA Trans 70:465–474
42. Singh S, Chana I, Buyya R (2015) Agri-Info: cloud based autonomic system for delivering agriculture as a service. arXiv preprint arXiv:1511.08986
43. Calheiros RN, Ranjan R, Beloglazov A, De Rose CA, Buyya R (2011) CloudSim: a toolkit for modeling and simulation of cloud computing environments and evaluation of resource provisioning algorithms. Softw: Pract Exper 41(1):23–50
44. Pimple N, Salunke T, Pawar U, Sangoi J (2020) Wireless security—an approach towards secured Wi-Fi connectivity. In: 2020 6th international conference on advanced computing and communication systems (ICACCS) (pp 872–876). IEEE
45. Kumkar V, Tiwari A, Tiwari P, Gupta A, Shrawne S (2012) Vulnerabilities of wireless security protocols (WEP and WPA2). Int J Adv Res Comput Eng Technol (IJARCET) 1(2):34–38
46. Lashkari AH, Danesh MMS, Samadi B (2009) A survey on wireless security protocols (WEP, WPA and WPA2/802.11 i). In: 2009 2nd IEEE international conference on computer science and information technology (pp 48–52). IEEE

47. Yin D, Cui K (2011) A research into the latent danger of WLAN. In: 2011 6th international conference on computer science & education (ICCSE) (pp 1085–1090). IEEE
48. Tsitroulis A, Lampoudis D, Tsekleves E (2014) Exposing WPA2 security protocol vulnerabilities. Int J Inf Comput Secur 6(1):93–107
49. Noor MM, Hassan WH (2013) Current threats of wireless networks. In: The Third international conference on digital information processing and communications (pp 704–713)
50. Prasad R, Rohokale V (2020) Mobile device cyber security. In: Cyber security: the lifeline of information and communication technology (pp 217–229). Springer, Cham
51. Fehér DJ, Sandor B (2018) Effects of the wpa2 Krack attack in real environment. In: 2018 IEEE 16th international symposium on intelligent systems and informatics (SISY) (pp 000239–000242). IEEE
52. Zou Y, Zhu J, Wang X, Hanzo L (2016) A survey on wireless security: Technical challenges, recent advances, and future trends. Proc IEEE 104(9):1727–1765
53. Alblwi S, Shujaee K (2017) A survey on wireless security protocol WPA2. In: Proceedings of the international conference on security and management (SAM) (pp 12–17). The Steering Committee of The World Congress in Computer Science, Computer Engineering and Applied Computing (WorldComp)
54. Čisar P, Čisar SM (2018) Ethical hacking of wireless networks in kali linux environment. Ann Faculty Eng Hunedoara 16(3):181–186
55. Buchanan C, Ramachandran V (2017) Kali Linux Wireless Penetration Testing Beginner's Guide: Master wireless testing techniques to survey and attack wireless networks with Kali Linux, including the KRACK attack. Packt Publishing Ltd
56. Nikolov LG (2018) Wireless network vulnerabilities estimation. Secur Fut 2(2):80–82
57. Fan L, Niu H, Yang X, Qin W, Bento CP, Ritsema CJ, Geissen V (2015) Factors affecting farmers' behaviour in pesticide use: Insights from a field study in northern China. Sci Total Environ 537:360–368
58. Wilson C, Tisdell C (2001) Why farmers continue to use pesticides despite environmental, health and sustainability costs. Ecol Econ 39(3):449–462
59. Jin J, Wang W, He R, Gong H (2017) Pesticide use and risk perceptions among small-scale farmers in Anqiu County, China. Int J Environ Res Public Health 14(1):29
60. Akter M, Fan L, Rahman MM, Geissen V, Ritsema CJ (2018) Vegetable farmers' behaviour and knowledge related to pesticide use and related health problems: A case study from Bangladesh. J Clean Prod 200:122–133
61. Rezaei R, Seidi M, Karbasioun M (2019) Pesticide exposure reduction: extending the theory of planned behavior to understand Iranian farmers' intention to apply personal protective equipment. Saf Sci 120:527–537
62. Pan D, He M, Kong F (2020) Risk attitude, risk perception, and farmers' pesticide application behavior in China: A moderation and mediation model. J Clean Prod 276:124241
63. Ian Fogg, 5G download speed is now faster than Wifi in seven leading 5G countries. https://bit.ly/2VfG8vI. Accessed 30 Nov 2020
64. Chung MA, Chang WH (2020) Low-cost, low-profile and miniaturized single-plane antenna design for an Internet of Thing device applications operating in 5G, 4G, V2X, DSRC, WiFi 6 band, WLAN, and WiMAX communication systems. Microw Optic Technol Lett 62(4):1765–1773
65. Khorov E, Kiryanov A, Lyakhov A, Bianchi G (2018) A tutorial on IEEE 802.11 ax high efficiency WLANs. IEEE Commun Surv Tutorials 21(1):197–216
66. Bellalta B (2016) IEEE 802.11 ax: High-efficiency WLANs. IEEE Wirel Commun 23(1):38–46
67. Afaqui MS, Garcia-Villegas E, Lopez-Aguilera E (2016) IEEE 802.11 ax: Challenges and requirements for future high efficiency WiFi. IEEE Wirel Commun 24(3):130–137
68. Afaqui MS, Garcia-Villegas E, Lopez-Aguilera E, Smith G, Camps D (2015) Evaluation of dynamic sensitivity control algorithm for IEEE 802.11 ax. In: 2015 IEEE wireless communications and networking conference (WCNC) (pp 1060–1065). IEEE

69. Deng DJ, Chen KC, Cheng RS (2014) IEEE 802.11 ax: Next generation wireless local area networks. In: 10th international conference on heterogeneous networking for quality, reliability, security and robustness (pp 77–82). IEEE
70. Deng DJ, Lien SY, Lee J, Chen KC (2016) On quality-of-service provisioning in IEEE 802.11 ax WLANs. IEEE Access 4:6086–6104
71. Deng DJ, Lin YP, Yang X, Zhu J, Li YB, Luo J, Chen KC (2017) IEEE 802.11 ax: Highly efficient WLANs for intelligent information infrastructure. IEEE Commun Mag 55(12):52–59.
72. Lee Y, Lee W, Shin G, Kim K (2017) Assessing the impact of dos attacks on iot gateway. In: Advanced multimedia and ubiquitous engineering (pp 252–257). Springer, Singapore
73. Butt SA, Diaz-Martinez JL, Jamal T, Ali A, De-La-Hoz-Franco E, Shoaib M (2019) IoT Smart health security threats. In: 2019 19th International conference on computational science and its applications (ICCSA) (pp. 26–31). IEEE
74. https://www.zdnet.com/article/d-link-routers-contain-remote-code-execution-vulnerability/. Accessed 15 Dec 2020
75. https://www.digital.security/en/blog/netis-routers-remote-code-execution-cve-2019-19356. Accessed 15 Dec 2020
76. https://nakedsecurity.sophos.com/2018/05/14/remote-code-execution-bug-found-in-gpon-routers-but-how-bad-is-it-really/. Accessed 15 Dec 2020
77. Wong H, Luo T, Man-in-the-Middle Attacks on MQTT-based IoT Using BERT based Adversarial Message Generation. KDD'20
78. https://www.zdnet.com/article/hacking-attacks-on-your-router-why-the-worst-is-yet-to-come/. Accessed 15 Dec 2020
79. Papp D, Tamás K, Buttyán L (2019) IoT Hacking–A Primer. Infocommun J 11(2):2–13
80. Morandi B, Manfrini L, Zibordi M, Noferini M, Fiori G, Grappadelli LC (2007) A low-cost device for accurate and continuous measurements of fruit diameter. HortScience 42(6):1380–1382
81. Link SO, Thiede ME, Bavel MV (1998) An improved strain-gauge device for continuous field measurement of stem and fruit diameter. J Exp Bot 49(326):1583–1587
82. Das S, Nayak S, Chakraborty B, Mitra S (2019) Continuous radial growth rate monitoring of horticultural crops using an optical mouse. Sens Actuators, a 297:111526
83. Thalheimer M (2016) A new optoelectronic sensor for monitoring fruit or stem radial growth. Comput Electron Agric 123:149–153
84. Dangare P, Mhizha T, Mashonjowa E (2018) Design, fabrication and testing of a low cost Trunk Diameter Variation (TDV) measurement system based on an ATmega 328/P microcontroller. Comput Electron Agric 148:197–206
85. Drew DM, Downes GM (2009) The use of precision dendrometers in research on daily stem size and wood property variation: a review. Dendrochronologia 27(2):159–172
86. Evans RG, Sadler EJ (2008) Methods and technologies to improve efficiency of water use. Water Resour Res 44(7)
87. Higgs KH, Jones HG (1984) A microcomputer-based system for continuous measurement and recording fruit diameter in relation to environmental factors. J Exp Bot 35(11):1646–1655
88. LANG, A. (1990) Xylem, phloem and transpiration flows in developing apple fruits. J Exp Bot 41(6):645–651
89. Gupta S, Mudgil A, Soni A (2012) Plant Growth monitoring system. Int J Eng Res Technol (IJERT) Mag 1(4)
90. Slamet W, Irham NM, Sutan MSA (2018) IoT based growth monitoring system of guava (Psidium guajava L.) Fruits. In: Proceedings of IOP Conference Series: Earth and Environmental Science (Vol 147)
91. Wu T, Lin Y, Zheng L, Guo Z, Xu J, Liang S, Liu Z, Lu Y, Shih TM, Chen Z (2018) Analyses of multi-color plant-growth light sources in achieving maximum photosynthesis efficiencies with enhanced color qualities. Opt Express 26(4):4135–4147
92. Othman MF, Shazali K (2012) Wireless sensor network applications: A study in environment monitoring system. Procedia Eng 41:1204–1210

93. Shinghal D, Noor A, Srivastava N, Singh R (2011) Intelligent humidity sensor for-wireless sensor network agricultural application. Int J Wirel Mob Netw (IJWMN) 3(1):118–128

94. Kiruthika M, ShwetaTripathi, MritunjayOjha, Kavita S (2015) Parameter Monitoring for Precision Agriculture. IJRSI, Volume II, Issue X, October 2015, ISSN 2321–2705

95. Wark T, Corke P, Sikka P, Klingbeil L, Guo Y, Crossman C, Valencia P, Swain D, Bishop-Hurley G (2007) Transforming agriculture through pervasive wireless sensor networks. IEEE Pervasive Comput 6(2):50–57

96. Awati JS, Patil VS, Awati SB (2012) Application of wireless sensor networks for agriculture parameters. Int J Agricult Sci 4(3):213

97. Awasthi A, Reddy SRN (2013) Monitoring for precision agriculture using wireless sensor network-a review. Global Journal of Computer Science and Technology

98. Hoy MB (2018) Alexa, Siri, Cortana, and more: an introduction to voice assistants. Med Ref Serv Q 37(1):81–88

99. Pu M, Zhong Y (2020) Rising concerns over agricultural production as COVID-19 spreads: Lessons from China. Global Food Security 26:100409

100. Chen Z, Chiu CL. Analyzing the Changes of Express Delivery Modules and Markets of Express Delivery Industry

101. Luo C, Wu L, Liu N (2020) Study based on contactless distribution patterns under the outbreak. In: IOP conference series: earth and environmental science (Vol 526, No 1, p 012204). IOP Publishing

102. Shi W, Cao J, Zhang Q, Li Y, Xu L (2016) Edge computing: Vision and challenges. IEEE Internet of Things Journal 3(5):637–646

103. Satyanarayanan M (2017) The emergence of edge computing. Computer 50(1):30–39

104. Shi W, Dustdar S (2016) The promise of edge computing. Computer 49(5):78–81

105. Abbas N, Zhang Y, Taherkordi A, Skeie T (2017) Mobile edge computing: A survey. IEEE Int Things J 5(1):450–465

106. Mao Y, You C, Zhang J, Huang K, Letaief KB (2017) A survey on mobile edge computing: The communication perspective. IEEE Commun Surv Tutorials 19(4):2322–2358

107. Yu W, Liang F, He X, Hatcher WG, Lu C, Lin J, Yang X (2017) A survey on the edge computing for the Internet of Things. IEEE Access 6:6900–6919

108. Hu YC, Patel M, Sabella D, Sprecher N, Young V (2015) Mobile edge computing—A key technology towards 5G. ETSI White Paper 11(11):1–16

109. Khan WZ, Ahmed E, Hakak S, Yaqoob I, Ahmed A (2019) Edge computing: A survey. Futur Gener Comput Syst 97:219–235

110. Ai Y, Peng M, Zhang K (2018) Edge computing technologies for Internet of Things: a primer. Digital Commun Netw 4(2):77–86

111. Chen B, Wan J, Celesti A, Li D, Abbas H, Zhang Q (2018) Edge computing in IoT-based manufacturing. IEEE Commun Mag 56(9):103–109

112. https://core.telegram.org/bots/faq#:~:text=When%20sending%20messages%20inside% 20a,messages%20per%20second%20or%20so. Accessed 12 Dec 2020

113. Sengupta A, Gill SS, Das A, De D (2021) Mobile Edge computing based internet of agricultural things: a systematic review and future directions. Springer Book, In press, Mobile Edge Computing

114. De Debashis (2016) Mobile cloud computing: architectures, algorithms and applications. CRC Press

115. Mukherjee, Anwesha, Payel Gupta, Debashis De (2014) Mobile cloud computing based energy efficient offloading strategies for femtocell network. Applications and Innovations in Mobile Computing (AIMoC), pp 28–35. IEEE, 2014

116. De D, Mukherjee A, Ray A, Roy DG, Mukherjee S (2016) Architecture of green sensor mobile cloud computing. IET Wirel Sens Syst 6(4):109–120

117. Popli S, Jha RK, Jain S (2018) A survey on energy efficient narrowband internet of things (NBIoT): architecture, application and challenges. IEEE Access 7:16739–16776

118. Klaina H, Vazquez Alejos A, Aghzout O, Falcone F (2018) Narrowband characterization of near-ground radio channel for wireless sensors networks at 5G-IoT bands. Sensors 18(8):2428

119. Aheleroff S, Xu X, Lu Y, Aristizabal M, Velásquez JP, Joa B, Valencia Y (2020) IoT-enabled smart appliances under industry 4.0: A case study. Adv Eng Inform 43:101043
120. Liu Y, Ma X, Shu L, Hancke GP, Abu-Mahfouz AM (2020) From Industry 4.0 to Agriculture 4.0: Current Status, Enabling Technologies, and Research Challenges. IEEE Transactions on Industrial Informatics
121. Tyagi S, Obaidat MS, Tanwar S, Kumar N, Lal M (2017) Sensor cloud based measurement to management system for precise irrigation. In: GLOBECOM 2017–2017 IEEE global communications conference (pp. 1–6). IEEE
122. Ojha T, Misra S, Raghuwanshi NS (2017) Sensing-cloud: Leveraging the benefits for agricultural applications. Comput Electron Agric 135:96–107
123. Kim K, Lee S, Yoo H, Kim D (2014) Agriculture sensor-cloud infrastructure and routing protocol in the physical sensor network layer. Int J Distrib Sens Netw 10(3):437535
124. Mishra, Moumita, Sayan Kumar Roy, Anwesha Mukherjee, Debashis De, Soumya K. Ghosh, Rajkumar Buyya (2019) An energy-aware multi-sensor geo-fog paradigm for mission critical applications. J Ambient Int Humanized Comput 1–19
125. Ferrández-Pastor FJ, García-Chamizo JM, Nieto-Hidalgo M, Mora-Pascual J, Mora-Martínez J (2016) Developing ubiquitous sensor network platform using internet of things: application in precision agriculture. Sensors 16(7):1141
126. Bhanumathi V, Kalaivanan K (2019) The role of geospatial technology with IoT for precision agriculture. In: Cloud computing for geospatial big data analytics (pp 225–250). Springer, Cham
127. Adebayo TB, Abegunrin TP, Awe GO, Are KS, Guo H, Onofua OE, Adegbola GA, Ojediran JO (2020) Geospatial mapping and suitability classification of groundwater quality for agriculture and domestic uses in a Precambrian basement complex. Groundwater for Sustainable Development, p 100497
128. Zhou Z, Chen X, Li E, Zeng L, Luo K, Zhang J (2019) Edge intelligence: Paving the last mile of artificial intelligence with edge computing. Proc IEEE 107(8):1738–1762
129. Roy DG, Das P, De D, Buyya R (2019) QoS-aware secure transaction framework for internet of things using blockchain mechanism. J Netw Comput Appl 144:59–78
130. Ahmed N, De D, Hussain MI (2018) A QoS-aware MAC protocol for IEEE 802.11 ah-based Internet of Things. In: 2018 fifteenth international conference on wireless and optical communications networks (WOCN) (pp 1–5). IEEE
131. Roy DG, De D, Alam MM, Chattopadhyay S (2016) Multi-cloud scenario based QoS enhancing virtual resource brokering. In: 2016 3rd international conference on recent advances in information technology (RAIT) (pp 576–581). IEEE
132. https://www.hivemq.com/blog/mqtt-essentials-part-6-mqtt-quality-of-service-levels/. Accessed 15 Dec 2020
133. Mistry I, Tanwar S, Tyagi S, Kumar N (2020) Blockchain for 5G-enabled IoT for industrial automation: A systematic review, solutions, and challenges. Mech Syst Sig Process 135:106382
134. Torky M, Hassanein AE (2020) Integrating blockchain and the internet of things in precision agriculture: Analysis, opportunities, and challenges. Computers and Electronics in Agriculture, p 105476
135. Ferrag MA, Shu L, Yang X, Derhab A, Maglaras L (2020) Security and Privacy for Green IoT-Based Agriculture: Review, Blockchain Solutions, and Challenges. IEEE Access 8:32031–32053
136. Vangala A, Das AK, Kumar N, Alazab M (2020) Smart secure sensing for IoT-based agriculture: Blockchain perspective. IEEE Sensors Journal
137. Bera B, Saha S, Das AK, Kumar N, Lorenz P, Alazab M (2020) Blockchain-envisioned secure data delivery and collection scheme for 5G-based IoT-enabled internet of drones environment. IEEE Trans Veh Technol 69(8):9097–9111
138. Niknejad N, Ismail W, Bahari M, Hendradi R, Salleh AZ (2020) Mapping the research trends on blockchain technology in food and agriculture industry: a bibliometric analysis. Environmental Technology & Innovation, p 101272

139. Lin YP, Petway JR, Anthony J, Mukhtar H, Liao SW, Chou CF, Ho YF (2017) Blockchain: The evolutionary next step for ICT e-agriculture. Environments 4(3):50
140. Kamilaris A, Fonts A, Prenafeta-Boldú FX (2019) The rise of blockchain technology in agriculture and food supply chains. Trends Food Sci Technol 91:640–652
141. Kamble SS, Gunasekaran A, Sharma R (2020) Modeling the blockchain enabled traceability in agriculture supply chain. Int J Inf Manage 52:101967
142. Li X, Wang D, Li M (2020) Convenience analysis of sustainable E-agriculture based on blockchain technology. J Clean Prod 271:122503
143. Vangala A, Das AK, Kumar N, Alazab M (2020) Smart secure sensing for IoT-based agriculture: Blockchain perspective. IEEE Sensors Journal
144. Ray PP, Dash D, De D (2019) Internet of things-based real-time model study on e-healthcare: Device, message service and dew computing. Comput Netw 149:226–239
145. Roy S, Sarkar D, De D (2020) DewMusic: crowdsourcing-based internet of music things in dew computing paradigm. Journal of Ambient Intelligence and Humanized Computing, pp 1–17
146. Ray PP, Dash D, De D (2019) Edge computing for Internet of Things: A survey, e-healthcare case study and future direction. J Netw Comp Appl 140:1–22
147. Rajakaruna, Archana, Ahsan Manzoor, Pawani Porambage, Madhusanka Liyanage, Mika Ylianttila, Andrei Gurtov (2018) Lightweight dew computing paradigm to manage heterogeneous wireless sensor networks with UAVs." arXiv preprint arXiv:1811.04283 (2018)
148. Abhishek B, Ranjit S, Shankar T, Eappen G, Sivasankar P, Rajesh A (2020) Hybrid PSO-HSA and PSO-GA algorithm for 3D path planning in autonomous UAVs. SN Appl Sci 2(11):1–16
149. Mirshamsi A, Godio S, Nobakhti A, Primatesta S, Dovis F, Guglieri G (2020) A 3D Path Planning Algorithm Based on PSO for Autonomous UAVs Navigation. In: International conference on bioinspired methods and their applications (pp 268–280). Springer, Cham
150. Sánchez-García J, Reina DG, Toral SL (2019) A distributed PSO-based exploration algorithm for a UAV network assisting a disaster scenario. Futur Gener Comput Syst 90:129–148
151. Ray A, De D (2016) An energy efficient sensor movement approach using multi-parameter reverse glowworm swarm optimization algorithm in mobile wireless sensor network. Simul Model Pract Theory 62:117–136
152. Chowdhury A, De D (2020) FIS-RGSO: Dynamic Fuzzy Inference System Based Reverse Glowworm Swarm Optimization of energy and coverage in green mobile wireless sensor networks. Comput Commun 163:12–34
153. Chowdhury A, De D (2020) MSLG-RGSO: Movement Score Based Limited Grid-Mobility Approach Using Reverse Glowworm Swarm Optimization Algorithm For Mobile Wireless Sensor Networks. Ad Hoc Networks, p 102191
154. Khan A, Aftab F, Zhang Z (2019) Self-organization based clustering scheme for FANETs using Glowworm Swarm Optimization. Phys Commun 36:100769
155. Goel U, Varshney S, Jain A, Maheshwari S, Shukla A (2018) Three dimensional path planning for UAVs in dynamic environment using glow-worm swarm optimization. Procedia Comput Sci 133:230–239
156. Pandey P, Shukla A, Tiwari R (2018) Three-dimensional path planning for unmanned aerial vehicles using glowworm swarm optimization algorithm. Int J Syst Assur Eng Manag 9(4):836–852

Aakashjit Bhattacharya is currently a Research Scholar at the Advanced Technology Development Centre of the Indian Institute of Technology Kharagpur, Kharagpur, West Bengal, India. He has earned his M. Tech in Computer Science and Engineering from the Maulana Abul Kalam Azad University of Technology (Formerly known as West Bengal University of Technology), West Bengal as a GATE CS Scholar in 2020. He has cleared GATE CS in 2018. He was a former full-time Development Engineer at Calsoft Pune, Inc. He also writes articles on topics related to Computer Science in GeeksforGeeks. He has earned his B. Tech degree from Sabita Devi Education Trusts Brainware Group of Institutions college affiliated under the Maulana Abul Kalam Azad University of Technology (Formerly known as West Bengal University of Technology), West Bengal in 2017. He was awarded Certificate of Merit, in 2013 for backing a State Rank 17th in the 12th National Cyber Olympiad conducted by Science Olympiad Foundation. He was also awarded Certificate of Merit, in 2013 for backing a State Rank 13th in the 15th National Science Olympiad conducted by Science Olympiad Foundation. He has backed an International Rank 10th in the final round of the International Informatics Olympiad conducted by Silver Zone in 2010.

Areas of Interest: Internet of Things, Edge Computing.

LinkedIn Profile: https://www.linkedin.com/in/aakashjit-bhattacharya-35b405133/

Prof. Debashis De earned his M. Tech from the University of Calcutta in 2002 and his Ph.D. (Engineering) from Jadavpur University in 2005. He is the Professor and Director in the Department of Computer Science and Engineering of the West Bengal University of Technology, India, and an Adjunct research fellow at the University of Western Australia, Australia. He is a senior member of the IEEE. Life Member of CSI and a member of the International Union of Radio Science. He was awarded the prestigious Boys cast Fellowship by the Department of Science and Technology, Government of India, to work at the Herriot-Watt University, Scotland, UK. He received the Endeavour Fellowship Award from 2008–2009 by DEST Australia to work at the University of Western Australia. He received the Young Scientist award both in 2005 at New Delhi and in 2011 at Istanbul, Turkey, from the International Union of Radio Science, Belgium. His research interests include mobile edge computing and IoT. He published in more than 300 peer-reviewed journals and 100 conference papers. He published eight research monographs in CRC, Springer, NOVA, Elsevier and five textbooks in Pesrson. His h index is 28, citation 4000.He is an Associate Editor of the journal IEEE ACCESS, Editor Hybrid computational intelligence.

Areas of Interest: Mobile Cloud Computing, Mobile crowd-sensing, IoT, Block-Chain, Computational Nanotechnology.

Email: dr.debashis.de@gmail.com.

Drones for Intelligent Agricultural Management

Subhranil Mustafi, Pritam Ghosh, Kunal Roy, Sanket Dan,
Kaushik Mukherjee, and Satyendra Nath Mandal

Abstract Agriculture is the primary source of livelihood to many countries around the globe and contributes to about 6.9% of the world's total economic production and worth about $5,084,800 million. Therefore, progressive growth in agriculture is very much needed for these countries. The production rate of crops in agriculture is affected by a number of factors such as temperature, humidity, rainfall, the onset of pests, etc. which are not under the direct control of the farmers. The management of diseases, pests and the fertility of the soil can be proclaimed by the application of different types of pesticides, insecticides, fertilizers, etc. through manual spraying by the crop scouts over the hectares of land, v hich affects the nervous system and most of the functionality of the human body parts. In this chapter, the concept of implementing drones for spraying pesticides, fertilizers, etc., has been proposed for intelligent agricultural management. It assures the deduction of negative effect on the farmers and helps to stimulate the idea of handling fertilizers and pesticides in the areas. The drone in the form of a quadcopter has been designed and implemented with the technology of spraying pesticides for instantaneous action as soon as the disease onset is confirmed either manually or technically. The practical implementation of this technology would help in covering large areas of agricultural land in a small quantum of time and also reduce the cost of spraying widely. A wide application is the use of the drone in precision agricultural management where with pre-defined trajectory and measurements, the farm productivity and management can be improved with least manual labour and most optimistic results.

Keywords Unmanned Aerial Vehicles (UAV) · Microcontroller · Quadcopter · Sprayer · Electronic Speed Controller

S. Mustafi (✉) · P. Ghosh · K. Roy · S. Dan · K. Mukherjee · S. Nath Mandal
Department of Information Technology, Kalyani Government Engineering College, Kalyani, Nadia 741235, India
e-mail: subhranilmustafi2011@gmail.com

© Springer Nature Switzerland AG 2021
P. Krause and F. Xhafa (eds.), *IoT-based Intelligent Modelling for Environmental and Ecological Engineering*, Lecture Notes on Data Engineering and Communications Technologies 67, https://doi.org/10.1007/978-3-030-71172-6_4

List of Acronyms

FCB	Flight Controller Board
PCB	Printed Circuit Board
IR	Infrared
PC	Personal Computer
IDE	Integrated Development Environment
USB	Universal Serial Bus
GPS	Global Positioning System
APM	Ardupilot Mega

1 Introduction

Agriculture is the most predominant economic structure in a country like India. It has been considered as the backbone of many economies for many years. It reserves more than 60% of the constitution in India as well. Therefore, it is very crucial to concentrate on the improvement of productivity and efficiency by boosting agricultural innovation and assure the farmers' safe cultivation. It is also very essential to improve the productive structure by offering innovative pesticides spraying technology and fertilizer. Spraying pesticides and sprinkling fertilizer are the most important segments of the Agricultural platform. World Health Organization (WHO) recently declared a status on pesticides cases happening in India, where they have estimated that almost 3 million farmers get affected every year by poisoning from pesticides. The pesticide is the foremost tool in agriculture as it helps to increase productivity but on the other hand, it is quite harmful to farmers. This chapter nurtures the idea of executing Agricultural Drone for Spraying Pesticides. This chapter will help out the consequence of spraying fertilizer manually. Automatic fertilizer will help to spray pesticides over a large area in short intervals compare to the manual or conventional spraying. This measurement of trajectory over an area in the agricultural field helps in increasing the farm productivity and economy. This is the primary concept of precision agriculture. The problem underlies the concept that the farmers are unable to reach specific areas for investigations of the soil and the crops, within the stipulated time and hence suffers from huge losses thereafter. The reaction to any kind of disruptions affect the crops nearby and hence the productivity of the respective land suffers at a great extent. Adaptive applications too play an important role in curbing the spread of diseases as it helps in optimal usage of resources as excessive usage of pesticides affect human health and that of the soil eventually. The advantage of using drones are:

1. Optimization on the usage of seeds, fertilizers, etc.
2. Quick and instantaneous reaction to the application of pesticides and insecticides to the area affected.
3. Saving time in the scouting of the crops and equivalent dedication of the respective time in validation treatment and actions to be taken.
4. Improvement in the variable-rate prescriptions of the amount of yield and measures.

The surveillance using the drones is an integrated part of the integrated agricultural management these days. The raw data collected by the drones get processed into comprehensible information with the help of dedicated algorithms. Some notable information applicable in the respective field by using the drones are:

1. Calculation of the height of the crops and density over a particular plot.
2. Calculation of vegetation indices such as the area of the leaf, efficiency of the treatment, detection of any kind of anomalies, phenology, etc.
3. Needs for the water and management of the drainage system through the lands.
4. Counting of plants and the statistics of the area.

The drone in the form of a quadcopter has a + or X shaped frame having brushless DC motors connected with ESCs at the four ends. The flight controller board is of Ardupilot (APM) having 8,000 mAh Li-Po battery and an FPV transmitter and receiver to patch and bind the remote controller with the quadcopter itself. A sprinkler system has been attached with the frame and programmed for the specificity of the region of spraying the pesticides [10]. The model proposed here is used to spray the pesticide content to the locations and areas which cannot be accessed by the humans instantaneously and easily. The sprayer system is used to spray the liquid as well of solid contents passing through the universal nozzle [13].

The chapter has been organised in the following sections. Section 1 gives an Introduction to the Proposed Model in Solving the Defined Problem followed by the Related Works and Background Study in Sect. 2. Section 3 contains the Usage, Applications and Benefits of using the drone technology in Precision Agricultural Management. Section 4 deals with the Methodology and Description having the Components of the Drone along with the Block Diagram of the Working Principle. The Hardware and Software Description has been discussed in Sects. 4.1 and 4.2. Section 5 discusses about the Spraying Mechanism. followed by Monitoring in Sect. 6, Results and Discussion in Sect. 7 and Conclusion in Sect. 8.

2 Related Works

Authors in [20] discussed the detailed implementation of the agriculturally based drones in view of the automatic mechanism of spraying pesticides and fertilizers in the crops. According to their proceedings and discussions about the reports generated by World Health Organizations stating about the deaths of about 2.2 million people regarding the pesticide poison, the adaptive precaution measures are stated for avoiding the harmful effects and eventually leading to the development of a cost-effective technology using PIC microcontroller for controlling the agricultural bots.

The author in [10] depicted the use of wireless flight controller boards in controlling a number of sensors for pressure, altitude, location measurements, etc. and in the usage of such programmable drones in autonomous and manual modes respectively.

Meivel et al. [13] has discussed the implementation of the drone in spraying pesticides where it is not easily accessible for the human beings. The usage of multispectral cameras in estimating the amount of greenfield and the edges of the crop field has also been discussed the authors of the respective paper. A few decades back, NASA developed solar-powered Pathfinder Plus Unmanned Aerial Vehicle (UAVs) which demonstrated the 3500 ha coffee plantation in Hawaii using the platform based on image collection [7, 8]. After that, another UAV, VIPtero [15] was developed for specific grapevine management using 63 multispectral image sin 10 min using MK-Okto / Hexa Titanium Rigger [1] for thermal and multispectral imagery. With the addition of the sensors for vision systems, thereby increasing the potential value to the UAVs [5], another mechanism that came into existence is the spraying mechanism as discussed in [17, 18]. All that is needed to incorporate such a system is to increase the payload of the UAV with increased power and RPM of the Propellers. This is a part of intelligent pest management and vector control where the precise application of chemicals and fertilizers at any instant of time is found to have immediate prevention of the crops from damages by diseases. The authors in [3] focussed on the research about the organization of spraying of pesticides which depends on the information fed back to the device, coming from different wireless sensor networks (WSN) deployed in the field. It resulted in the short delay in the control loop for the analysis of the area of incoming information of the respective WSN and re-routing according to the shortest possible route thereafter [4]. Reference [24] focussed on the development of the integrated Aerial Automated Pesticide Sprayer (AAPS) quadcopter, used for spraying pesticides using GPS on low altitude area. A flexible cost-effective sprayer drone, "Freyr" was developed as instantiated in [22], which can be controlled by an android application. With all such advancement in the technology, excessive usage of pesticides eventually harms the crops aggressively. To curb such excessive spraying of pesticides, an electrostatic sprayer has been designed under the technology of electrostatic spraying using Hexa rotor UAV [27]. The method of particle image velocimetry has been used to measure the spraying and droplet movement under the deposition and downfall over the crops at the different RPMs of the rotor fixed in an octocopter using double pulsed laser [16]. Several types of water sensitive sprayers are used to study the droplet coverage and spraying deposition over the fields [25].

3 Use of Drone in Precision Agricultural Management

Farming and management are facing severe challenges due to the decreased manual labour, primarily because of the depopulation in the rural areas, urbanization, increasing trend toward the consumption of the animal proteins completely and adulteration of foods due to the cost-effectiveness and imbalanced demand and supply chain. To improve such adverse condition with such decreased labour, the use of machines is required. Drones play a vital role in this area where with defined trajectory and measurement of the amount of action to be performed in a specific area, several applications such as a spray of pesticides, calculation of the presence of water due to the distribution of heatwave pattern from thermal scanners and most importantly the prediction of the early onset of disease with the help of hyperspectral cameras, can easily be accomplished with decreased cost, labour and optimised time. The general steps followed during the capturing of images are:

- **Analysis of the area**: The area requiring the management has to be analyzed by demarcating the territory and uploading the GPS information to the drones navigation system.
- **Uploading the data**: After capturing the data through the sensors and cameras, the same data has to be analysed through numerous softwares for which it is primarily required to be uploaded to a database server.
- **Result and Actuation**: After uploading and the analysis of the data, the required result has to be depicted in a simple and hassle-free format, for which photogrammetry can be extensively used.

The applications served while using drones in intelligent agricultural management are:

- **Irrigation Monitoring**: Drones, included with thermal, hyperspectral or other multispectral cameras serve an important purpose in managing water efficiency in an area, finds potential pool and leaks through the calculation of the vegetation index, wet and dry balance of the land, etc.
- **Crop health monitoring and surveillance**: It is an important task to estimate the health of the crops by sensing the different amount of green light through near infrared spectroscopy. The presence of blights and spots can be detected and be applied with the treatment instantaneously. This has a wider benefit for the farmer in claiming the crop insurance schemes.
- **Assessing the Crop damage**: Apart from identifying the presence of weeds and fungi in the crops, the amount of chemicals required to fight such diseases is also looked for to minimize the amount of cost in applying the pesticides and insecticides.
- **Planting of seeds**: The invention of drone planting has not only helped reducing the cost by about 85% but also in specific application of nutrients in the soil depending on the presence of such nutrients by previous analysis.

The advantages of using the technology of drones in intelligent agricultural management are depicted below:

- **Increased Production**: Comprehensive plan in irrigation, crop health monitoring, presence of fungi or weeds with the use of drone technology help in the increased production of crops and thereby the economy of the farmers.
- **Minimizing Risk for Farmers and Adaptive Techniques**: Often in terrain and hilly areas, it becomes difficult for farmers to climb and spray since the areas may contain power lines adjacent or wild animals. Drones with adaptive techniques built in serve the purpose thereby minimizing the risk for the farmers.
- **Minimizing the wastage of resources**: The drones help in minimizing the use of resources such as pesticides, insecticides, etc.
- **The usefulness of the shreds of evidence for insurance claims**: The data captured through the drones are trustworthy and serve as an important document for the insurance authorities in granting the claims to the farmers.

4   Methodology and Description

Suitability, Adaptability and Scalability are the major factors taken into consideration for the respective chapter focussed hereafter. Suitability deals with the type of drones necessary in solving the purpose of intelligent agricultural management. Adaptability deals with the variability in the quantity of pesticides or fertilizers in a particular area and scalability deals with the lifetime and flight time of the drone used. Flight time tends to around 51 min in single charging and lifetime depends on the integrated components and on average scale it varies from 6–7 years under proper care. However, it is generally advised in checking the synchronization of all motors with that of Electronic Speed Controllers (ESCs) every time before the flight.

The block diagram of the quadcopter has been shown in Fig. 1, where the flight controller board (FCB) has been connected to four brushless (Direct Current) DC motor and (Electronic Speed Controller) ESCs along with the battery for calibration of the motors. The ESCs are used to determine and allow the flow of a controlled amount of power into the brushless motors to control its speed. The receiver is used for binding and patching with the transmitter from the remote controller and (Global Positioning System) GPS for locating the programmed location for implementation. Telemetry has been attached with the components of the receiver and transmitter for the transfer and receipt of the signals in the respective frequency with that of the controller. Ardupilot Mega (APM) flight controller board has been used as it provides easy mobility and stability in balancing the drones in comparison to other boards. The GPS module provides the present location of the drone in association with the home location for landing information.

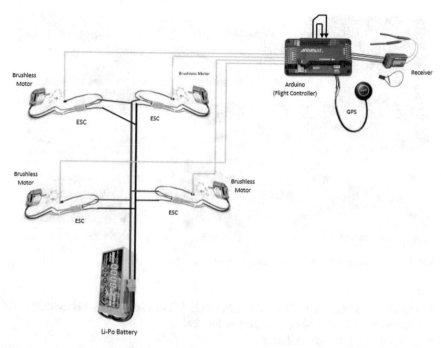

Fig. 1 Circuit diagram for quadcopter with APM as FCB

4.1 Drone Hardware and Description

Figure 2 contains all the necessary components required to configure a drone in the form of a quadcopter. Ardupilot (APM) 2.8 is considered as the Flight Controller Board (FCB) for the drone since it has the complete open source Ardupilot system for easy configuration of the specific task. A 1-L empty tank in the form of a bottle has been annexed in an upright position along with a horizontal discharge tube facilitating the method of spraying.

4.1.1 Flight Controller: Ardupilot Flight Controller (APM)

1. Arduino Compatible.
2. Includes 3-axis gyro, accelerometer and magnetometer, along with a high-performance barometer.
3. Onboard 4 Megabyte Data flash chip for automatic data logging.
4. One of the first open-source autopilot systems to use Invensense's 6 DoF Accelerometer/Gyro MPU-6000.
5. Upgradation of Barometric Pressure Sensor from Measurement Specialities to MS5611-01BA03.

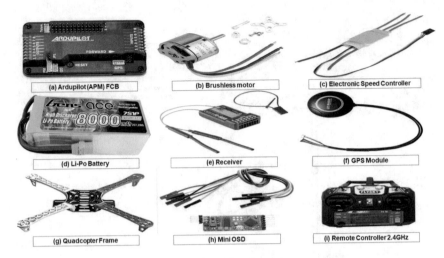

(a) Ardupilot (APM) FCB	(b) Brushless motor	(c) Electronic Speed Controller
(d) Li-Po Battery	(e) Receiver	(f) GPS Module
(g) Quadcopter Frame	(h) Mini OSD	(i) Remote Controller 2.4GHz

Fig. 2 Major components for setting up a drone (quadcopter)

6. Optional off-board GPS, MediaTek MT3329 V2 or uBlox LEA-6H module.
7. Pre soldered header pins, as shown in Fig. 2a.
8. Costs about $35 (source Internet).

4.1.2 BLDC Brushless Motor A2212

1. Type: Outrunner Motor.
2. Operating Voltage: 18–36VDC.
3. Maximum Efficiency: 80%.
4. Maximum Efficiency Current: 4–10 A (>75%).
5. Soldered Bullet connectors for easy connections (Fig. 2b).
6. About $5 each. So, total $20 (source Internet).

4.1.3 ESC (Electronic Speed Controller)

1. Controls the electronic speed and the BLDC Motor.
2. The signal gets broken into 3 parts after coming out from micro-controller and reaches the BLDC Motor.
3. Equal number of ESCs and BLDC Motors used.
4. Used for Optimal Stabilization and is controlled independently (Fig. 2c).
5. About $5 each. So, sums up to $20 (source Internet).

4.1.4 Li-Po Battery

1. Used for driving power in most of the electric modules.
2. Energy Storage and Discharge Ratio is high.
3. Generally combined under the formation of 3SP1, where 3 cells are connected in Series and 1 in Parallel. Also, available in Single Cell format of 3.7 V.
4. Respective Battery Configuration: 12 V, 8000 mAh (Fig. 2d).
5. Costs about $90 (source Internet).

4.1.5 Transmitter and Receiver

1. The Transmitter side produces 2.4 GHz signal.
2. Receiver contains 6 independent channels for receiving the signal.
3. Sending the signal further to the microcontroller for getting processed.
4. Works on 5-V Power Supply where the Current Consumption is less than 40 mA (Fig. 2e).
5. About $50 (source Internet).

4.1.6 GPS Module

1. Ready for the connection to the PIXHAWK FC with attachment of 6-pin connector.
2. Type of Receiver is 72-channel Ublox M8 engine.
3. Galileo-ready architecture of code E1B/C (NEO-M8N).
4. Updation of Navigation at Rate1 Single Global Navigation Satellite System (GNSS): up 18 Hz.
5. Cold Acquisition starts: 26 s.
6. Onboard Compass (Fig. 2f).
7. Costs $20 (source Internet).

4.1.7 Frame

Glass Fibre is being used for making the respective Q450 Quadcopter Frame to incorporate the properties of toughness and durability. Polyamide-Nylon are used for making the arms to prevent the breakage form hard landing at worst cases. The support ridges on the arm improves the stability and provides faster forward flight (Fig. 2g). This frame is also a budget-friendly and costs about $12 (source Internet).

4.1.8   Propellers

The Propellers (Orange HD 8038(8X3.8)) made of Carbon Nylon Black is of high-quality designed especially for multi-copters. The propellers designed are light, durable and have a 15° angle design at the tail for avoiding whirlpooling during the flight. They also cost low with an easily available retail price of $2 (source Internet) for a pair, here we have used two pairs as it is a quadcopter and therefore sums to $4. The number of pairs of propellor used might vary with kind of frame i.e. if it is a quadcopter or a hexacopter.

4.1.9   Monitor Display Screen

The Display Screen module of the 7-in. Monitor has been designed especially for First-Person View (FPV) and other outdoor purposes. Image resolution depends on the manufacturer of the screens and are generally highly defined having resolution 1366 X 720 pixels and supply voltage range of 7–12 V. It serves to be an important feature for operators and users.

4.1.10   Video Transmitter

1. Choice of 48 channels for getting the best transmission quality.
2. A, B, E and F frequency bands compatibility.
3. Lighter weight and smaller size.
4. 48 channels compatibility to all the FPV 5.8 GHz receivers.
5. 5.8 G with power 600 mW and 48 Channels wireless FPV transmitter
6. Wireless transmitter having power 600 mW super small 200 mA current.
7. 600 mW transmission power having assured 5 KM distance in an open area, 5–8 KM availability if being worked with the bigger gain antenna.
8. Costs just about $23 (source Internet).

4.1.11   Mini OSD

The Mini On-Screen Display (OSD) for APM 2.6 and 2.8 is an Arduino based OSD board used in the respective functionalities. It has been tailored for use with the Ardupilot Mega under the MAVlink protocol and has been designed to be as small as possible (Fig. 2h). Costs about $15 (source Internet).

4.1.12   Antenna

Weight of the Antenna is about 8 g but includes heavy-duty heat-sink made of Poly-Olefin to protect the Printed-Circuit Board from damage. The Pagoda II has quite a few optimized axial ratio measurement <1.3 and SWR of 1.2. It will work great as either TX or RX antenna and for best performance we recommend using it as a matched set. Costs about $4 (source Internet).

4.1.13 Remote Controller

Many embedded devices use IR and RF remote controls. Most cars now have a radio frequency (RF) remote key free on board (FOB). Several Wireless keyboards and mice use Radio Frequency links gained at 24 MHz or 2.4 GHz. Instead of Infrared, Node MCU might also be used as less complicated Wi-Fi module is built in the node MCU [11] (Fig. 2i). It also costs much less at just $4 (source Internet) than many other modules.

The most considerable competitor in the market to Ardupilot APM is Raspberry Pi. But we are not considering Raspberry Pi in this chapter for some of its drawbacks:

1. The software code for the drone is needed to be written from the scratch, which might contain various bugs and internal errors, compile-time errors and many more. On the other hand, APM along with Mission Planner provides pre-built and intensively tested error-free software. The functionality of each port can be changed easily using the GUI (Graphical User Interface) of the mission planner instead of changing lines and lines of code which gives the user much more freedom.
2. Raspberry Pi requires a 5.1-V power supply which is not directly available from the battery packs needed to power the brushless motor and are required to be passed through a converter before using. While APM requires a power of 12–16 V which is directly available from the Li-Po battery used.
3. The more powerful processor of the Raspberry Pi produces more amount of heat for the same task done on both of the modules.

4.2 Software Description

Along with the hardware components, the software components are also required for engaging the full functionality of the drones. Several IDEs such as Arduino IDE, Raspbian OS, Cygnus, PlutoX, DroneKit are present but amongst them, Arduino IDE has been used due to the features of open-sourcing, robustness, scalability and mobility of the platform in comparison to others. The calibration and transceiving of the bandwidth of the channel has been done with the help of the mission planner, which has been explained thereafter.

4.2.1 Arduino IDE

This is a computing platform which is open-source and based on a simple software written on the board of the microcontroller. It is interactive in nature in writing, compiling and deploying the codes to the respective modules and sensors such as light, sound, heart rate, ultrasonic, etc. for the respective functionalities being deisgnated for the proper use of them. It has an integrated development environment with simple functionalities and syntaxes where a novice may be able to develop a simple code and deploy the same in the respective sensor. The main code, popularly known as Sketch, being generated in the Arduino IDE, develops a run time code known as Hex and is deployed in the respective module for the written purpose mentioned in the code. Many downgraded versions of the Arduino are also available with the same capabilities but at a much lower price which makes it cost-efficient to use. Arduino has also an alternative namely Raspberry pi but its usage is not recommended due to some of its flaws like the power usage. Raspberry pi sucks a great amount of power to run its processor from the battery which reduces the total flight time and is also less economic. It also lacks an inbuilt analog to digital converter and faces problems of overheating [12].

4.3 Mission Planner

This is an integrated development environment for calibrating all sorts of sensors and synchronizing with the motor. The test signal is passed between the transmitter and receiver after the patching and binding between the drone and the remote controller. The flight controller board is connected to the PC with the help of a USB cable through a communication port and armed to synchronize all the individual module of the quadcopter. Use of mission planner is recommended as it provides the user with a GUI based interface which is easy to understand as well as use. Mission planner supports a wide variety of development boards and also provides the software needed to be installed in the APM for its usage all by itself. It gives an interface for easy viewing of the level at which the drone is placed or alerts the user for any fault in the systems or the hardware of the drone. This software is publicly available for free for users to use it and hence makes it the best software to be used in this process.

5 Spraying Mechanism

For the spraying of pesticides on crops, the spraying arrangement has been used which helps in spraying the fertilizers or pesticides on the crops. For this a tank has been used for storing the pesticide and the sprayer has been technically connected to the tank which helps in spraying the pesticides. There is a nozzle being connected which when switched ON, starts the motor placed in the tank to pump the pesticide

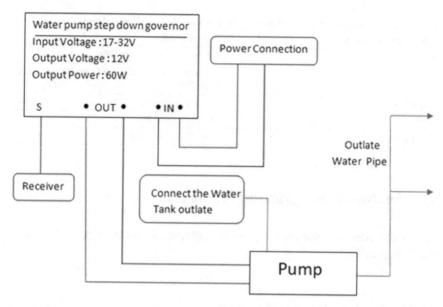

Fig. 3 Block diagram of spraying mechanism

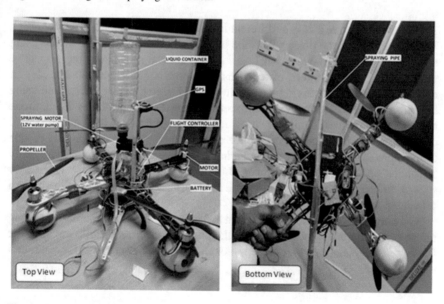

Fig. 4 The spraying drone

through the pipe with the power generated by the battery. And to avoid the wastage of pesticide a uniform pressure has been applied to designate in the respective point of interest [6] (Figs. 3 and 4).

Table 1 Coordination between nozzle type and spraying time for different drones

Spraying type and speed	Type of nozzle
1.15 ha/h	Fan having flat surface [26]
0.3–0.81/m	Fan having a flat surface with centrifugal action force [19]
850 ml/min	Electric centrifugal [25]
4.45 m/s	Not elaborated [9]
0.6–1 lit/min	Centrifugal action [8]

5.1 Analysis of the Sprinkling Mechanism

Table 1 describes the speed and type of the spraying versus the type of the nozzle used for spraying.

6 Monitoring Crops and Lands

The ability to observe a crop at different indices can be programmed in UAVs [21]. The coordinates of the place where the image has been captured are stored using the help of the GPS module and later be used for the analysis of the respective vegetation index. The criteria of reflectance, absorptance and transmittance play an important factor in identifying the quality of land from the images captured using high-resolution RGB camera or multispectral cameras in other cases. For a drone having a flight time of 50 min and speed of 40 km/h, the distance that can be covered in the ideal case is around 34 km, which can be expressed as a square field of side 6 km (approx.). With charging at regular intervals, several hectares of land can easily be studied with these proposed drones. But, for the aforementioned case, it perfectly suits the criteria for deployment in a small farming enterprise which has already been calculated and proved mathematically.

Since, a payload has also been attached to the body of the drone, merely increasing the power of the battery will not serve the purpose because, the arm length of the frame, propeller length, power of the motor also needs to be optimised for maximum utilization.

7 Results and Applications

As the spraying time depends on the quantity of the pesticide available in the tank of fertilizer, the increase in the quantity of pesticide available in the tank leads to the increase in the weight of the tank. With the increase in th weight, the corresponding

Fig. 5 Pictorial representation of the measurement of crop land

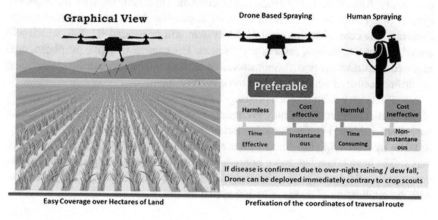

Fig. 6 Comparison of the evaluation of drone based activity

capacity in handling the weight should be increased. This can be done by taking higher ratings of the BLDC motors. If the flight timing of the quadcopter is required to be increased, then a higher-rated Lithium-Polymer Battery should be selected. This requires less capital cost. It is also safe to human beings as it avoids the direct contact of human beings to pesticides. It also reduces the time required and the cost per area of spraying the pesticide on the crops.

Figure 5 depicts the pictorial overview of the usage of agricultural drones in working over the respective field/plot.

The evaluation of the data in place of capturing the images in the field by the drone has been depicted in the following Fig. 6. It has been found for agricultural drones that for a fly of about 50 min, at a height of 50–100 m the area covered will be about 12 Sq. km [2]. The cost would be around €1,300 for public use and lifetime of the drone would be around 2–4 year, based on the extent of usage. Such drones are highly efficient and once programmed, they would perform their duty where visual interception can be observed through the FPV (First Person View) module attached as a component to the circuitry.

The drones will help the farmers in providing the information based on the raw data captured and machines that react on the onset of the issues created which cannot be solved by the drones alone, if being integrated into the proposed technology, will help widely in developing the intelligent agricultural management.

The tremendous acceptance of drones in agricultural management, according to FAO, 2018 [23] has been proved in the US with the sale being risen to 117% of the previous demand of the current year. The improvement in the commercialization for a medium farming enterprise will be around $32.4 bn with high demand in optimizing the time scalability for human sprayers.

The proposed system is suitable for choice in the small and medium farming enterprise as it may be utilized as a ground for testing the efficacy of the drones, practically. Also, a variety of crops can be cultivated at each cycle with the proposed system being able to optimize the cost of deploying the system, experimentally. Presently, the cost of deployment of the system will be around €1,700, which will contain drones, actuators, sensors and camera. The camera will give the photograph required for further process, the actuators and sensors will be very helpful in providing the desired sense of action based on the course of inputs.

Situations may arise where a prolonged dewfall or rainfall increases the humidity of that particular area which may eventually lead to the onset of disease if it stays for more than 6–7 hrs [14]. On confirmation of the onset of diseases, instant action can be taken by employing the drone with the required pesticides and insecticides which may rather have taken few more hours to call for the scouts, arrange them with the necessary pesticides and set to leave for action. This time can be smoothly reduced with the application of drone in action at that instant.

8   Conclusion

Drones dedicated to agricultural use have the ability to uphold the agricultural yield. It can also promote and transform the agricultural industry. Efficient management of large farms with the help of constant monitoring and dedicated application as measures can be instantaneous. The use of the respective components has been proved to be optimized for carrying a litre of pesticides at a particular time. However, the payload can be improved with the increase in the power rating of the brushless DC motors. The chapter and its work can be extrapolated with the implantation of new seeds in the agricultural land with the help of drones. This would help in reducing the labour cost and the time required for such implantation. However, with the concept proposed in this chapter, the technology has been tested and proven to be an efficient one in carrying out its purpose.

Funding Source This work was supported by the Department of Higher Education, Science & Technology and Bio-Technology, Government of West Bengal [Memo No. 33(Sanc.)/ST/P/S&T/6G- 41/2017 dated 12/06/2018]

Acknowledgements The authors would like to express sincere gratitude to Department of Higher Education, Science & Technology and Bio-Technology, West Bengal for funding the research work (Memo No. 33(Sanc.)/ST/P/S&T/6G- 41/2017 dated 12/06/2018), where this technology can be applied. The authors would also thank Dr A Bandyopadhyay, Former Director, ICAR-DGR, Junagadh, Dr. Sourabh Kumar Das, Principal, Kalyani Government Engineering College for their valuable knowledge transfer, Nairita Ghosh and Rupam Saha for their involvement in implementing the technology and Bidhan Chandra Krishi Viswavidyalaya for allowing us to access their agricultural fields.

References

1. Bendig J, Bolten A, Bareth G (2012) Introducing a low-cost mini-UAV for thermal-and multispectral-imaging. Int Arch Photogramm Remote Sens Spat Inf Sci 39:345–349
2. CEMA: Flyer drones FFA, CEMA, European agricultural machinery association (2016) https://www.cema-agri.org/index.php?option=com_content&view=article&id=490:drones&catid=18:news-publications&idU=1&acm=_62
3. Costa FG, Ueyama J, Braun T, Pessin G, Osório FS, Vargas PA (2012) The use of unmanned aerial vehicles and wireless sensor network in agricultural applications. In: 2012 IEEE international geoscience and remote sensing symposium. IEEE, pp 5045–5048
4. Faiçal BS, Costa FG, Pessin G, Ueyama J, Freitas H, Colombo A, Fini PH, Villas L, Osório FS, Vargas PA et al (2014) The use of unmanned aerial vehicles and wireless sensor networks for spraying pesticides. J Syst Arch 60(4):393–404
5. Gupte S, Mohandas PIT, Conrad JM (2012) A survey of quadrotor unmanned aerial vehicles. In: 2012 proceedings of IEEE Southeastcon. IEEE, pp 1–6
6. Harsh Vardhan PDPR, Deephak S, Aditya PT, Arul S (2014) Int J Res Eng Technol (IJRET) 3(4):856–861
7. Herwitz S, Johnson L, Arvesen J, Higgins R, Leung J, Dunagan S (2002) Precision agriculture as a commercial application for solar-powered unmanned aerial vehicles. In: 1st UAV conference, p 3404
8. Herwitz S, Johnson L, Dunagan S, Higgins R, Sullivan D, Zheng J, Lobitz B, Leung J, Gallmeyer B, Aoyagi M et al (2004) Imaging from an unmanned aerial vehicle: agricultural surveillance and decision support. Comput Electron Agric 44(1):49–61
9. Kabra TS, Kardile AV, Deeksha M, Mane DB, Bhosale PR, Belekar AM (2017) Design, development & optimization of a quad-copter for agricultural applications. Int Res J Eng Technol 4
10. Korlahalli KB, Hangal MA, Jitpuri N, Rego PF, Raykar SM (2020) An automatically controlled drone based aerial pesticide sprayer. http://www.kscst.iisc.ernet.in/spp/39_series/SPP39S/02_Exhibition_Projects/147_39S_BE_0564.pdf
11. Kurkute SR, Thenge S, Hirve S, Gosavi D (2018) Int J Adv Res Comput Commun Eng 7(1):139–140
12. McFadden C (2018) Raspberry pi and arduino: what's the difference and which is best for your project? https://rb.gy/qbmbme
13. Meivel S, Maguteeswaran R, Gandhiraj N, Srinivasan G (2016) Quadcopter UAV based fertilizer and pesticides spraying system. Int Acad Res J Eng Sci. http://acrpub.com/article/publishedarticles/24102016IARJES343.pdf

14. Mustafi S, Ghosh P, Dan S, Mukherjee K, Roy K, Mandal SN (2020) IOT based leaf wetness sensors. In: International conference / IETE zonal seminar on IOT in present wireless revolution (IOTWR)
15. Primicerio J, Di Gennaro SF, Fiorillo E, Genesio L, Lugato E, Matese A, Vaccari FP (2012) A flexible unmanned aerial vehicle for precision agriculture. Precis Agric 13(4):517–523
16. Qing T, Ruirui Z, Liping C, Min X, Tongchuan Y, Bin Z (2017) Droplets movement and deposition of an eight-rotor agricultural UAV in downwash flow field. Int J Agric Biol Eng 10(3):47–56
17. Sarghini F, De Vivo A (2017) Analysis of preliminary design requirements of a heavy lift multirotor drone for agricultural use. Chem Eng Trans 58:625–630
18. Sarghini F, De Vivo A (2017) Interference analysis of an heavy lift multirotor drone flow field and transported spraying system. Chem Eng Trans 58:631–636
19. Shilin W, Jianli S, Xiongkui H, Le S, Xiaonan W, Changling W, Zhichong W, Yun L (2017) Performances evaluation of four typical unmanned aerial vehicles used for pesticide application in china. Int J Agric Biol Eng 10(4):22–31
20. Shivaji CP, Tanaji JK, Satish NA, Mone PP (2017) Agriculture drone for spraying fertilizer and pesticides. Int J Res Trends Innov (IJRTI) 2(6):34–36
21. Simelli I, Tsagaris A (2015) The use of unmanned aerial systems (UAS) in agriculture. In: HAICTA, pp 730–736
22. Spoorthi S, Shadaksharappa B, Suraj S, Manasa V (2017) Freyr drone: pesticide/fertilizers spraying drone-an agricultural approach. In: 2017 2nd international conference on computing and communications technologies (ICCCT). IEEE, pp 252–255
23. Sylvester G (2018) E-agriculture in action: drones for agriculture. http://www.fao.org/documents/card/en/c/I8494EN/
24. Vardhan PH, Dheepak S, Aditya P, Arul S (2014) Development of automated aerial pesticide sprayer. Int J Eng Sci Res Technol 3(4):458–62
25. Xinyu X, Kang T, Weicai Q, Yubin L, Huihui Z (2014) Drift and deposition of ultra-low altitude and low volume application in paddy field. Int J Agric Biol Eng 7(4):23–28
26. Yallappa D, Veerangouda M, Maski D, Palled V, Bheemanna M (2017) Development and evaluation of drone mounted sprayer for pesticide applications to crops. In: 2017 IEEE global humanitarian technology conference (GHTC). IEEE, pp 1–7
27. Yanliang Z, Qi L, Wei Z (2017) Design and test of a six-rotor unmanned aerial vehicle (UAV) electrostatic spraying system for crop protection. Int J Agric Biol Eng 10(6):68–76

Subhranil Mustafi completed his M.Tech in Information Technology from Kalyani Government Engineering College in 2020. He qualified Graduate Aptitude Test in Engineering (GATE) in 2018. He is a Gold Medalist in Information Technology from MAKAUT, WB, India. He has authored more than 10 research papers and articles in 2020 and currently working in the domain of animal identification. He received the Best Student Project Award for the final year Research Project in B.Tech in Animal Identification. His research interest lies in Digital Image Processing (extensively in the area of Biometric Identification), Internet of Things and Data Science.

Pritam Ghosh did his Masters in Information Technology from Kalyani Government Engineering College and is currently working as a software developer in a MNC. His research interests lie in Machine Learning and Computer Vision and has contributed extensively in the ares of Smart Farming, Intelligent Livestock Management and Animal Biometrics.

Kunal Roy completed his M.Tech in Information Technology from Kalyani Government Engineering College and is currently the Junior Research Fellow of WBDSTBT Project for Plant Disease Identification. His contribution lies in the area of Image Processing extensively for the purpose of Crop Disease Identiifcation.

Sanket Dan completed his M.Tech in Information Technology from Kalyani Government Engineering College and is currently the Rajiv Gandhi National Fellow for pursuing Ph.D. in Computer Science and Engineering. He has an Intellectual Property Certification for copyrighting Individual Pig Identification using Biometrics.

Kaushik Mukherjee completed his M.Tech in Information Technology from Kalyani Government Engineering College and is currently the Junior Research Fellow of DST-SERB Project, Government of India, for Plant Disease Identification. His contribution lies in the area of Machine Learning for the purpose of Crop Disease Identiifcation.

Satyendra Nath Mandal completed his Ph.D. from Maulana Abul Kalam Azad University of Technology, Kolkata, India and is currently an Assistant Professor in the Department of Information Technology of Kalyani Government Engineering College. He has authored more than 100 Research Papers including Conferences and Journals and is currently a Principal Investigator in a number of Government Funded Research Projects.

Multi-Modal Sensor Nodes in Experimental Scalable Agricultural IoT Application Scenarios

Dimitrios Loukatos and Konstantinos G. Arvanitis

Abstract During the last years, a wide variety of credit card-sized computer systems have appeared, systems characterized by continuously diminishing cost and size and plenty of features. The boost in the electronics industry had also a strong impact on the accompanying sensing, acting and transmitting modules. On the other hand, the efficient monitoring and control, which is of vital importance for any agricultural process, demands diverse resources to be allocated, in terms of communication bandwidth, distance coverage, energy consumption and processing power. Apparently, this flourishing in electronics is beneficial for any similar process but the high availability of cutting-edge components does not guarantee their efficient interoperation as well. Indeed, the task of providing a wide set of heterogeneous monitoring features while maintaining in parallel the overall processing, bandwidth and energy needs at a minimum level is a real challenge. As the candidate technologies being available are exhibiting complementary characteristics, this work is willing to highlight both the importance and the feasibility of synergy actions between the various corresponding components participating to form an efficient set of multi-modal network nodes. In this regard, a pilot system is implemented and evaluated. This system is mainly using LoRa (or ZigBee) and Wi-Fi interfaces to provide both simple and composite information flows and to drive various actuator modules. A key factor is to keep active only the necessary components for the tasks to be performed at a specific moment. For instance, the nodes should be smart enough to decide which radio interface to use for each specific data transmit action. The extent to which popular IP-based services can be supported by non-standard radios (e.g., the LoRa), in case of emergency, is also a challenging issue being examined.

Keywords Multi-modal nodes · WSAN · IoT · Diversity · Energy optimization

D. Loukatos (✉) · K. G. Arvanitis
Agricultural University of Athens, Iera Odos, Athens 75, 11855, Greece
e-mail: dlouka@aua.gr

K. G. Arvanitis
e-mail: karvan@aua.gr

© Springer Nature Switzerland AG 2021
P. Krause and F. Xhafa (eds.), *IoT-based Intelligent Modelling for Environmental and Ecological Engineering*, Lecture Notes on Data Engineering and Communications Technologies 67, https://doi.org/10.1007/978-3-030-71172-6_5

List of Acronyms with Explanation

FAO　　Food and Agriculture Organization
GPIO　General-Purpose Input Output pins
IoT　　Internet of Things
MQTT　Message Queuing Telemetry Transport protocol
LiPo　Lithium Polymer technology batteries
LoRa　Long Range communication protocol
NDVI　Normalized Difference Vegetation Index
RSSI　Received Signal Strength Indicator
SSH　　Secure Shell Protocol
TCP　　Transmission Control Protocol
UDP　　User Datagram Protocol
UN　　United Nations
VNC　　Virtual Network Computing
WSAN　Wireless Sensor and Actuator Network
WSN　　Wireless Sensor Network

1　Introduction

According to recent United Nations (UN) projections, the world population is about to increase to 9.7 billion by 2050 [1]. In order to tackle such a growth, the agricultural production should be raised up to 60 percent during the twenty-first century, according to the estimates of the Food and Agriculture Organization (FAO) [2]. This raise in production should be carried out against the deteriorating environmental conditions of our era, which directly impact the quantity and quality of the crops. For this reason, the only way to succeed is the agricultural practices to become more productive and "climate-smart", by successfully exploiting a variety of existing and emerging technologies [3]. Thankfully, the progress in Information and Communication Technologies (ICT) puts the basis for significant improvements in agri-production [4]. The vast booming in the mobile phone market made available a plethora of amazing computer-related features (i.e., in terms of computation, communication and sensing) that had never been experienced before at such price levels. Apparently, this fact favored any relevant application field, with the agricultural sector not to be an exception. In this context, the term "Agriculture 4.0" [5] is willing to express the beneficial synergy between the diverse cutting-edge technologies in the agricultural industry.

　　The modern wireless sensor networks (WSNs), often enriched with actuator elements and thus forming the so-called wireless sensor/actuator networks (WSANs) [6] provide the potential for sensing and remotely controlling several environmental parameters and agricultural operations. The WSNs typically are trying to optimize the efficiency of the network and to minimize the energy use [7], but the harsh environmental conditions and the energy limitations, especially in rural agricultural areas, are posing further challenges in finding an effective combination of solutions. The bouquet of candidate technologies to be used are of high heterogeneity, and thus they are difficult to combine, either due to the multi-level nature of the field operations or due to the large variety of electronic components being available at the market. Furthermore, the proposed solutions should be cost effective in order to be affordable by small to medium-scale farmers as well. In this regard, this chapter intends to highlight, via the necessary paradigm, the importance, the feasibility and the challenges of a synergy between different technologies, in terms of networking, processing, sensing and acting, to better support the agricultural processes. Indeed, data can be collected and decisions can be made in more efficient manner, by systems able to dynamically adapt their active set of resources to the current needs. By this way, increased production, by healthier animals and plants, at reduced costs, will be possible. The selection of cost effective and easy-to-find, to combine and to program components has the additional benefit of creating an ecosystem suitable for experimentation orchestrated by the students of agricultural engineering and future professionals [8].

The rest of this chapter is organized as follows. Apart this short introduction, Section two further highlights the requirements and the challenges of developing a sensor node network suitable for serving many of the diverse needs of modern agriculture. Section three provides the necessary architecture details of a pilot multi-modal WSAN system for agricultural purposes that has been selected to test the feasibility of having various states of operation for its nodes. Section four highlights important issues related to the implementation of the proposed sensor nodes and their performance evaluation methodology. Section five is dedicated to corresponding experimental measurements and useful discussion results. Finally, Section six summarizes the main points of this research and also presents issues left open for future investigation.

2 IoT Wireless Technologies in Agriculture

The nature itself is exhibiting an extraordinary degree of diversity and both sustainability and evolution are heavily based on this mechanism [9]. Any ecosystem is characterized by a wide variety of species (e.g., animals, plants) that had to coexist and develop the necessary potential to better adapt to a competitive environment. In agriculture, diversity, which is more often used as a synonym of biodiversity, consists both a reality and a necessity that should be taken into account, for better production quality and preservation of resources [9]. Apparently, this dynamic directly affects the technological solutions that are adopted by everyday agricultural practice. In this regard, some applications are more straightforward, like the deployment of smart irrigation systems, based on fusion of soil moisture and weather data, for crop management [10, 11]. Classical security surveillance infrastructures, using video cameras guided by motion detection triggers remain very important, as the theft of plant production (i.e., medical marijuana) or of farm machinery or of animals is a considerable issue [12–14]. Other application cases are less apparent but equally important. For instance, the need for alarm cameras can be remarkably beneficial for inspecting the troughs used in order the sheep to drink water, offering additional and more accurate information than the water level indicators, or even for monitoring cows or ewes about to give birth [15, 16]. The insect population can be effectively monitored and the proper/optimal spraying action to be taken by a fusion of techniques involving cameras, weather sensors and artificial intelligence techniques [17, 18]. Furthermore, health condition or aggressive behaviors like the tail-biting among pigs can be tackled effectively by fusion of data provided by audio and video sensors [19–21]. Similarly, in marine fish farms, the authors in [22] highlight the large variety of technical challenges experienced by various researchers for maintaining and improving a fishery unit. The water quality can be monitored and even predicted using wide variety of sensors in conjunction with visual data and further processing, involving from simple methods to neural networks, alarm actions should be triggered as well. The presence of predators, the hunger cycle of fishes and the waste management of their remaining food are also a critical factors requiring from conventional sensing elements to acoustic radars and sophisticated vision techniques [23, 24].

Despite the advances in technology of the WSNs, there is a considerable gap between theoretical research and practice in real-world conditions [25, 26]. To tackle this issue, more effort should be made to understand the idiosyncrasies of the underlying components and to practice with experimental implementations. Several studies are assessing the diverse technologies and reporting on the factors that should be taken into account while selecting a specific WSN solution [27–30], as many researchers have worked on different IoT-based agricultural projects and technologies to improve the quality and increase productivity [31]. The important points of these approaches can be enriched with further requirements to form a set objectives during the WSN nodes deployment:

- The ability to monitor and/or control of biological and physical parameters related with the agricultural operations (both for plants and livestock).
- The ability to monitor and/or control of environmental conditions such as temperature, humidity, gas emissions, etc. as well as meteorological data.
- The incorporation of mechanisms that provide security surveillance (and secure data transfer) and control of the farm premises.
- Robust operation (tolerance to nodes' failures) and longevity under the harsh conditions of the farm. Ability for easy repairs and maintenance.
- Ability to alter their state/cooperation schema in order to adapt to the rapidly varying conditions of the agricultural field.
- Fast response and autonomy while delivering the supported operations, towards minimization of the intervention with the humans.
- Energy autonomy, usually via exploiting solar panels, and also minimization of the environmental impact, in terms of energy needs and pollution.
- Cooperation with pre-existing or complementary and often heterogeneous systems, i.e., for providing easy visualization of the processes of interest.

- The use low-cost and easy-to-find equipment, based on widespread protocols and methods and allowing the fast deployment of applications and repairs.
- Scalability and reusability of equipment and that can be easily modified and exploited by non-experienced personnel and used in future projects.

In agriculture, the candidate network technologies being available for the deployment of sensors networks should be carefully examined, in conjunction with the idiosyncrasies of the specific application case, as each solution has both advantages and disadvantages. Technically speaking, some of the above requirements might look quite diverse or even contradicting. For instance, the rapid development of information technology, allows the formation of 5G cellular networks, which are able to provide high speeds, low delay and large capacity for many demanding IoT application cases. Despite the increased quality characteristics that the cellular technology is offering, it has considerable billing costs, requires the presence of permanent infrastructures (i.e., bearers and base stations) nearby and is energy-demanding [32]. The satellite links are offering excellent coverage to remote locations but they are quite expensive, involve energy consuming terminals, are exhibiting increased latency and are more sensitive to the weather conditions. For these reasons, both cellular and satellite technologies do not always comprise a total solution for data communication but are rather challenging options to be assessed among the others. Indeed, alternative technological approaches may also have a slice of the pie, as they are less costly and probably suit better in cases exhibiting moderate or scarce data traffic or delay-tolerance. Protocols like the ZigBee [33] (for shorter distances) or the LoRa [34] (for longer distances) can cover the latter needs at reduced energy and money costs. Apparently, wired networks are also an option for high-quality communication but with serious limitations. More specifically, apart from the lack of mobility, the cable solutions for interconnecting scattered livestock/crop farm facilities, which quite often are in wide isolated areas, may have high installation and maintenance costs. The cables are sensitive to the harsh environmental conditions dominating the rural areas and can be damaged by the farming machines or even be chewed by animals like birds or rats [15]. In all cases, the need for secure communications is increasing the packet delays and the energy spent, due to the extra processing and payload bytes. Furthermore, the demand for long-range and noise-resilient communication, at low power consumption, is incurring lower data traffic rates [35]. The trivial Wi-Fi radios are still in use, for communication cases requiring high data rate, like the real-time video delivery, at the cost of limiting the transmission range. The diversity of the agricultural processes demands matching (and thus diverse) sensor elements in order to be optimally monitored, which also means that different data manipulation techniques should be applied, in terms of sampling or processing (e.g., the monitoring of the soil humidity once per hour is adequate while a full NDVI [36] map calculation, by a UAV, may require extended processing of large imagery data many times per second). Different sensing and processing potential signifies completely different energy consumption and connectivity characteristics (e.g., the energy and the bandwidth demands of a video streaming camera are completely different from the ones of a simple luminosity or temperature sensor). The need for decision making and action taking (e.g., watering the crops, providing ventilation, turning a light on, etc.) makes things even more complicated, as the sensor nodes can also be used to control action elements, like electric water pumps, solenoid valves, ventilators or security alarms, components that occasionally demand high amperage to operate.

The increase of the IoT sensor applications generates increased amounts of aggregate data that may push the network and central systems to the limit. In this regard, the Edge Computing [37] tries to move the computation from the data centers towards the edges of the network, in order to improve response times and save bandwidth, by exploiting smart things, mobile phones or network gateways to perform tasks and provide services on behalf of the cloud [38, 39]. Towards this direction, the modern technology is offering a variety of credit card-sized systems with sufficient computational power to support edge intelligence actions as in [40]. The edge computing functionality can be supported either by the sensor/actuator nodes or by the network nodes nearby. The need for autonomous operation is in-line with this edge computing policy, as many decisions have to be taken on-site (i.e., locally). In general, decisions and processing at the edges, especially in cases of scarce traffic, signifies peaks and periods of inactivity (i.e., increased occasional demands in computational power, bandwidth and energy by the participating nodes), and thus the idea of selectively activating/deactivating the corresponding processing or networking modules for energy saving purposes is quite appealing.

In recapitulating, all these diverse requirements demand a sophisticated node architecture in order to maximize efficiency and, in parallel, to minimize complexity and the implementation, installation, operation and maintenance costs. Instead of having completely separate equipment to cover each operational need, sensor nodes with more than one network interfaces, able for different functions, can be implemented to reduce redundancies and save money and space. More specifically, the corresponding nodes may selectively activate/deactivate their appropriate subsystems to optimally respond to the current needs. The term "multi-modal" expresses exactly this ability of the nodes to selectively alter their mode of operation. Indeed, the idea of multi-modal operation is very appealing to the research community and many scientific teams have contributed in it, as reported in [41], especially as this multi-modality may also refer to nodes with diverse sensor elements that can significantly attribute to the information fusion process [42]. The extent to which this aggregation of features can function in a satisfactory degree is not always straightforward to answer without further experimentation.

3   Sensor Nodes Architecture Issues

The proposed analysis intends to indicate the complementarity of strengths and weaknesses characterizing the existing technologies and to investigate methods for a beneficial synergy between them. For this reason, in order to highlight critical issues like the radio coverage, the bandwidth availability, the computational efficiency or the power consumption of the WSN (or the WSAN) nodes under dynamically varying conditions, it is necessary to adopt a flexible model, to implement a set of pilot nodes according to it and to perform series of tests, using an evaluation methodology that matches the diverse profiles of the agricultural applications demands described in Section two.

An Experimental WSAN Nodes Model

The proposed model for WSAN node testing comprises of characteristic radio links, of sensing and action elements, all of low-cost, as this model tries to be in-line with the limited budget of the small or medium size agricultural premises. The design being followed favors the hiring of generic components that allow for fast and easy deployment, modifications and testing.

In terms of communication needs, as a basis, the experimental setup involved Wi-Fi radios, for supporting conventional IP and monitoring functionality. In addition to this, LoRa radio transceivers for long-range, low-energy, low-rate communications were present. As an alternative, modules based on the IEEE 802.15.4 protocol [43] (like the ZigBee) were also included. The latter low-energy modules are exhibiting comparatively short-range characteristics but they are capable for multi-hop operation, which is beneficial for many application cases.

Two processing units were also present on each node: a raspberry pi (model 2B or 3B) [44] for the high-complexity operations (e.g., for real-time video capturing and delivery or image recognition) and an arduino uno unit [45] for the trivial and low-complexity operations (e.g., temperature, humidity, proximity monitoring). Indicative actuators were also hosted on this composite sensor layout. Typically, they were managed by the arduino, via its GPIO (Generic Purpose Input Output) pins, using the necessary driving circuits.

The monitoring data traffic, originating from the nodes, is travelling towards the sink node for further processing. This sink node can send configuration change or monitoring requests towards the sensor nodes. The sink node is also known as "gateway" or "coordinator" and, apart from Wi-Fi, LoRa or ZigBee interface modules, it can be equipped with an Ethernet (or 3G/4G/5G) interface, in order to communicate with the cloud hosting additional services. Via a simple implementation, based on the lightweight MQTT protocol [46] and its publisher and subscriber entities, commands and data can be exchanged between the cloud and the gateway node, in order the experimental testbed to be accessible from everywhere. These arrangements are depicted in Fig. 1.

Power Management Issues

In accordance with the design arrangements of the experimental WSAN nodes, the following types of operation are possible: (a) Data of low-rate can be exchanged via the low-energy links (e.g., the LoRa ones) between the sensor nodes and the gateway, carrying managerial commands or monitoring parameter values like temperature, humidity or luminosity indications. (b) Commands sent while in mode (a) activate/deactivate high-rate traffic generation, like the video traffic, which is delivered via fast but energy consuming links (i.e., the Wi-Fi ones). (c) low-rate signals, via

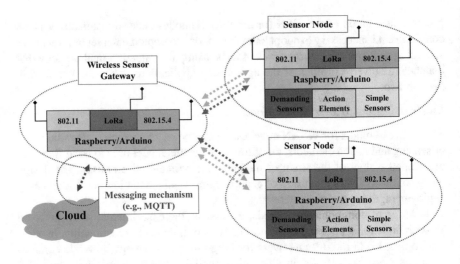

Fig. 1 Multi-modal nodes exhibiting diverse elements and their coordinating infrastructure

the low-rate links, can be sent to the sensor nodes, in order to turn on the actuator units via the necessary driver circuicity. All these types of operations may coexist, but the activity of type (c) is the most important.

It must be noted that the radio transceiver characteristics is not the only factor influencing the power consumption of a WSAN node. The aggregate power consumption of a node relies also on the nature of the sensing elements it is hosting. Most of these elements consume small amounts of energy, but there are exceptions. For instance, some gas-sensing modules require several mA to operate, a camera is also a greedy component. The total number of sensing elements a WSAN node has should also be taken into account. The adoption of small solar panels or of equivalent energy harvesting techniques drastically improves the power autonomy of the sensor nodes. The existence of actuator elements, which usually are "greedy" modules, in terms of power consumption, is a dominant factor for selecting the energy-saving policy to be followed. The overall node setup, with respect to energy management issues, is shown in Fig. 2.

Typically, the necessary modules can be turned on and off via system managerial commands, like linux shell commands (on the raspberry pi) or library-specific application interface commands (on both the raspberry pi and the arduino). Modules like a USB hub or a video display adapter or a radio transceiver chip or an analog-to-digital converter module can be managed this way. Furthermore, many processors are capable for entering into sleep mode, during which a minimal set of modules/operations is active, and thus, the power consumption is minimized. The latter feature is supported by the arduino microcontroller but not by the raspberry pi unit. At the absence of this feature, a problem common in less sophisticated equipment, the modules can be activated/deactivated less "politely" by turning them on/off via direct control of their power supply pins. This control method is possible grace to

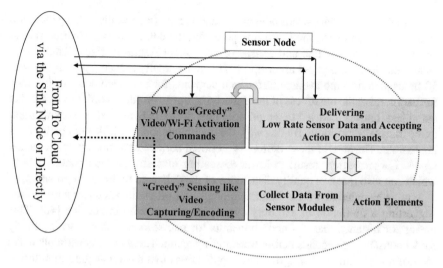

Fig. 2 Detailed WSAN node architecture example with respect to energy management issues

relays, either conventional or digital ones (that are often classified as driving circuits). For very modest powering cases (of the order of a few mA), the GPIO pins of the microcontroller can be used to power up these additional modules as well.

While, in many (cloud-based) cases, the commands to make a node to enter into a specific state of operation are coming via the central gateway/coordinator node, the decision a node to alter its operational status can also be generated by the gateway/coordinator or even by the node itself, in-line with the edge computing approach, according to criteria and thresholds formed via simple or more complicated algorithms that are running locally. For instance, in a livestock monitoring application, the excessive values of the moving average of a sound sensor readings may automatically fire up the video delivery process. The radio signal strength deteriorations may also lead to transmit power level shifts or even to handover decision towards another candidate radio interface. Further analysis of this mechanism is beyond the scope of this work.

Supporting IP-Based Video Content Over LoRa

As discussed above, for IP-based functionality and video traffic delivery, the most suitable solution is the occasional activation of the high-speed Wi-Fi radio link (which is quite greedy in terms of energy), along with the video camera device, as long as the activity of interest is happening. For instance, when an elementary motion or sound sensing unit intercepts a potential alarm (i.e., an intruder, a predator, aggressive animal behavior or even excessive water consumption), more detailed information (e.g., containing images or video content) should be provided via the network infrastructures to allow the proper decisions and actions. At the absence of a Wi-Fi link or an equivalent cellular solution, the existing LoRa transceiver module should be an alternative. The LoRa link capacity resources are limited to a few kbps while the video traffic is very bandwidth-demanding and bursty. In order the video generation characteristics to match the LoRa link potential, the content to be delivered should be of very low frame rate and consist of very small frames in size. Compression techniques like the mpeg-4 algorithm [47] can drastically reduce (typically by 40–50 times or more) the average traffic rate demands for the video delivery but they further increase the burstiness of the underlying traffic profile. In order to tackle this burstiness, buffers are used to temporarily host the extra bytes, before passing them to the transceiver module. By this way, the communication with slower components is possible. The size of these buffers is a critical parameter to prevent from packet losses. Larger buffers result in lower losses but increase the inter-packet delays, which can also result in quality distortions of the delivered real-time application.

It must be noted that the communication using LoRa radios is under a fair policy suggesting a limitation of 1% duty cycle per each transmitting device [48]. This means for instance that if a node transmits for one second it should wait politely for 99 consecutive seconds before transmitting again. These extra constraints make the LoRa ratio suitable for agricultural applications that have matching monitoring parameters needs, i.e., that have a scarce traffic activity potential, while seems quite "unorthodox" the LoRa radios to be used for (voice/video) streaming applications. Nevertheless, in case of emergencies, the hiring of the LoRa protocol for the transmission of more bandwidth-demanding, real-time applications, like surveillance video content might be justified. Apparently, this option is expected to provide results of very poor quality but it is worth mentioning as, in some cases, there is no radio link alternative. The challenging issue of the delivery of imaging content over LoRa links, especially in rural areas, is drawing the attention of other research teams [49, 50] as well.

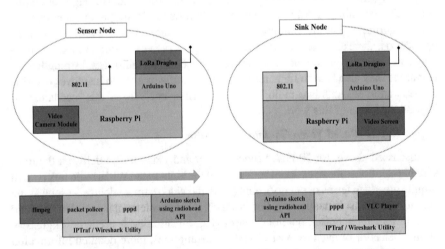

Fig. 3 Architecture arrangements for testing the (video) IP functionality over LoRa links

In order to test the latter case and to make, in parallel, the minimal hardware and software module modifications, the configuration depicted in Fig. 3 has been selected for testing the IP-style video delivery over LoRa links.

At the sensor node end, a raspberry pi unit, equipped with a camera module (either the built-in or a USB one), was configured to run the ffmpeg software package [51] for parametric generation and delivery of mpeg-4 video traffic. In parallel, the pppd software [52] was used to provide IP functionality over a serial link. The serial link was initially implemented using a USB connection between the raspberry pi and an arduino uno unit. The arduino uno unit was equipped with a LoRa dragino shield [53] and was running specially-written software which encapsulated the USB serial data (containing the pppd frames with the ffmpeg payload) into corresponding LoRa frames, using the RadioHead library [54].

At the sink (gateway) node, the hardware setup was identical, except the absence of a camera module. More specifically, an arduino uno unit, equipped with a LoRa shied was receiving the LoRa frames and extracting the LoRa payload. The latter payload was delivered throughout the USB (serial) connection towards the receiving raspberry pi unit that was running the pairing pppd service and an mpeg-4 video client (e.g., like the VLC client application [55]).

The extent to which other conventional services based on the Internet Protocol could function over LoRa links seems to be also a challenging issue. Indeed, the pppd mechanism, being used, can provide the necessary abstraction from the LoRa implementation details to test similar services. Nevertheless, the various bottlenecks that the LoRa protocol is implying, due to its nature, would limit any potential service to very poor data rate performance, especially a TCP-based one.

4 Implementation Details and Evaluation Techniques

This section provides further details on how the separate parts of the experimental WSAN multi-modal nodes have been interconnected and programmed. It also highlights the materials and the methods being followed to assess the most characteristic performance issues of these nodes. In order to implement and test the proposed pilot sensor and actuator nodes, a wide set of hardware components have been combined, as indicates the list below:

- Arduino uno and arduino mega microcontroller boards.
- Raspberry pi (model 3) boards.
- Esp8266 (WeMos D1 R2) boards [56].
- Wi-Fi radio modules (TP Link TL-WN722N).
- LoRa dragino shields for arduino boards at 868 MHz radio frequency.
- Digi XBee and Waveshare ZigBee, IEEE802.15.4-based modules.
- Raspberry pi camera module and Logitech C170 web camera.
- Several low-cost sensor components (e.g., elements for measuring temperature, humidity, luminosity, etc.)

- Indicative actuator components (relays, motor drivers, speakers, etc.)
- Components for direct or indirect energy measurements and some discrete electronics (e.g., capacitors and resistors).
- Solar panels, batteries and voltage regulating and charging equipment.
- Box enclosure units and connecting cables and wires.

The role, the strengths and the weaknesses of these important modules is further explained through the following description, via prototype configuration deployment and experimentation. In all cases, the overall cost is kept low, typically varying from 50€ to 250€, for wider feasibility purposes.

The Diversity of the Participating Modules

The multi-modal sensor nodes are comprised of modules of diverse characteristics, in terms of performance and operational requirements. A wide variety of actuator components may coexist with the basic sensing elements. The power needs of all these modules, the time granularity of their operation, their tolerance to environmental conditions, their communication interface and many other characteristics may vary considerably. For on–off operations (e.g., for solenoid valves or constant lighting), relays of low-driving current, either conventional or solid-state ones are needed. In case of variable operation (e.g., ventilation of controllable flow or adjustable lighting), the driving circuits may involve more composite power control methods. The information to be gathered in a farm is acquired by a variety of sensors (e.g., for measuring air/soil humidity, luminosity, distance, position or temperature) connected on the microcontroller unit, via its GPIO pins. This information can be simple on–off (e.g., corresponding to a position switch) or more complex that needs special protocols (e.g., the I2C or analog to digital conversions). The left part of Fig. 4 depicts exactly this diversity of components being examined.

In response to this diversity, the sensor node may follow a quite simple arrangement of components or a more sophisticated one. The right part of Fig. 4 depicts a node setup, able for simple operations, involving an arduino unit, a gas sensor, a simple relay and a ZigBee transceiver. A more composite sensor example is depicted in the left part of Fig. 5. According to it, low-power elements (an arduino unit, an XBee module, plus an array of sensoria for luminosity, temperature, humidity, proximity and sound) coexist with high-power units (a raspberry pi with Wi-Fi and a camera). In the right part of Fig. 5, the raspberry pi, the arduino and the LoRa radio components are clearly shown.

Two types of batteries have been used, the sealed acid-lead batteries and the LiPo ones. The first type is a very common option for home security systems, the second one is similar to the powering unit of mobile phones and tablet devices. The sealed-lead batteries are more tolerant to misuse while the LiPo ones are more expensive and require meticulous handling (e.g., are more flammable) and sophisticated circuits for charging them properly (e.g., efficient switching regulators). The matching solar panel dimensions depended on the expected average consumption of the node and varied from 1 W at 6 V to 15 W at 12 W (or larger units).

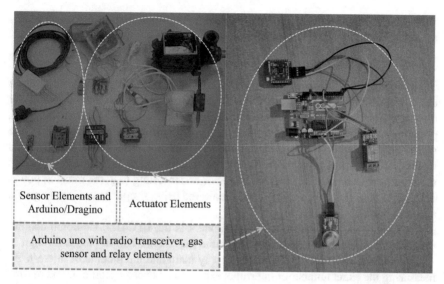

Fig. 4 The diversity of sensor and actuator elements and an indicative implementation example

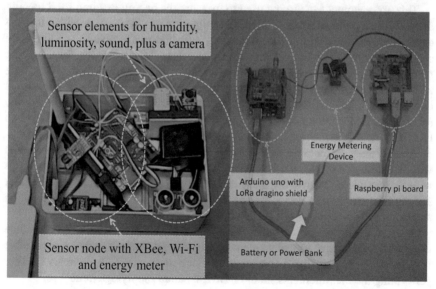

Fig. 5 Sensor implementation example (left) and energy measuring equipment details (right)

The low-level software needed to support this functionality comprises of vendor-specific parts as well as custom implementations (usually in C language-based, python or linux Bash environments). Network programming techniques were also applied. More specifically, linux wireless tools packages, containing commands like ifconfig or iwconfig, and parallel threads/processes communicating mainly via

UDP sockets have been exploited to implement the corner-stone components of this functionality.

Methodology and Equipment for Energy Measurements

Many operational parameters (e.g., radio coverage, transmit power level) can be directly acquired/adjusted using the software application interfaces provided by the component manufacturers. Added to this, the experimental deployment being presented also involves modules dedicated in measuring the performance of the radio interfaces and of the other components of interest, during their activity. This methodology is mostly based on the approach described in [57]. Special software modules have been written so as the behavior under testing (e.g. the transmit power or the packet rate) to be controlled and monitored using simple commands (e.g., via a smart phone).

For energy measurements, a key component is the LTC4150 Coulomb counter [58], which generates an interrupt every time that 0.0001707Ah passes through it. This module is connected between the supply main and the module of interest. By measuring the exact number of interrupts experienced over a specific period of time the overall consumption is calculated. The role of the LTC4150 meter is shown in both parts of Fig. 5 (at the right in greater detail).

This arrangement can also be used to measure the energy required for a specific fast-varying repetitive action (e.g., the transmission of a packet), provided that the aggregate quantities during inactivity (i.e., with the radio module at idling) are subtracted from the corresponding aggregate values during activity, for a specific number of repetitions, N. Finally, the exact amount of energy spent for a single activity of interest is calculated through division of the latter difference by the N quantity [59]. This mechanism, installed in a system consisting of an arduino uno (or a WeMos, after minor connection modifications), a LoRa dragino shield and an LTC4150 module to provide LoRa radio-specific measurements, is depicted in Fig. 6.

Alternatively, to measure the consumption of a module (or of the overall node) with fine time granularity, the voltage drops over a small, in value, resistor, in series with the module of interest are measured. An ADS1015 [60] differential amplifier and digitizer component provides these readings to an arduino unit for visualization and storage. This arrangement is shown in Fig. 7.

Apart from the resistor, a comparatively large capacitor, in parallel with the module under testing, is used as a memory element, in order not to miss the impact of possible very sudden peaks in current consumption. For better results, the modules used for performing the measurements should have a separate supplying circuit.

Arrangements for Testing Video over LoRa Links

The hardware arrangements for testing the IP-style video delivery over LoRa radio links are depicted in Fig. 8. More specifically, a camera module, a raspberry pi and an arduino unit (or a WeMos device, after minor connection modifications for the basic signals, due to the slightly different pinout of this board compared with the arduino uno/mega pinout) equipped with a LoRa dragino shield.

As the video traffic generated by the ffmpeg application is expected to be bursty, apart from the limited LoRa link capacity, special care should be taken to counterbalance this behavior as well and avoid losses. Towards this direction, the combination of two sets of methods was necessary:

- The first set was hiring traffic shaping techniques via the suitable tc [61] linux package running on the raspberry pi. More specifically, in-line with the study in [62], the qdisc method has been properly parameterized to make the video traffic profile more "smooth" before the delivery via the USB and LoRa interfaces.
- The second set of methods involved modification of the serial buffer size characteristics of the microcontroller hosting the LoRa transceiver module. More specifically, from within the arduino IDE parameters, this buffer size was set to values near the maximum to avoid losses caused by overflows and finally microcontrollers with larger amount of memory were also hired.

Detailed variants of the matching software configuration are depicted in Fig. 9. The pppd connection is first established, a traffic policy mechanism is applied to the corresponding ppp0 interface and finally, the video flow is generated in the form of mpeg-4 UDP packets. Low-level software, running on the unit hosting the dragino shield, intercepted the video frames and encapsulated them into LoRa frames, using the RadioHead library.

The two raspberry pi nodes occasionally had their Wi-Fi interface active (and in some cases their ethernet interface) to provide an out-of-band link, for monitoring purposes. Indeed, SSH or VNC sessions were opened through these Wi-Fi interfaces,

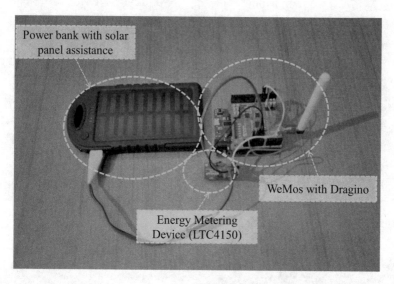

Fig. 6 Arrangements for testing the energy requirements of the packet delivery over LoRa links

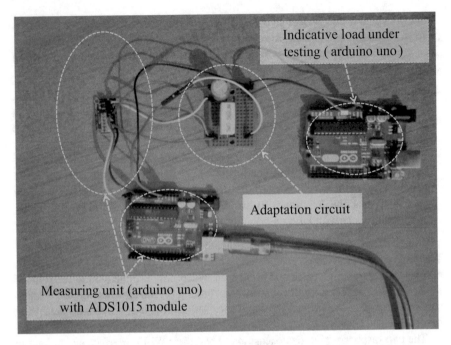

Fig. 7 Alternative measuring circuicity for fast-varying power (current) measurements

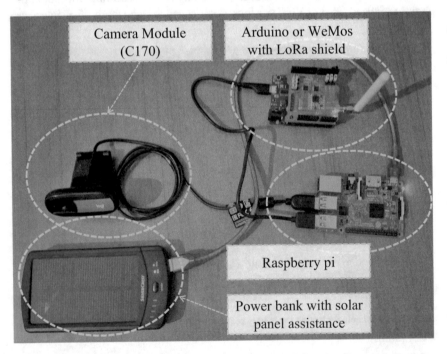

Fig. 8 Hardware implementation details for testing the (video) IP functionality over LoRa links

```
pppd -d /dev/ttyACM0 9600 noauth nodetach

killall pppd

pppd -d /dev/ttyACM0 38400 noauth nodetach noccp

tc qdisc del dev ppp0 root

tc qdisc add dev ppp0 root tbf rate 16kbit latency 5000ms burst 350

tc qdisc add dev ppp0 root tbf rate 15kbit latency 5000ms burst 500

tc qdisc add dev ppp0 root tbf rate 14kbit latency 8000ms burst 500

tc qdisc add dev ppp0 root tbf rate 15kbit latency 10000ms burst 250 peakrate 25kbit
minburst 250

ffmpeg -i /dev/video0 -s 320x240 -r 2 -vcodec mpeg4 -q 25 -f mpegts
udp://192.168.5.101:12345?pkt_size=195

ffmpeg -i /dev/video0 -s 160x120 -r 4 -vcodec mpeg4 -q 25 -f mpegts
udp://192.168.5.101:12345?pkt_size=195

ffmpeg -i /dev/video0 -s 320x240 -r 2 -vcodec mpeg4 -q 25 -f mpegts
udp://127.0.0.1:12345?pkt_size=192

ffmpeg -i /dev/video0 -s 320x200 -r 3 -vcodec mpeg4 -q 25 -f mpegts
udp://192.168.5.101:12345?pkt_size=200

frame= 443 fps=2.8 q=25.0 Lsize=   254kB time=00:02:40.00 bitrate=  13.0kbits/s dup=0
drop=1067 speed=  1x
```

Fig. 9 Examples of linux shell commands (on the raspberry pi unit) to establish the proper connection for generating and delivering video streaming over LoRa, in three steps

for coordinating the overall node activity. A set of monitoring tools, like the IPTraf [63], were used to verify that the traffic of interest was properly generated and delivered, according to methods described in [64]. Further experiments involved the Wireshark [65] utility package, a continuously evolving tool that provides detailed information and statistics referring to the packets circulating on the network.

5   Results and Discussion

This section provides characteristic results revealing the idiosyncrasies of the diverse components comprising the multi-modal nodes of interest. Several issues, such as the of radio coverage, the bandwidth requirements, the power consumption and the response times, are assessed.

Using the software provided by the manufacturers of the radio components, in conjunction with custom modules being developed, several radio performance metrics were gathered and analyzed. Figure 10, depicts the characteristic differences in performance behavior (i.e., the RSSI [66] over distance metric) between LoRa and Wi-Fi radios. The vertical axis values are in dBm while the horizontal axis values

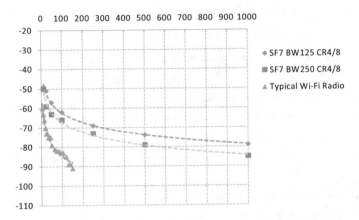

Fig. 10 Different performance behaviors (RSSI over distance) between LoRa and Wi-Fi radios

are in meters. By modifying LoRa radio link parameters, like the spreading factor—SF, the coding rate—CR and the bandwidthBW, two different configurations were created, both adjusted at the 10dBm transmit power level, and compared with the performance of a Wi-Fi radio having the same transmit power and a matching type of antenna. Due to the superior modulation techniques the LoRa technology is incorporating, the LoRa link was capable for much longer distance communication, i.e., more than 1000 m, while the Wi-Fi did not exceed the 150 m limit. On the other hand, as indicated by performance tools like the IPTraff, the Wireshark and custom ones, the Wi-Fi was capable for rates of the order of Mbps, while potential of the LoRa link was limited to a few kbps. These results justify the decision to use LoRa links for the transmission of elementary sensing data and/or some metadata and Wi-Fi links (or other fast links like the cellular ones) for the transmission of bandwidth-greedy content, like images or video.

In terms of power consumption, not only the differences among the various modules are worth mentioning but even the behavior of each one of them for different operational parameters settings. In this concept, the left part of Fig. 11 depicts the power consumption, in mA, of a LoRa dragino unit, as a function of the transmit power, in dBm, (the consumption of the hosting arduino at idling has been subtracted), while the right part of Fig. 11 highlights the drastic changes in the energy required (in mJ) for the transmission of a specific length payload inside a LoRa frame, for different coding rate (CR) arrangements.

Figure 12 depicts the changes in the overall power consumption, of a multi-modal composite node, due to the activation of the Wi-Fi unit and (later on) of the camera module. The vertical axis values are in mA while the horizontal axis values are in seconds. The involvement of the Wi-Fi radio was incurring an increase of 100 mA, while the video activity (i.e., including the camera sensor, the video packet encoding and the delivery over the Wi-Fi link) was incurring a 90 mA shift, approximately. The non-activity phase (i.e., at the level of 275 mA) was dominated by the consumption of the raspberry pi itself.

The power consumption of the raspberry pi unit at idling can be trimmed down to drop below the 150 mA border (by deactivating additional greedy modules like the USB hub, if not needed), but, even then, it remains be quite significant. On the other hand, the power consumption of the arduino uno unit was 50 mA approximately. Slight modifications of the arduino uno board, or the hiring of an arduino nano or micro unit dropped the consumption bellow the 5 mA level. The power consumption of the LoRa (dragino) module at idling was also bellow 5 mA, and the one of the ZigBee at about 50 mA, via USB, and 30 mA, directly. Furthermore, the LoRa was experiencing more sudden peaks in its power consumption during transmissions, compared with the Wi-Fi and the ZigBee modules. The latter had almost negligible changes from its idling level. The arrays of sensing elements participating in our experiments typically consumed a few mA, in total. The consumption of the relays and similar switching/driving components, without load, varied according to the type of the module, starting from 10 mA. The loads typically consumed for several tenths of mA to a few amperes.

After this analysis, it is apparent that the aggregate consumption of a WSAN node is not negligible, and thus, the hiring of a duty cycle (dc) policy, with the low-energy processor to activate/deactivate the rest of the modules, would provide considerable power efficiency improvements. The hiring of an even "loose" sleep policy (i.e., via the SLEEP_MODE_PWR_DOWN mode) for the low processor unit (i.e., the arduino) can further decrease the power consumption. Figure 13 depicts the estimated savings in energy (as a percentage of the always-active case—vertical axis) over the extra load during the period of activity (in mA – horizontal axis), for different duty cycle arrangements. The total off-state consumption is assumed to be at the 100 mA level. As extracted by the inspection of the graphs, the lower the duty cycle is, the better the energy savings. Even modest duty cycle values can save a lot of energy, provided that the loads at the period of activity are quite high.

The total power consumption, in case that a high-power component (raspberry pi) is activated/deactivated by a low-power component (arduino), through external control of its power supply via relay, is shown in Fig. 14. The vertical axis values

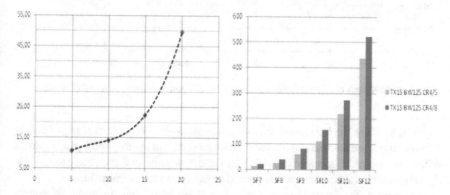

Fig. 11 Consumption of LoRa transceiver modules for various operation settings, in mA and mJ

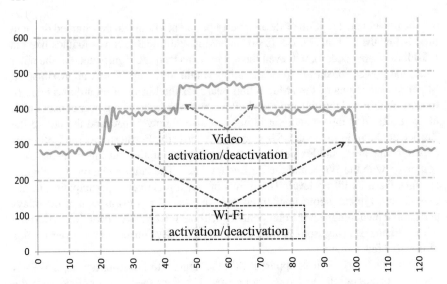

Fig. 12 Sensor's consumption, in mA, during the various phases of the high activity state

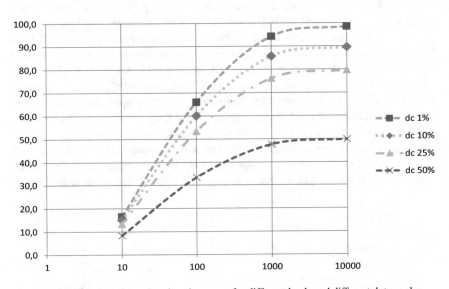

Fig. 13 WSAN node estimated savings in energy for different loads and different duty cycles

are in mA, while the horizontal axis values in seconds. The system is experiencing intense amperage fluctuations, as the various modules on the raspberry pi are started. After a period of 10 s, approximately, the system is stabilized and ready to use. On the other hand, the response to the powering off command is immediate.

Apparently, the latter control method is more beneficial, in terms of energy savings, for high-demand components that do not support sleep mode. Indeed, the

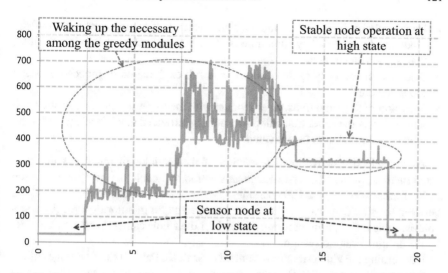

Fig. 14 Consumption while waking up "greedy" modules, using an external relay mechanism

improvement in performance is apparent by inspecting/comparing the behaviors described in Figs. 12 and 14. On the other hand, by this method, the amperage fluctuations are more intense and the time until stable operation is longer, compared with the 1–3 ms required according to the techniques behind the behavior depicted in Fig. 12.

Concerning the characteristics of the batteries involved in the experiments, the smaller and lighter LiPo batteries, are exhibiting smaller internal resistor and almost linear relation between their voltage and their remaining capacity. The latter feature simplifies the extraction of node's health status data (e.g., the remaining time of activity can be estimated directly based on these voltage measurements). The use of solar panel assisted power banks in our experiments is indicative and partial. This is because, many of these devices have an automatic shut-off operation, typically when the draining current becomes lower 50 mA or 100 mA. This feature makes them unsuitable for supplying modules with a short-time consumption average below this threshold. Alternatively, a resistor, in parallel, can be used to increase the total consumption beyond this limit, but this non-optimal constant increase should be taken into account during the energy estimations.

The whole involvement in the implementation and the testing process made apparent the importance of a set of good practices to be followed, for reducing the energy consumption of the multi-modal sensor nodes:

- Operation of the modules at the lowest possible supply voltage option (e.g., at 3.3 V instead of 5 V).
- Usage of high efficiency components for voltage regulating and power supplying and actuator driving issues (i.e., MOSFET instead of JFET transistors).

- Usage only of the necessary modules and features. If components with no sleep mode are necessary, they should be switched on and off by the node's modules that support sleep mode.
- Selection of low duty cycle operation and full exploitation of processors' sleep mode features.
- Selection of the radio transceiver module with the lower power consumption and adjustment at the lower transmit power setting, provided that it is adequate for the type of information to be delivered.

As the selective activation/deactivation of the modules of interest is the indicated method, in the IP-over-LoRa encapsulation approach, one-two bytes of each LoRa frame were reserved for passing module activation/deactivation commands between the sink (gateway) node and the sensors and for monitoring data. A noticeable example is the ability of turning on and off the whole camera/raspberry module, by the arduino/dragino system.

Concerning the video streaming content over LoRa links, at the first stages of the testing process, dummy UDP packets were generated from the sensor node towards the sink node, via a custom application, and were delivered through the LoRa radios, while the classic ping command was used for testing the connectivity between the sensor and the sink node. The LoRa link capacity was limited to a few kbps, even using the faster transceiver arrangements [34, 35], and, to tackle the burstiness of the video traffic, an extra mechanism had to be involved. By activating a linux traffic shaping mechanism these losses were reduced. Nevertheless, as the handling of these bursts is not an easy task, significant losses (near to 15%) were still experienced, due to the limited buffer size of the serial port of the underlying arduino uno units.

Things became slightly better by increasing the incoming serial buffer size (RX) of the arduino that was connected to the sensor node that generated the video traffic and by decreasing its outgoing buffer size (the opposite settings were applied to the arduino of the sink (receiver) node), arrangements possible via the arduino IDE environment. The replacement of the arduino uno unit by an arduino mega unit (that has 4 times more RAM) and, in turn, by a WeMos unit (which has even better specifications) made the losses to drop below the 5% level.

System also performed slightly better with video frames of smaller size (of about 200 bytes), values bellow the max packet frame of the LoRa protocol, but in any case the total video traffic rate being achieved could not exceed the 9kbps border, for the faster LoRa configuration settings. Figure 15 depicts (in the top part) the packet rate of the video traffic over the LoRa links, as it was intercepted at the sinks' end, by the Wireshark application. The bottom part of Fig. 15 depicts the packet size distribution of this traffic stream. The protocol headers, during the encapsulation process generated a non-negligible overhead of about 50 bytes over the about 200 bytes of the video payload.

Concluding, the delivery of video over LoRa, by using generic hardware and software methods, was really achieved, but the content was of very poor quality. It must be noted that any TCP-based application (like SSH), that typically require a fast acknowledgement mechanism to work, experienced very long delays. More

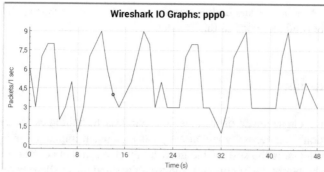

```
=====================================================================================
Packet Lengths:
Topic / Item        Count       Average      Min val      Max val     Rate (ms)   Percent
-------------------------------------------------------------------------------------
Packet Lengths      243         232,00       72           242         0,0050      100%
 0-19               0           -            -            -           0,0000      0,00%
 20-39              0           -            -            -           0,0000      0,00%
 40-79              1           72,00        72           72          0,0000      0,41%
 80-159             5           84,00        82           92          0,0001      2,06%
 160-319            237         235,80       202          242         0,0049      97,53%
 320-639            0           -            -            -           0,0000      0,00%
 640-1279           0           -            -            -           0,0000      0,00%
 1280-2559          0           -            -            -           0,0000      0,00%
 2560-5119          0           -            -            -           0,0000      0,00%
 5120 and greater   0           -            -            -           0,0000      0,00%
-------------------------------------------------------------------------------------
```

Fig. 15 Video packet delivery characteristics during transmission over the LoRa radio link

sophisticated shaping techniques involving linux kernel modification might provide slightly better results but always with respect to the 9kbps border.

In recapitulating, the modern computer systems being available to compose sensor and actuator nodes are advanced enough to provide a variety of configuration options in response to the wide range of the operational requirements of the agricultural sector, involving activation/deactivation of critical modules. This fact may lead to a more efficient operation, usually having as priorities the optimization of the radio coverage, the processing power, the response times and of the energy consumption. The performance measurements being conducted reveal this potential for diverse behavior settings. In some cases it is preferred to completely powering off a "greedy" module instead of handling it "gently". The nature/quality of the participating actuator and energy harvesting/regulating equipment also has a strong impact on the overall performance. The quite "unorthodox" solution of hiring low-energy, long-range and low-cost radio modules to carry conventional IP services has also been assessed but delivered very poor quality results.

The selection of low-cost and easy-to-find components that offer sufficient support and allow for fast application development, for implementing both the network nodes and the measurement equipment, had an additional benefit. More specifically, it comprised an educationally-friendly environment for the students of agricultural engineering participating in the experiments, who acquired hands-on experiences and enriched their technological background as future professionals. The latter benefit was verified by methods and findings similar to the ones described in [8], but further analysis is beyond the scope of this chapter.

6 Conclusions

The effective engagement of various IoT wireless technologies in the agricultural operations is a necessity and already has significant impact on many production sectors, towards the sustainable growth objective. In this context, this chapter intended to highlight the importance, the feasibility and the challenges of a synergy between diverse emerging technologies in terms of networking, processing, sensing and acting, thus aiming at agricultural production of improved qualitative and quantitative characteristics, at reduced cost levels. More specifically, a set of possible implementation and configuration arrangements involving multi-modal sensor nodes were studied and relevant performance evaluation results were presented. The modern systems used to compose sensor and actuator nodes are advanced enough to respond to the diverse set of the operational requirements, the agricultural tasks are imposing. Data can be collected and decisions can be made in a faster, more accurate and less energy-consuming manner, by systems which are able to dynamically adapt their set of active resources to the current needs. This technique leads in more compact layouts and also saves energy and money. The whole approach is also beneficial for the university students getting involved who developed better engineering background.

Future plans include conducting similar experiments with a wider set of more compact and efficient hardware modules, in terms of sensing, communicating, acting and processing. Added to that, more sophisticated multi-hop scenarios and decision algorithms of increased edge intelligence will be studied, in order to leave less job for the cloud and further assist the growth of self-adaptive autonomous interconnected systems that will be able to make efficient decisions and participate in larger digital ecosystems of smart agriculture.

References

1. United Nations (2019) Population. Online at: https://www.un.org/en/sections/issues-depth/population/index.html. Accessed 30 Apr 2020
2. FAO (2017) The Future of Food and Agriculture—Trends and Challenges. Online at: https://www.fao.org/3/a-i6583e.pdf. Accessed 30 Apr 2020
3. Symeonaki EG, Arvanitis KG, Piromalis DD (2019) Cloud computing for IoT applications in Climate-Smart agriculture: a review on the trends and challenges toward sustainability. In: Theodoridis A, Ragkos A, Salampasis M (eds) Innovative approaches and applications for sustainable rural development. HAICTA 2017. Springer Earth System Sciences. Springer, Cham, pp 147–167
4. O'Grady M, O'Hare G (2017) "Modelling the smart farm" in Information Processing in Agriculture, 4(3). https://doi.org/10.1016/j.inpa.2017.05.001

5. Bonneau V, Copigneaux B, Probst L, Pedersen B (2017) Industry 4.0 in agriculture: focus on IoT aspects. Eur Comm Online at: https://ec.europa.eu/growth/tools-databases/dem/monitor/sites/default/files/DTM_Agriculture%204.0%20IoT%20v1.pdf. Accessed 30 Apr 2020
6. Akyildiz IF, Kasimoglu IH (2004) Wireless sensor and actor networks: research challenges. Ad Hoc Netw. 2(4):351–367
7. Welsh M, Mainland G (2004) 'Programming Sensor Networks Using Abstract Regions', In: Proceedings of USENIX NSDI Conf
8. Loukatos D, Arvanitis KG (2019) "Extending Smart Phone Based Techniques to Provide AI Flavored Interaction with DIY Robots, over Wi-Fi and LoRa interfaces", MDPI – Education Sciences,9(3):224–241. https://doi.org/10.3390/educsci9030224
9. FAO (2020) Biodiversity for food and agriculture and ecosystem services – thematic study for the State of the World's biodiversity for food and agriculture. Rome. https://doi.org/10.4060/cb0649en
10. Math RKM, Dharwadkar NV (2018) "IoT Based Low-cost Weather Station and Monitoring System for Precision Agriculture in India", In: Second International conference on I-SMAC (IoT in Social, Mobile, Analytics and Cloud) (I-SMAC 2018), pp 81–86
11. Ramani JG, LakshmiPriya A, Madhusudan S, Kishore PJR, Madhisha M, Preethi U (2020) "Solar Powered Automatic Irrigation Monitoring System," In: Sixth International Conference on advanced computing and communication systems (ICACCS), Coimbatore, India, pp 293–297. https://doi.org/10.1109/ICACCS48705.2020.9074220
12. Cousins D (2017) "Ultimate guide to farm security kit". Online at: https://www.fwi.co.uk/machinery/ultimate-guide-farm-security-kit. Accessed 20 August 2020
13. Dunn C (2019) "How Surveillance Solutions Ensure Cannabis Compliance". Online at: https://www.securitymagazine.com/articles/89937-how-surveillance-solutions-ensure-cannabis-complianceteafy. Accessed 20 August 2020
14. Weinstock D, Harler C (2014) "Emerging Markets: From Cattle to Deere: Security technology can help protect a farm's valuable livestock and expensive equipment". Online at: https://www.securityinfowatch.com/video-surveillance/article/12138944/security-technology-can-help-protect-a-farms-valuable-livestock-and-expensive-equipment. Accessed 20 August 2020
15. Patmore S (2017) "Cameras key to cut sheep water runs". Online at: https://www.countryman.com.au/countryman/livestock/cameras-key-to-cut-sheep-water-runs-ng-b88687389z Accessed 20 August 2020
16. SIBI (2019) "New on-farm technology for sheep producers", Sheep Industry Business Innovation project (SIBI). Online at: https://www.agric.wa.gov.au/feeding-nutrition/new-farm-technology-sheep-producers. Accessed 20 August 2020
17. Pontikakos C, Tsiligiridis T, Yialouris C, Kontodimas D (2012) Pest management control of olive fruit fly (Bactrocera oleae) based on a location-aware agro-environmental system. Comput Electron Agric 87:39–50. https://doi.org/10.1016/j.compag.2012.05.001
18. Shaked B et al (2018) 2018) Electronic traps for detection and population monitoring of adult fruit flies (Diptera: Tephritidae. J Appl Entomol 142:43–51. https://doi.org/10.1111/jen.12422
19. Matthews S, Miller A, Clapp J, Ploetz T, Kyriazakis I (2016) Early detection of health and welfare compromises through automated detection of behavioural changes in pigs. Vet J 217:43–51. https://doi.org/10.1016/j.tvjl.2016.09.005
20. Tscharke M, Banhazi TM (2016) A brief review of the application of machine vision in livestock behaviour analysis. J Agric Inform 7(1):23–42
21. Tzanidakis C, Simitzis P, Arvanitis K, Panagakis P (2019) Precision livestock farming (PLF) techniques in Pig husbandry. In: 11th National Conference of the Hellenic society of agricultural engineer, Volos, Thessaly, Greece
22. Hassan D-S, Hasan M, Li D (2016) Information fusion in aquaculture: a state of art review. Front Agric Sci Eng 3. https://doi.org/10.15302/J-FASE-2016111
23. Garcia M, Sendra S, Lloret G, Lloret J (2011) Monitoring and control sensor system for fish feeding in marine fish farms. IET Commun 5(12):1682–1690
24. Lloret J, Garcia M, Sendra S et al (2015) An underwater wireless group-based sensor network for marine fish farms sustainability monitoring. Telecommun Syst 60:67–84. https://doi.org/10.1007/s11235-014-9922-3

25. Oteafy S, Hassanein H (2014) "Future Directions in Sensor Networks", pp 115–121. https://doi.org/10.1002/9781118761977.ch10
26. EPRS (2016) European Parliament Research Service—EPRS, "Precision agriculture and the future of farming in Europe". https://www.europarl.europa.eu/RegData/etudes/STUD/2016/581892/EPRS_STU(2016)581892_EN.pdf. Accessed 25 August 2020
27. Bhakta I, Phadikar S, Majumder K (2019) State-of-the-art technologies in precision agriculture: a systematic review. J Sci Food Agric 99:4878–4888. https://doi.org/10.1002/jsfa.9693
28. Jawad HM, Nordin R, Gharghan SK, Jawad AM, Ismail M (2017) Energy-efficient wireless sensor networks for precision agriculture: a review. Sensors 17:1781
29. Symeonaki E, Arvanitis K, Piromalis D (2020) A context-aware middleware cloud approach for integrating precision farming facilities into the IoT toward agriculture 4.0. Appl Sci 10(3):813. https://doi.org/10.3390/app10030813
30. Suresh P, Daniel JV, Parthasarathy V, Aswathy RH (2014) "A state of the art review on the Internet of Things (IoT) history, technology and fields of deployment", In: International Conference on science engineering and management research (ICSEMR) 2014:1–8
31. Farooq MS, Riaz S, Abid A, Umer T, Zikria YB (2020) Role of IoT technology in agriculture: a systematic literature review. Electronics 9(2):319. https://doi.org/10.3390/electronics9020319
32. Han C et al (2011) Green radio: radio techniques to enable energy-efficient wireless networks. IEEE Commun Mag 49(6):46–54. https://doi.org/10.1109/MCOM.2011.5783984
33. ZigBee (2020), ZigBee protocol description on Wikipedia. https://en.wikipedia.org/wiki/Zigbee. Accessed 20 August 2020
34. Semtech (2020). LoRa FAQs. https://docplayer.net/9473940-Lora-faqswww-semtech-com-1-of-4-semtech-semtech-corporation-lora-faq.html. Accessed 25 August 2020
35. Sanchez-Iborra R, Sanchez-Gomez J, Ballesta-Viñas J, Cano MD, Skarmeta AF (2018) "Performance Evaluation of LoRa Considering Scenario Conditions." MDPI Sensors 18(3). https://doi.org/10.3390/s18030772
36. NDVI (2020) Normalized difference vegetation index description on Wikipedia. https://en.wikipedia.org/wiki/Normalized_difference_vegetation_index. Accessed 20 August 2020
37. Edge Computing (2020) Edge computing description on Wikipedia. https://en.wikipedia.org/wiki/Edge_computing. Accessed 20 August 2020
38. Garcia LP, Montresor A et al (2015) Edge-centric computing: vision and challenges. ACM SIGCOMM Comput Commun Rev 45(5):37–42. https://doi.org/10.1145/2831347.2831354
39. Merenda M, Porcaro C, Iero D (2020) Edge machine learning for AI-Enabled IoT devices: a review. Sensors. 20(9):2533. https://doi.org/10.3390/s20092533
40. Xhafa F, Kilic B, Krause P (2020) Evaluation of IoT stream processing at edge computing layer for semantic data enrichment. Futur Gener Comput Syst 105:730–736. https://doi.org/10.1016/j.future.2019.12.031
41. Aranda J, Mendez D, Carrillo H (2020) MultiModal wireless sensor networks for monitoring applications: a review. J Circuits Syst Comput 29:1–34. https://doi.org/10.1142/S0218126620300032
42. Lahat D, Adali T, Jutten C (2015) Multimodal data fusion: an overview of methods, challenges and prospects. Proc IEEE 103. https://doi.org/10.1109/JPROC.2015.2460697
43. IEEE 802.15.4–2015 (2016) IEEE standard for low-rate wireless networks, in IEEE Std 802.15.4–2015 (Revision of IEEE Std 802.15.4–2011), pp 1–709. https://doi.org/10.1109/IEEESTD.2016.7460875
44. Raspberry (2020) Raspberry Pi 3 Model B board description on the official Raspberry site. Retrieved in April of 2020 from the site: https://www.raspberrypi.org/products/raspberry-pi-3-model-b/
45. Arduino (2020) Arduino Uno board description on the official Arduino site. Retrieved in April 2020 from the site: https://store.arduino.cc/arduino-uno-rev3
46. MQTT (2020) The MQ telemetry transport protocol (Wikipedia). Retrieved in April 2020 from the site: https://en.wikipedia.org/wiki/MQTT
47. Sikora T (1997) The MPEG-4 video standard verification model. IEEE Trans Circuits Syst Video Technol 7(1):19–31. https://doi.org/10.1109/76.554415

48. ETSI (2012) "ETSI EN 300 220–1: Electromagnetic compatibility and radio spectrum matters (ERM); short range devices (SRD); radio equipment to be used in the 25 MHz to 1000 MHz frequency range with power levels ranging up to 500 mw; part 1 (2012)". https://www.etsi.org/deliver/etsi_en/300200_300299/30022001/02.04.01_40/en_30022001v020401o.pdf. Accessed April 2020

49. Jebril A, Sali A, Ismail A, Rasid M (2018) Overcoming limitations of LoRa physical layer in image transmission. Sensors 18:3257. https://doi.org/10.3390/s18103257

50. Fan C, Ding Q (2018) A novel wireless visual sensor network protocol based on LoRa modulation. Int J Distrib Sens Netw 14:155014771876598. https://doi.org/10.1177/155014 7718765980

51. ffmpeg (2020) The ffmpeg package documentation page. https://ffmpeg.org/ffmpeg.html. Accessed 30 April 2020

52. pppd (2020) The Point-to-Point Protocol (Wikipedia). https://en.wikipedia.org/wiki/Point-to-Point_Protocol. Accessed 30 April 2020

53. Dragino (2020) The LoRa Dragino shield for arduino. https://www.dragino.com/products/module/item/102-lora-shield.html. Accessed 25 April 2020

54. RadioHead (2020) The RadioHead library to support LoRa modules. https://www.airspayce.com/mikem/arduino/RadioHead/. Accessed 25 April 2020

55. VLC (2020) The VLC media player software. https://wiki.videolan.org/VLC_media_player/. Accessed 30 April 2020

56. WeMos (2019) The WeMos D1 R2 board. https://wiki.wemos.cc/products:d1:d1. Accessed 25 April 2020

57. Loukatos D, Manolopoulos I, Arvaniti E, Arvanitis K, Sigrimis N (2018) Experimental testbed for monitoring the energy requirements of LPWAN equipped sensor nodes. In: Proceedings of 6th IFAC Conference on Bio-Robotics, Beijing, China. https://doi.org/10.1016/j.ifacol.2018.08.196

58. LTC4150 (2019) The LTC4150 coulomb meter. https://www.sparkfun.com/products/12052. Accessed 25 April 2020

59. Wassie D, Loukatos D, Sarakis L, Kontovasilis K, Skianis C (2012) "On the energy requirements of vertical handover operations: Measurement-based results for the IEEE 802.21 framework". In: Proceedings of 17th IEEE International Workshop on computer aided modeling and design of communication links and networks (CAMAD 2012), pp 145–149 https://doi.org/10.1109/CAMAD.2012.6335316

60. ADS1015 (2020) The ADS1015 Analog-to-Digital converter circuit. https://cdn-shop.adafruit.com/datasheets/ads1015.pdf Accessed 25 April 2020

61. tc (2020) The linux traffic control package. https://wiki.debian.org/TrafficControl. Accessed 25 April 2020

62. Zoi S, Loukatos D, Sarakis L, Stathopoulos P, Mitrou N (2003) "Extending an Open MPEG-4 Video Streaming Platform to Exploit a Differentiated Services Network" in high-speed networks and multimedia communications (HSNMC 2003). Lecture Notes in Computer Science, vol 2720. Springer. https://doi.org/10.1007/978-3-540-45076-4_39

63. IPTraf (2020) The IPTraf IP network monitoring software. Retrieved in April 2020 from the web site: https://iptraf.seul.org/

64. Loukatos D, Sarakis L, Kontovasilis K, Skianis C, Kormentzas G (2007) Tools and practices for measurement-based network performance evaluation. 1–5. https://doi.org/10.1109/PIMRC.2007.4394164

65. Wireshark (2020) The wireshark network monitoring software. https://www.wireshark.org/. Accessed 30 April 2020

66. RSSI (2019) Received signal strength indicator – RSSI. https://en.wikipedia.org/wiki/Received_signal_strength_indication. Accessed 30 April 2020

Dr. Dimitrios Loukatos received the diploma in Electrical and Computer Engineering and the Ph.D. degree in Telecommunications and Computing, both from the National Technical University of Athens (NTUA), Greece. He currently is staff member of the Agricultural Engineering Laboratory of the Agricultural University of Athens, Greece. He has worked as a research associate at the NTUA and the National Centre for Scientific Research 'Demokritos'. He also worked as a senior engineer for the Institute of Geodynamics of the National Observatory of Athens, enhancing their seismic sensor network. His research interests include hardware/software platforms evaluation and optimization, management and applications of wired and/or wireless (sensor) networks, artificial intelligence, human–computer interaction, robotics and applications. His most recent work is focused on the area of IoT and Physical Computing. His research work has been published in many national and international scientific conference proceedings, book chapters and journals.

Dr. Konstantinos G. Arvanitis received the B.Sc. and Ph.D. degrees from the National Technical University of Athens, Department of Electrical and Computer Engineering, in 1986 and 1994, respectively. He currently serves as a Professor of "Automation in Agriculture" in Agricultural University of Athens, Department of Natural Resources Management and Agricultural Engineering, Laboratory of Farm Machinery, where he also served as Head/Deputy Head of the Section of Farm Structures and Farm Machinery. He published over 285 research papers in refereed journals, international scientific conferences and book chapters. His research work has received significant international recognition (over 3200 citations, h-index = 26, g-index = 53). His main research interests are: Electrification and Automation in Agriculture, Advanced Process Control, Wireless Sensor Networks, ICT and Artificial Intelligence Applications in Agriculture, Remote Sensing, Optimization Techniques, Energy Management and Control of Autonomous Micro-Grids, Internet of Things and Cloud Computing.

Design Architecture of Intelligent Agri-Infrastructure Incorporating IoT and Cloud: Link Budget and Socio-Economic Impact

Mobasshir Mahbub

Abstract With the geometrical development of mankind, that means the growth of the populace, the traditional or antiquated harvesting techniques are getting unfit to adapt to the growth with a satisfactory level. Subsequently, advanced harvesting techniques are much necessary to approach the need for nourishment of these growing numbers of populace. In nearby years, smart farming frameworks aided by embedded electronics and the Internet of Things (IoT) getting fascination and ubiquity among people to upgrade crop production for people. This work has endorsed a harvesting framework dependent on the embedded electronics, IoT, and wireless sensor systems for farm fields. This chapter incorporates the portrayal of frameworks with the associated electronic equipment and circuitry of the framework, utilized network protocols and smart remote monitoring frameworks for PCs and Smartphones, link budget and analysis, etc. Later it incorporates socio-economic impacts of intelligent farming and finally, the work concludes with conclusion including the portrayal of the future extents of relevant advancements in smart farming.

1 Introduction

To enhance agricultural production with fewer assets and work endeavors, significant developments have been made all through mankind's history. By and by, the higher growth of the population never lets the demand and supply coordinate during all these times. As per the forecasted figures, in 2050, the total population is expected to reach 9.8 billion, an expansion of around 25% from the present figure. Nearly the whole referenced ascent of the population is forecasted to happen among the developing nations. On the other hand, the pattern of urbanization is forecasted to proceed at a quickened pace, with about 70% of the total population anticipated being urban until 2050 (right now 49%). Moreover, the income rate will be the several times

M. Mahbub (✉)
Department of Electrical and Electronic Engineering, Ahsanullah University of Science and Technology, Dhaka, Bangladesh
e-mail: mbsrmhb@gmail.com

© Springer Nature Switzerland AG 2021
P. Krause and F. Xhafa (eds.), *IoT-based Intelligent Modelling for Environmental and Ecological Engineering*, Lecture Notes on Data Engineering and Communications Technologies 67, https://doi.org/10.1007/978-3-030-71172-6_6

129

multiplication of what they are currently, which will drive the food demand further, particularly in developing nations. Thus, these countries will be increasingly cautious about their eating regimen and food quality; henceforth, consumer preferences can move from wheat and grains to vegetables and, later, to meat. To take care of this bigger, increasingly urban, and more extravagant population, food production should double by 2050 [1, 2]. Especially, the present figure of 2.1 billion tons of yearly oat production should reach around 3 billion tons, and the yearly meat production should increase by more than 200 million tons to satisfy the demand of 470 million tons [3, 4].

We know that agricultural production is one of the crucial facts of mankind. In the whole world, around 60% of peoples are engaged in cultivation. The progressing headway in communication and data advancements has empowered farmers to accumulate an enormous proportion of site-explicit data for the farm fields. Farmers still relying upon the conventional cultivation systems such as the manual circulation of seeds and wrinkling, two harvests in a year strategy, informal structures of cultivation. The monsoons are erratic, and the lopsidedness of availability of water throughout the year represents a significant issue. Also, the farmers are depending upon the regular procedures for watering the farm field, fertilizing the field, splashing pesticides without decisively observing the definite condition of the farm field. This prompts insufficient growth and low productivity of yields. The execution of logical methodologies in the field of agribusiness can upgrade the productivity of harvests, in light of the improved efficiency in the cultivation [5].

To guarantee better farming and crop production sufficient watering, treating and pesticides are compulsory. It is a lot simpler if we make a mechanized framework to do these errands automatically. The endorsed frameworks in this work consist of farm field moisture observing and mechanized watering, computerized insect identification and pesticide splashing, pH measuring and fertilizing with quad-copters, intruder alert system for the farm field, etc.

The [6] agricultural production has accomplished a huge improvement. These upgrades have presented a robotized framework in which the development of yields can be monitored and devices are constrained by using WSN (Wireless Sensor Network). The key activity of WSN is to detect the information from a remote place and transmit the detected information over wireless networks that can be checked by the recipient. The WSN [7] technology can be used in farming particularly for managing dispersed information assortment from cultivating environments and significantly for coordinating farmers with continuous data of the cultivating field. The prescribed framework has additionally executed WSN furnished with all the dedicated sensing equipment for observing a large farming environment [8, 9].

In this work, the further sections will include some relevant literature review (Sect. 2), system architectures for smart farming including a description of used devices, circuit designing, networking layers and protocols, link budget (Sects. 3 and 4), socio-economic impacts (Sect. 5) and conclusion (Sect. 6).

2 Related Works and Literature

This first topical cluster is entrenched, with one line of inquiry concentrated on various parts of precision technology appropriation on the farm, analyzing both monetary and conducts perspectives. This writing focuses on singular adoption determinants portrayed in works of Barnes et al. [10], Kernecker et al. [11], Leonard et al. [12], just as expansion and open mediations to animate appropriation introduced by Kutter et al. [13]. A different line of inquiry looks at precision agribusiness's deployment on the farm and how it influences cultivating rehearses portrayed in the article of Hansen [14] and post-adoption procedures of adaptation by Higgins et al. [15] through ideas, for example, 'tinkering' and 'arrays'. The latter point has likewise been dissected from a beyond-farm level viewpoint, looking at the more extensive systems and advancement frameworks in which innovation is formed and where co-evolution between the innovation and more extensive social and institutional conditions described in the work of Eastwood et al. [16]. A bunch of expansions on assortment of techniques, ranging from modeling approaches of the expenses and advantages of precision cultivation introduced in the work of Schimmelpfennig et al. [17]. Quantitative or econometric methodologies for testing the impacts of various factors on adoption, for example, farm size and specialization, farmers' age, training, and so on., are described in the works of Annosi et al. [18]. Lowenberg-DeBoer et al. [19] in their work, featured the circumstances of both adopters and non-adopters of precision agriculture, and represented less quantifiable perspectives, for example, material utiliztion possibilities and social elements of information regarding the farming. It ought to be noticed that while most research in industrial contexts focused on the appropriation of precision agri-technologies, research in Africa rather focused on the adoption (or non-adoption) of market data frameworks developed by Wyche et al. [20]. The literature on the African perspective looks at agribusiness explicit decision support mechanisms as well as the role of nonexclusive innovations, for example, mobile phones, in the case of access to information and commodities' costs are introduced in the work of Baumüller [21].

Gutiérrez et al. [22] proposed an IoT-aided irrigation system framework including a WSN associated with a gateway and a remote server. Other than the proficient utilization of water, this project additionally adopted soil moisture sensing systems and temperature sensors. Smart sensing web [23] was proposed to quantify soil moisture by the sensing system and controlled them from a Web-aided framework to screen pH, temperature, humidity, and so on. Cambra et al. [24] proposed video analytics-based compost controlling in agribusiness production. An IoT-aided savvy security and monitoring framework is created in [25]. A soil moisture observing system is introduced in [26], which operates based on an energy effective algorithm. Dan et al. [27] observed atmosphere conditions utilizing a nursery monitoring framework with short-extend ZigBee technology. Ojha et al. [28] talked about the arrangement procedures of WSN for agribusiness and farming and investigated the system concerning advancements, principles, and equipment. Notwithstanding, these solutions mainly considered short-extend systems deployed in a territory where network connectivity over the Internet is accessible nearby. As farms and agribusiness essentially available in rural territories, interfacing the short-extend WSN from long distances and solving the farming issues still challenging. Further, maintaining QoS for such applications to satisfy distinctive system prerequisites is significant.

Contribution of This Work: This work has prescribed a system that contains the necessary components to make agricultural tasks automated. The work first prescribed embedded systems to scientifically sense the environment and automate tasks like watering, pH monitoring of soil, CO_2 monitoring, detecting the presence of animals near the field which can harm crops, etc. Moreover, the work has designed a quadcopter based on embedded electronics through which pesticides and fertilizer spraying can be done. The work also performed a link budget and link-level analysis of the communication modules to show the empirical performance of the system in a farming field (empirically). Finally, it has discussed the socio-economic impact of the automated farming system. According to the author's best of the knowledge and as per reviewed articles there is no such article or work in which all these topics are included. Hence, this work extends the literature and development regarding the intelligent farming concept and might lead the way towards further research and development.

3 System Architecture, Software Architecture, and Frameworks

3.1 System Description

The number of sensor nodes to be deployed in a farm field is relying upon the total area of the field. Each of the sensor nodes is equipped with the sensing devices mentioned earlier. Every sensor will gather information from the farm environment and the information will be handled by the relevant microcontroller unit. The DHT11 temperature and humidity sensor will distinguish the humidity and temperature of the field, PIR sensor will recognize the movement from objects that emanate infrared signal to forestall the trespassing of people or animals or birds in the field and if something identified, it will trigger the alert utilizing the buzzer, MQ-135 gas sensor is utilized to identify and measure the amount of CO_2 present in the field, the barometric pressure sensor is utilized to quantify the pressure in the field to foresee the climate, for example, the water identification sensor is utilized to identify the rainfall, the soil moisture sensor is utilized to measure the amount of moisture in the soil of the farm field, pH sensor is utilized to quantify the pH estimation of the soil of the farm field. At that point sensed information from sensor nodes will be transmitted to the central hub or node utilizing the nRF24L01 wireless transceiver. Utilizing the nRF24L01 module as a receiver the central node will assemble every sensor node information and the corresponding sensor data as well. A specific threshold is set in the central node's MCU for every sensor. The central node ceaselessly compares the assembled information of every sensor from every sensor node and when the reading goes above or underneath the defined threshold, the central node will transmit the corresponding information by an SMS to the client's cell phone employing GSM module and the client can likewise capable of observing the sensor data of every sensor node through the different web pages assigned for every sensor node utilizing a cell phone or PCs with the assistance of ESP8266 Wi-Fi module. The central node is designated with the capability of doing some computerized errands, for example, depending upon the soil moisture data it can begin the robotized water pumping system to provide water in the farm field. It is also capable to inform the client with pH sensor data when fertilization is required to the field. The pesticide spraying and fertilizing will be maintained by the farming-copter (farming purpose quad-copter). The depiction of the farming-copter will be given in the forthcoming section of the chapter.

3.2 System Framework

The IoT based smart farming structure comprises of five interconnected sub-systems. These are—detection, analysis, communication, representation, and execution sub-system. The detecting or sensing sub-system is interconnected with the analysis sub-system where crude information of every sensor manipulated individually, prepared and analyzed utilizing various algorithms actualized in MCU for representation and execution sub-system [29]. Here the representation procedure of the manipulated sensed information is legitimately relied upon the communication sub-system, to visualize the information through the cell phone or PC's, one must be connected to the modules of the communication sub-system relying upon the information visualization procedure (If the client (generally a farmer) needs to show the information through the internet browser, then it certainly needs ESP8266 IoT module [30], if "he/she" needs to get SMS in a cell phone than the GSM module [31] required, if "he/she" attempts to visualize information over Arduino 'Sketch' IDE's 'Serial Monitor' than must need to rely upon the nRF24L01 module [32]). The framework is wisely capable of making decisions and execution depending on the manipulated sensed information through the execution sub-system which is the fundamental objective of the prescribed framework. Figure 1 shows the IoT based smart farming system.

3.3 Functional Framework

The prescribed IoT based system of the smart farming framework is divided into some functionality layers. The layers are shown in Fig. 2.

The functional framework represents how farmers can access multiple functionalities with the aid of the support layer from the application layer. The implementation

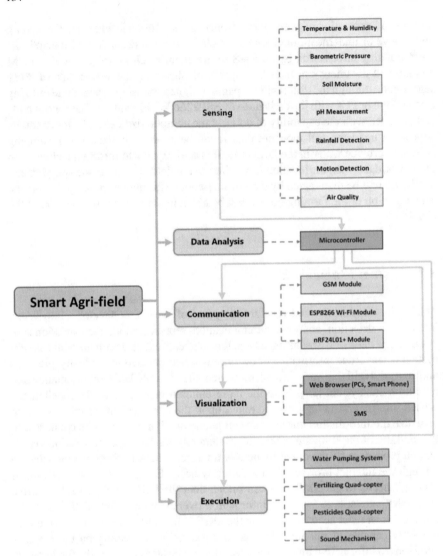

Fig. 1 IoT based smart agri-field system framework

layer contains the fundamental operations which are significant for any IoT based farming. The data acquisition layer establishes a connection with the session layer through IoT relevant protocols, for example, HTTP.

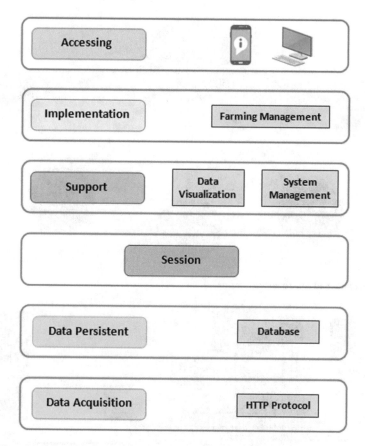

Fig. 2 Functional framework

3.4 Circuit Design

The smart agri-field system consists of two different circuits:

(i)   Central Node and
(ii) Wireless Sensor Nodes.

Figure 3 shows an illustration of central and sensor nodes.

Wireless Sensor Nodes—The sensor nodes are consists of Arduino Mega as MCU (Microcontroller Unit), nRF24L01 + PA/LNA wireless transceiver module, DHT11 temperature and humidity sensor, PIR motion sensor, MQ-135 gas sensor, barometric pressure sensor, rainfall/water detection sensor, pH sensor, soil moisture sensor, and a buzzer. Figure 4 shows the circuit design of each wireless sensor nodes deployed in the farming field with certain distant keeping in mind the range of the nRF24L01 wireless transceiver module.

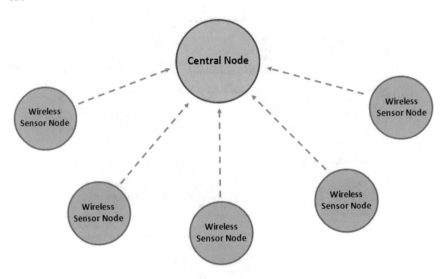

Fig. 3 Central and sensor nodes

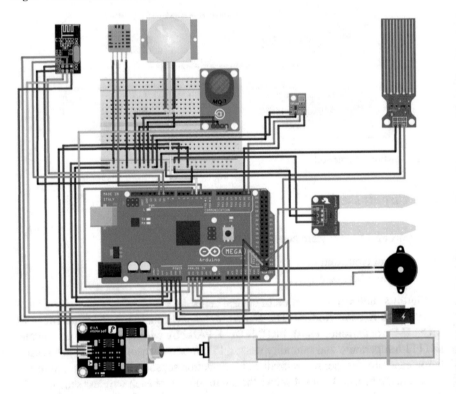

Fig. 4 Wireless sensor node

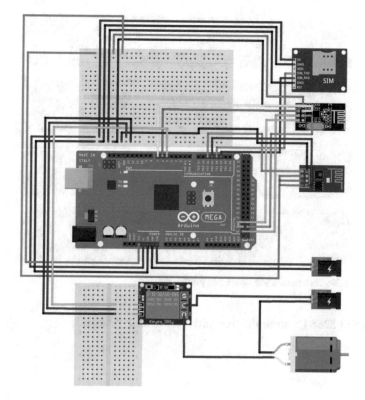

Fig. 5 Central node

Central Node—The central node consists of an Arduino Mega as a microcontroller, GSM module, ESP8266 Wi-Fi module, and nRF24L01 + PA/LNA wireless transceiver module and pumping system. Figure 5 shows the corresponding circuitry of the central node.

3.5 Communication Networks and Protocols

IoT farming system is formed depending on several kinds of long and short-range networking systems and modules for communications. A few IoT network technologies have utilized to construct farm field observing gadgets and sensors [33]. Communication protocols are the foundation of IoT based farming frameworks and applications [34]. They are utilized to serve the purpose of exchange of every single agrarian data over the network.

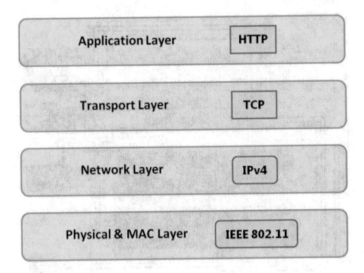

Fig. 6 ESP8266 communication layers and protocols

3.5.1 ESP8266 Communication and Networking Protocols

Application Layer

Transport Layer

Network Layer

Physical and MAC Layer.

Figure 6 shows communication layers and protocols.

3.5.2 GSM SIM900A Module's Communication Protocols and Networking

GSM SMS Protocols—To realize how the SMS goes from the gadget (in this circumstance the GSM module) to the SMSC please investigate Fig. 7. Observing this figure one can understand the conventions utilized and the GSM network components which are taking care of the communication procedure. The communication procedure can be described as; the GSM module transmits the SMS (Short Message Service) to the BTS (Base Transceiver Station) over a radio connection. Afterward, the message goes through the backbone transmission system of the service provider. The MSC (Mobile Switching Center), the HLR (Home Location Register) and alternatively the VLR (Visitor Location Register) play out all the necessary specialized functionalities to determine the specific Short Message Service Center (SMSC) that will preserve

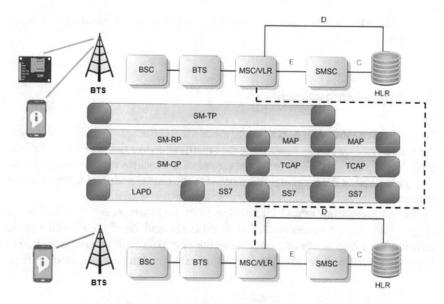

Fig. 7 SMS Network and Protocols

and forward the SMS determining the present status (accessibility) of the recipient [35]. The figure underneath (Fig. 7) visualizes the GSM SMS transmission system and protocols for SMS going from the GSM module to the cell phone.

3.6 Agri-Robotics

Distinctive Agribots have been intended to fill the needs of smart farming which are limiting the extra endeavors of farmers by expanding the speed of work through advanced innovations. Agribots perform fundamental work like weeding, splashing, and planting, etc. Each of these robots is controlled by using IoT to upgrade crop productivity and compelling asset usage [36].

3.7 Physical Implementation

Various sensors, different sorts of actuators, and microcontrollers are executed physically to monitor assorted farming applications. Different network hardware is also actualized at the physical layer. At this layer, the whole ecological states of the farm field are detected and thereafter incite as indicated by the predefined rules. The microcontroller is the prime controller here and plays out the supervisory role including networking relevant tasks and some other functionality that is executed by sensors and actuators [37, 38].

3.8 Data Communication

By using ESP8266 in AP (Access Point) mode users can access the ESP module with the smartphone by simply connecting the smartphone over the Wi-Fi with the ESP module using assigned SSID name and password to visualize the data in web pages.

Figure 8 shows the communication between the central node and wireless sensor nodes and data visualization using the central node.

The corresponding sensed information from the sensor nodes can likewise be logged into the datasheets of different databases and distributed cloud storage utilizing the cloud servers by interfacing the ESP8266 with an access point that is providing the web access. The logged information can be envisioned utilizing

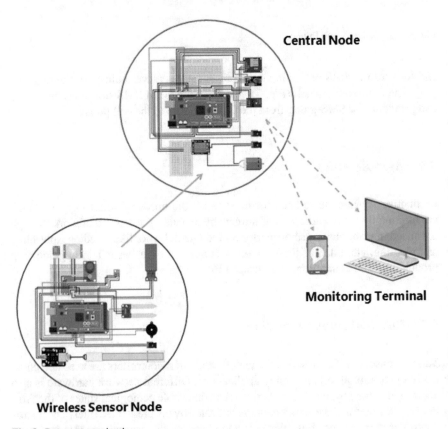

Fig. 8 Data communication

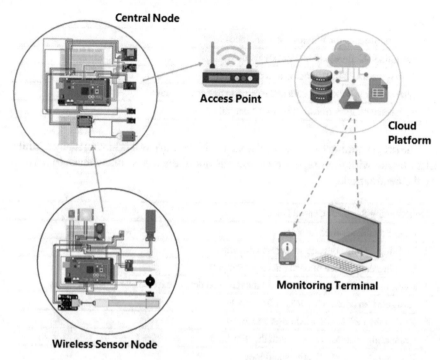

Fig. 9 Data communication with the cloud

cell phones, PCs from anyplace by accessing the cloud server. Figure 9 shows communication with the cloud.

3.9 Code Architecture

To obtain the desired output from the system it is required to program the system effectively. The mentioned central node and wireless sensor nodes MCU are needed to be programmed separately to make them do their assigned work. At first, focusing on the programming of the wireless sensor nodes. As the sensor nodes are containing the same types of equipment so that the programming of each of them will be the same [39].

Programming Steps for Wireless Sensor Nodes
1. Including the required headers
2. Defining the pins of each sensor and modules
3. Defining the purpose of each pin (Input/output)
4. Defining variables for containing data of each sensor

(continued)

(continued)

Programming Steps for Wireless Sensor Nodes
5. Assigning an address to nRF24L01 module
6. Setting the nRF24L01 module as a transmitter
7. Each sensor is collecting data and storing to assigned variable
8. Each sensor data is transmitting to the central node

The programming for the central node is a bit lengthy as it is having many modules that operate with their separate commands and there are so many tasks to perform by the central node.

Programming Steps for Central Node
1. Including the required headers
2. Defining the pins of each sensor and modules
3. Defining the purpose of each pin (Input/output)
4. Defining variables that will contain the received data of each sensor of sensor nodes
5. Assigning an address to nRF24L01 module
6. Setting the nRF24L01 module as a receiver
7. Comparing each sensor data with the threshold
8. If sensor data satisfy the threshold loop
9. Send SMS to the user's mobile phone via GSM module
10. Depending on the threshold start the certain automated task
11. Setup and design of webpage for the ESP8266 module
12. Cloud storage configuration (if want to store and visualize data using cloud storage)

3.10 Data Visualization and Storing

The information gathered by every sensor from each of the sensor nodes can be preserved and represented via web pages arranged for every sensor node by accessing to the ESP8266 module (access point) of the central hub or node. To visualize the information client needs to initially turn on "his/her" cell phone or PC's Wi-Fi and afterward from the Wi-Fi settings the client needs to discover SSID appointed and enter the passcode for the ESP module to get access. From that point onward, the client needs to open up the internet browser and need to enter the IP address of the ESP module in the URL box. Finally, the internet browser will divert the client to the principle page of the data representation, from which "he/she" can get access to the pages of every sensor node to screen every sensor data. Figure 10 shows the principle access page and Fig. 11 shows the page for each of the sensor nodes for observing

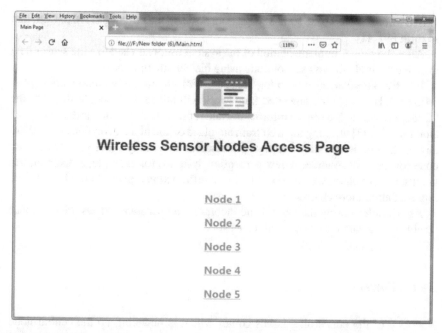

Fig. 10 Main page

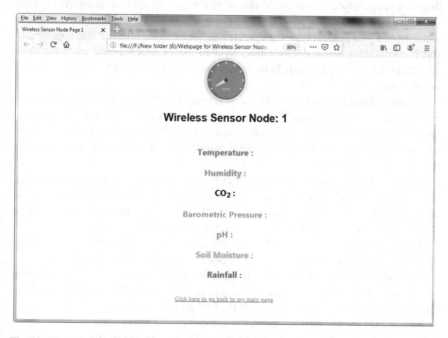

Fig. 11 The web page for the wireless sensor node

sensed information. These sorts of pages can be designed utilizing HTML, CSS, and so forth [40, 41].

After the proper implementation of the system user can visualize the sensor data from the desired wireless sensor node using his/her smartphone or PC.

It is likewise conceivable to log the sensed information in a cloud storage platform [42]. Here the recommended framework will utilize the "Google sheet" of the "Google Drive" to log sensed information and represent information preserved in the cloud server [43] utilizing the web from anyplace on the planet. Anybody can utilize any of the distributed cloud storage frameworks by simply modifying the source code for the ESP module. A few perception systems, for example, a diagram, bar graphs, and pie charts can also be used in the information representation to make the representation more charming.

After implementing the system and incorporating necessary codes the sheet will be able to log data for corresponding sensors.

3.11 Power

Now one of the concerning issues comes which is powering up the central node and the wireless sensor nodes. Batteries can be used to power up the system, but there are some major drawbacks to power up the system with batteries. As batteries are required to be recharged and they have a shorter life that is why the usage of batteries as a power source for the system will not be much efficient and appropriate. But incorporating Solar Photo-voltaic System the overall farming system can be operated with more efficiency. In this case, the "Off-grid solar PV" system can be a better source of power [44].

Off-grid Solar PV System—An Off-grid system is not connected to the electricity grid and therefore requires battery storage. An Off-grid solar system must be designed appropriately so that it will generate enough power throughout the year and have enough battery capacity to meet the requirements, even in the depths of winter when there is less sunlight [45].

Modern Off-grid solar systems use multi-mode inverters or chargers to manage batteries, solar, and back-up power sources such as a generator. The inverter or charger is the central energy management unit and can be either AC coupled with a solar inverter, or DC coupled with a solar charge controller, or both.

The figure below will show the Off-grid solar PV system to the smart farming system (central node and wireless sensor node) connection diagram (Fig. 12).

Here one point to be noted that, the pumping system that means the water pump will be connected to the main 220 V AC system. Only the MCU and relevant micro-electronic circuits will be connected to the Off-grid PV system.

This is how the overall smart agri-system or farming system can be made highly effective and efficient by incorporating Solar PV systems rather than depending on the batteries for power.

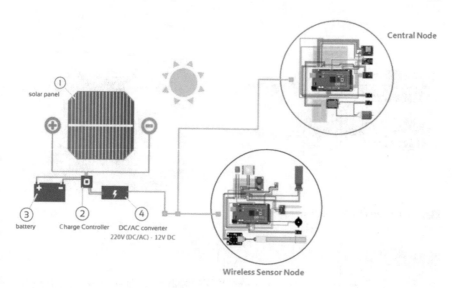

Fig. 12 An Off-grid system is powering the farming system

3.12   Agri-Copter

The prescribed farming system will use the quad-copter specially designed for agricultural purposes to spray pesticides and fertilizer over the farming field [46].

Design—The farming agri-copter is based on Arduino as a controller, nRF24L01P wireless transceiver module for wireless communication and control, four BLDC motors (Brushless DC), four Electronic Speed Controller (ESC) and Li-Po (Lithium-Polymer) battery.

Here one thing must be kept in mind that in case of a quad-copter among four motors, two motors are needed to be rotated in the clockwise direction (CW) and the other two are required to be rotated in the counter-clockwise direction (CCW) which is mandatory for flight control. Figure 13 shows the CW and CCW rotation configuration of the BLDC motor of quad-copter.

A quad-copter frame is required to set up all circuits into the frame. Any kind of quad-copter frame from different manufacturers can be used to build the agri-copter.

Figure 14 visualizes the circuit diagram of the agri-copter.

Now the copter required a dedicated controller to control its flight. A flight controller for the copter can easily be built using Arduino (MCU), nRF24L01P wireless transceiver (Communication), joystick module (Flight control), and push buttons (To start and stop the flow of pesticides or fertilizer from the bucket). The flight controller unit is shown in Fig. 15.

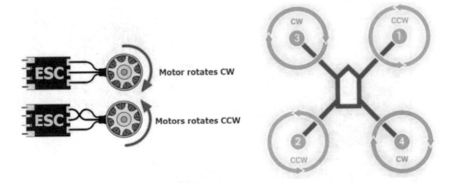

Fig. 13 CW and CCW rotation

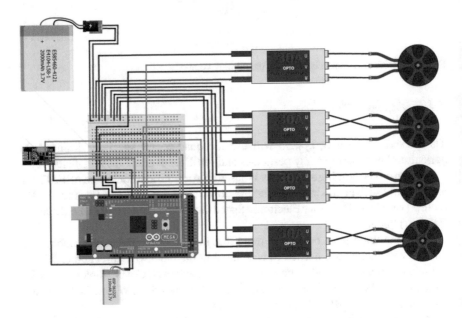

Fig. 14 Circuit of the agri-copter

3.13 Link Budget and Link-Level Analysis

The link budget and analysis approach introduced in this work are based upon the communication devices used in this work for communication purposes such as wireless sensor node to central node (nRF24L01 + PA/LNA based), central node to access point (ESP8266 Module based), and central node to user (GSM Module based) communication. The communication frequencies for those mentioned devices are

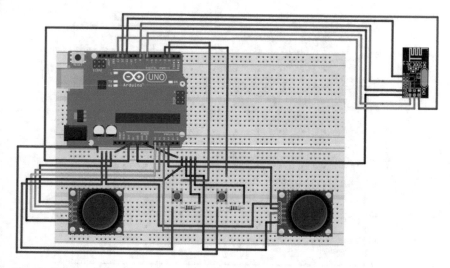

Fig. 15 Flight controller of the agri-copter

900 MHz and 2.4 GHz correspondingly and the considered wireless communication link distance (transmitter–receiver separation) is up to 1000 m.

3.13.1 Path Loss Models for Vegetated Environments

Radio wave propagation in vegetated environments is classified into three unique classifications: direct, reflected, and horizontal waves. The direct and reflected waves traveling over the vegetated area are subjected to an expanded loss because of absorption and scattering from the foliage. The horizontal waves propagate from the transmitter through the tree crowns, over the highest point of the vegetation and afterward back down through the vegetation to the recipient.

Experimental propagation loss models for horizontal propagation in vegetated regions known as exponential decay (EXD) models [47]. These models combined with the adjusted exponential decay (MED) models are shown in the equation below (Eq. 1) [48].

$$PL_{MED(COST235;In-leaf)}[dB] = Af^B D^C \tag{1}$$

where A, B, and C are fitted quantities for amplitude, frequency, and distance dependency respectively. f indicates the frequency in MHz and D is the transmitter–receiver separation distance in meters. The Fig. 16 below shows the loss (propagation loss or path loss) of the propagated signals for corresponding parameters in a vegetated environment.

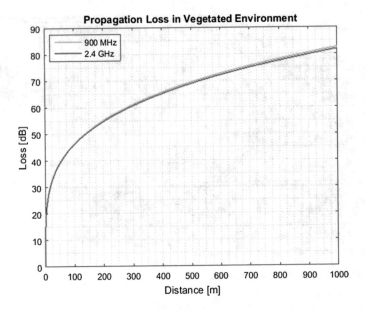

Fig. 16 Propagation characteristics of a vegetated environment

3.13.2 RSSI (Received Signal Strength Indicator) Measurement Approach

The RSSI measurement approach is based on transmitted power, gain of the transmitter and receiver, and the loss determined by the corresponding signal propagation model [49] and stated in Eq. (2).

$$RSSI[dBm] = P_T + G_T + G_R - PL \tag{2}$$

where P_T is the transmitted power in dBm (30 dBm for GSM and ESP8266 module, and 20 dBm for nRF module), G_T is the transmitter gain in dBi (6 dBi for GSM and ESP8266 module, 3.0 dBi for nRF module), G_R is the receiver gain in dBi (6 dBi for GSM and ESP8266 module, 3.0 dBi for nRF module), PL is the path loss or propagation of the considered channel model. Figures 17, 18 and 19 show the RSSI measurements for mentioned the modules.

The link budget is included to provide a better insight into the performance of the prescribed system after implementing it in a farm field. Though it is an empirical analysis it will provide sufficient knowledge and aid in designing and deploying the networking and communication infrastructure of the system.

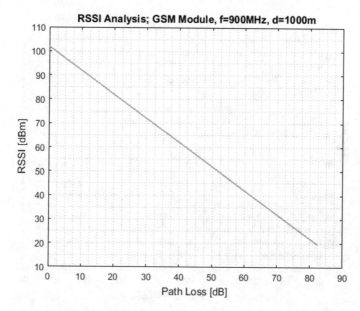

Fig. 17 RSSI analysis of GSM module

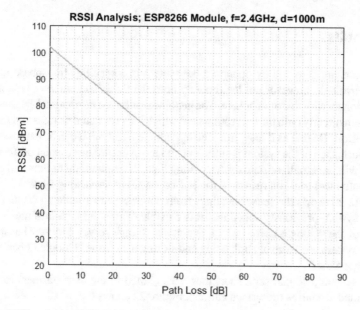

Fig. 18 RSSI analysis of ESP8266 module

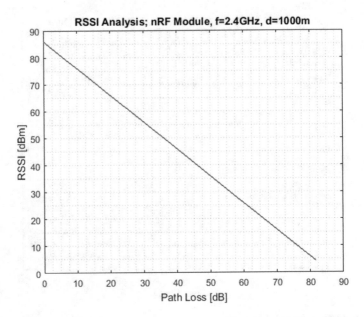

Fig. 19 RSSI analysis of nRF module

4 Results

The results are acquired by operating the system in a specific atmosphere relevant to the farm field however not legitimately onto the field. The outcomes are almost proportionate with that atmosphere. The genuine results can be obtained by executing the system directly onto the harvesting field which is a protracted process and requires a huge effort. Be that as it may, from the hypothetical test run it can be expressed that the performance of the system is sufficient and it can ensure more exact performance when it will be actualized onto the harvesting field. The upcoming figures will picture the hypothetical trial run results of the IoT based smart harvesting system.

Figure 20 shows the sensor readings from wireless sensor node 1 (WSN-1).

The system previously stated that it is capable of logging data into the "Google Sheet" of the "Google Drive" which is a cloud storage server. Figure 21 visualizes that the system is effectively uploading and storing data to the "Google Drive" within a "Google Sheet".

The system also can send short messages (SMS) to the user (farmer) for some pre-defined scenarios (described earlier). Figure 22 shows the SMS.

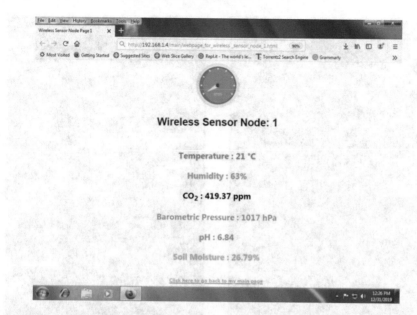

Fig. 20 Sensor readings from WSN-1

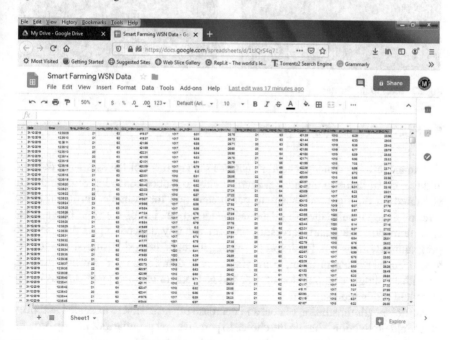

Fig. 21 Stored data in the Google Sheet of Google Drive

Fig. 22 SMS to the farmer

5 Socio-Economic Impacts

The smart farming concept ensures cautious management of the estimated demand and release of items to the market immediately to reduce squander. Smart farming is fixated on the right enhancement parameters and for example, fertilizer, moisture, or material substance to guarantee better production of the right harvest that is in demand. The smart harvesting system actually depends upon the use of a programmed framework for the proper management of the agri-business. The control system manages sensor input, conveying remote data for proper decision making, notwithstanding the automation of gear and gadgets for responding to various issues and yield production support.

Sufficient administration of cost and decrease of waste lead to extended authority over yield production. Having the choice to perceive any peculiarities in crop growth, one will have the choice to limit the dangers of losing yield.

The effectiveness of agri-business is increased by adopting a robotized framework. By utilizing smart gadgets, one can mechanize different processes over the production cycle, for example, irrigation system, fertilizing, or effective pest control.

At the basic level, smart farming depends intensely on proper data management is made conceivable by incorporating new technological improvements. It offers unlimited opportunities in expanding benefits while diminishing the ecological adverse impacts of harvesting. This accomplished by intently coordinating the utilization of farm inputs, for example, pesticides, irrigation, and fertilizers with a measure of actual conditions in explicit territories of the field [50].

Smart farming, which is otherwise called site-explicit management, is dealing with the harvesting at a spatial scale which is typically littler than an individual field. In many cases, the harvest's condition changes generously, ranging from one section then onto the next. Modifying input levels and the executives' works according to the proper requirements for the particular local environment, reduce the farming cost, reduce unfavorable environmental impacts, and improve farming can be achieved. The very first improvements of smart farming took place in the late 1980s in the wheat and soybean production frameworks of the Midwest.

As indicated by recent works, smart farming is expected to enhance nourishment strength, while simultaneously guaranteeing that sustainability is maintained. These practices include the utilization of innovations to improve the ratio between agrarian information (irrigation, fertilizers, soil, vitality, pesticides, and so on.) and crop productivity (typically food). Smart farming comprises the utilization of sensors to distinguish harvest needs exactly. The farmer at that point intercedes in a calculated manner to expand the profitability of crops while lessening asset wastage [51].

Another promising and incredible aspect of smart farming is its capacity to lessen the sector's negative effect on the environment. As per statistics, agribusiness is responsible for at least 10% of the worldwide greenhouse gas substance emanations. Further, the abuse of pesticides and fertilizers, as well as soil erosion, cause concerns with respect to the atmosphere. In this manner, smart farming is truly helping control these worries.

According to a market insight report by BIS Research, the worldwide smart harvesting market is expected to reach $23.14 billion by 2022, ascending at a compound yearly development rate of 19.3% from 2017 to 2022. The market development is essentially ascribed to the expanding interest for higher harvest yield, the developing penetration of ICT in farming, and the expanding requirement for climate-smart harvesting.

In the upcoming years, smart harvesting is anticipated to make a huge effect on the agri-economy by overcoming any issues among miniature and large-scale organizations. The pattern isn't just appropriate in developed nations—developing nations have understood its massive significance also. In nations, for example, China and Japan, large-scale deployment of the internet of things (IoT) frameworks have prompted a quick adoption of smart farming frameworks. The administrations of a few nations have likewise understood the requirement for, and the merits of these technologies, and thus, their initiatives to enhance smart farming strategies are expected to drive the development of the market further. However, such progressive change in harvesting rehearses come with opportunities as well as specific difficulties which are generating some restrictions in the development of the market. The knowledge and awareness about more up to date agribusiness innovations are yet to spread broadly, particularly in rising nations [52].

Regionally, North America is at the bleeding edge of the worldwide smart farming market, with higher market penetration in the United States. Nonetheless, Mexico is expected to have the most elevated market growth in the upcoming five years.

The Asia–Pacific is anticipated to exhibit the quickest market growth from 2017 to 2022. The area presents a huge scope for market development, owing to the expanding urban populace size, developing business sector entrance of web in farm management, and favorable government ventures. Also, the presence of financially propelling nations, for example, India and China are expected to make the locale an essential part of the development of smart farming in the forthcoming years.

A shift in the worldwide mature demographic has set off the adoption of computerized systems in harvesting rehearses. Automation and control frameworks producers have observed a definite surge in their deals because of this significant change in the farming business [53].

In recent years, agrarian robots have likewise been adopted into farming tasks as they treat soil and harvests specifically according to their requirements and minimize the requirement for manual labor. UAVs produced the highest revenue among every horticultural robot used in smart farming. Most of the robot deployment was done for harvest management.

Key players working right now increase their item dispatch exercises over ongoing years to create public awareness about their current and new items and advancements and contend with their rivals' item portfolio.

Partnerships and joint venture strategies have additionally been utilized for greater expansion of the smart farming market. With the expanding growth in the worldwide market, organizations working in this industry are now constrained to think of collaborative processes with other agrarian OEMs to continue in the strongly competitive market [54].

6 Conclusion

Throughout the world, scientists are investigating technological advancements to upgrade crop productivity in a way that supplements existing services by implementing IoT innovations. To encourage the development of agricultural profitability, this work has presented a smart harvesting framework or system dependent on embedded electronic equipment, IoT, and wireless sensor networks. This work prior introduced an overview of progressing research on smart agribusiness and talked about conceivable outcomes of smart farming through a literature survey. A while later presented the endorsed smart farming framework by portraying the framework architecture including core equipment or devices of the framework, communication protocols utilized by the wireless networking modules present at the framework to utilize the features of IoT, additionally depicted the functional architecture of the system. Moreover, the work introduced the point by point structure of the recommended IoT based smart farming framework, included circuit diagrams of the embedded architecture, depicted the working principles of the endorsed framework with the essential description of core components. Finally, the work included the link budget to provide a better insight into the performance of the system after implementing it in the farm field. Further researches can be performed to enhance the farming system such as the incorporation of agri-robots, video analytics-based crop monitoring, AI (Artificial Intelligence) and ML (Machine Learning) based harvesting analysis, utilization of advanced computation units (STM, Raspberry Pi) for sophisticated tasks of farming, etc. The smart farming these days is one of the vital R&D (Research and development) topics of the world as farming is one of the crucial bases of mankind. As the world population is expanding exponentially upgraded farming facilities are exceptionally important to supply a satisfactory level of nourishment to that expanded population. That is why analysts and researchers can more and more emphasis on the advancement of smart farming. This work might be assistive to the researchers to pursue better knowledge on IoT aided farming as this work has emphasized on the previously mentioned points.

References

1. Majumdar J, Naraseeyappa S, Ankalaki S (2017) Analysis of agriculture data using data mining techniques: application of big data. J Big Data 4:20. https://doi.org/10.1186/s40537-017-0077-4
2. Rosenstock TS, Rohrbach D, Nowak A, Girvetz E (2019) An Introduction to the Climate-Smart agriculture chapters. In: Rosenstock T, Nowak A, Girvetz E (eds) The Climate-Smart agriculture chapters. Springer, Cham
3. Koshy SS, Sunnam VS, Rajgarhia P et al (2018) Application of the internet of things (IoT) for smart farming: a case study on groundnut and castor pest and disease forewarning. CSIT 6:311–318. https://doi.org/10.1007/s40012-018-0213-0
4. Stagnari F, Maggio A, Galieni A et al (2017) Multiple benefits of legumes for agriculture sustainability: an overview. Chem Biol Technol Agric 4:2. https://doi.org/10.1186/s40538-016-0085-1
5. Yoon C et al (2018) Implement Smart Farm with IoT Technology. In: 2018 20th International Conference on advanced communication technology (ICACT), 749–752, Chuncheon-si Gangwon-do, South Korea
6. Wu Q, Liang Y, Li Y, Liang Y (2017) Research on intelligent acquisition of smart agricultural big data. In: 2017 25th International Conference on geoinformatics, IEEE, pp 1–7
7. Baldovino RG et al (2019) Implementation of a low-power wireless sensor network for smart farm applications. In: 2018 IEEE 10th International Conference on Humanoid, Nanotechnology, Information Technology, Communication and Control, Environment and Management (HNICEM), Baguio City, Philippines
8. Tamoghna O, Sudip M, Narendra SG (2015) Wireless sensor networks for agriculture: the state of the art in practice and future challenges. Comput Electron Agric 118:66–84. https://doi.org/10.1016/j.compag.2015.08.011
9. Hu Z et al (2019) Application of non-orthogonal multiple access in wireless sensor networks for smart agriculture. IEEE Access 7:87582–87592
10. Barnes A et al (2019) Influencing factors and incentives on the intention to adopt precision agricultural technologies within arable farming systems. Environ Sci Policy 93:66–74

11. Kernecker M, Knierim A, Wurbs A, Kraus T, Borges F (2019) Experience versus expectation: farmers' perceptions of smart farming technologies for cropping systems across Europe. Precis Agric https://doi.org/10.1007/s11119-019-09651-z
12. Leonard E et al (2017) Accelerating precision agriculture to decision agriculture: enabling digital agriculture in Australia. Cotton Research and Development Corporation
13. Kutter T, Tiemann S, Siebert R, Fountas S (2011) The role of communication and cooperation in the adoption of precision farming. Precis Agric 12:2–17
14. Hansen BG (2015) Robotic milking-farmer experiences and adoption rate in Jæren. Norway J Rural Stud 41:109–117
15. Higgins V, Bryant M, Howell A, Battersby J (2017) Ordering adoption: materiality, knowledge and farmer engagement with precision agriculture technologies. J Rural Stud 55:193–202
16. Eastwood C, Ayre M, Nettle R, Dela Rue B (2019) Making sense in the cloud: farm advisory services in a smart farming future. Njas - Wageningen J Life Sci https://doi.org/10.1016/j.njas.2019.04.004
17. Schimmelpfennig D, Ebel R (2016) Sequential adoption and cost savings from precision agriculture. J Agric Resource Econ 41:97–115 https://doi.org/10.22004/ag.econ.230776
18. Annosi MC, Brunetta F, Monti A, Nat F (2019) Is the trend your friend? An analysis of technology 4.0 investment decisions in agricultural SMEs. Comput Ind 109:59–71
19. Lowenberg-DeBoer J, Erickson B (2019) Setting the record straight on precision agriculture adoption. Agron J 111:1552. https://doi.org/10.2134/agronj2018.12.0779
20. Wyche S, Steinfield C (2016) Why don't farmers use cell phones to access market prices? technology affordances and barriers to market information services adoption in rural Kenya. Inf Technol Dev 22:320–333. https://doi.org/10.1080/02681102.2015.1048184
21. Baumüller H (2017) The little we know: an exploratory literature review on the utility of mobile phone-enabled services for smallholder farmers. J Int Dev 30:134–154
22. Gutiérrez J et al (2014) Automated irrigation system using a wireless sensor network and GPRS module. IEEE Trans Instrum Meas 63(1):166–176
23. Moghaddam M et al (2010) A wireless soil moisture smart sensor Web using physics-based optimal control: concept and initial demonstrations. IEEE J Sel Topics Appl Earth Observ Remote Sens 3(4):522–535
24. Cambra C et al (2014) Deployment and performance study of an ad hoc network protocol for intelligent video sensing in precision agriculture. In: Proceedings of International Conference on Ad Hoc Network and Wireless. pp 165–175
25. Baranwal T et al (2016) Development of IoT based smart security and monitoring devices for agriculture. In: Proceedings of IEEE 6th International Conference-Cloud System and Big Data Engineering. pp 597–602
26. Lerdsuwan P et al (2017) An energy-efficient transmission framework for IoT monitoring systems in precision agriculture. In: Proceedings of International Conference on Information Science and Applications. pp 714–721
27. Dan L et al (2015) Intelligent agriculture greenhouse environment monitoring system based on IoT technology. In: Proceedings of IEEE Intelligent Transportation, Big Data Smart City (ICITBS). pp 487–490
28. Ojha T et al (2015) Wireless sensor networks for agriculture: the state-of-the-art in practice and future challenges. Comput Electron Agricult 118:66–84
29. Dachyar M et al (2019) Knowledge growth and development: internet of things (IoT) research, 2006–2018. Elsevier–Helion 5(8):1–14
30. Shun WG, Muda WMW, Hassan WHW, Annuar AZ (2020) Wireless sensor network for temperature and humidity monitoring systems based on NodeMCU ESP8266. In: Anbar M, Abdullah N, Manickam S (eds) Advances in Cyber Security. ACeS 2019. Communications in Computer and Information Science, vol 1132. Springer, Singapore
31. Obikoya GD (2014) Design, construction, and implementation of a remote fuel-level monitoring system. J Wireless Com Network 2014:76. https://doi.org/10.1186/1687-1499-2014-76

32. Wu G et al (2019) Application and design of wireless community alarm system based on nRF24L01 module. In: 2019 Chinese Control and Decision Conference (CCDC). Nanchang, China
33. Navulur S et al (2017) Agricultural management through wireless sensors and Internet of Things. Int J Electr Comput Eng 7(6):3492
34. Sarawi SA et al (2017) Internet of Things (IoT) communication protocols. In: Proceedings of 8th International Conference on Infocomm Technology (ICIT). pp 685–690
35. Zhen Y et al (2009) On user data protocol of SMS in remote monitoring system. In: 2009 International Conference on environmental science and information application technology. Wuhan, China
36. Milella A, Reina G, Nielsen M (2019) A multi-sensor robotic platform for ground mapping and estimation beyond the visible spectrum. Precision Agric 20:423–444. https://doi.org/10.1007/s11119-018-9605-2
37. Strobel D et al (2014) Microcontrollers as (In) Security Devices for Pervasive Computing Applications. Proc IEEE 102(8):1157–1173
38. Zhu J et al (2019) Efficient actuator failure avoidance mobile charging for wireless sensor and actuator networks. IEEE Access 7:104197–104209
39. Novák M et al (2018) Use of the Arduino platform in teaching programming. In: 2018 IV International Conference on information technologies in engineering education (Inforino). Moscow, Russia
40. Sorn D et al (2014) Web page template design using interactive genetic algorithm. In: 2013 International computer science and engineering conference (ICSEC), Nakorn Pathom, Thailand
41. Carter B (2015) HTML architecture, a novel development system (HANDS): an approach for web development. In: 2014 Annual Global Online Conference on information and computer technology. Louisville, KY, USA
42. Qiang W (2019) Performance and security in cloud computing. J Supercomput 75:1–3. https://doi.org/10.1007/s11227-018-2671-4
43. Park DS (2018) Future computing with IoT and cloud computing. J Supercomput 74:6401–6407. https://doi.org/10.1007/s11227-018-2652-7
44. Ivanova IY et al (2018) Comparative analysis of approaches to consider rationale of use of solar panel plants for power supply of off-grid consumers. In: 2018 International Ural Conference on Green Energy (UralCon). Chelyabinsk, Russia
45. Hemmati R (2019) Stochastic energy investment in off-grid renewable energy hub for autonomous building. IET Renew Power Gener 13(12):2232–2239
46. Bnhamdoon OAA, Mohamad NHH, Akmeliawati R (2020) Identification of a quadcopter autopilot system via Box-Jenkins structure. Int J Dynam Control https://doi.org/10.1007/s40435-019-00605-x
47. Li Z, Wang N, Hong T (2013) RF propagation patterns at 915 MHZ and 2.4 GHZ bands for in-field wireless sensor networks. Trans ASABE 56:787–796
48. Shutimarrungson N, Wuttidittachotti P (2019) Realistic propagation effects on wireless sensor networks for landslide management. J Wireless Com Network 2019:94. https://doi.org/10.1186/s13638-019-1412-6
49. Yamamoto B, Wong A, Agcanas PJ, Jones K, Gaspar D, Andrade R, Trimble AZ (2019) Received signal strength indication (RSSI) of 2.4 GHz and 5 GHz wireless local area network systems projected over land and sea for near-shore maritime robot operations. J Mar Sci Eng 7(9):290
50. Therond O, Debril T, Duru M, Magrini MB, Plumecocq G, Sarthou JP (2019) Socio-economic characterisation of agriculture models. In: Bergez JE, Audouin E, Therond O (eds) Agroecological transitions: from theory to practice in local participatory design. Springer, Cham
51. McCarthy N, Lipper L, Zilberman D (2018) Economics of climate smart agriculture: an overview. In: Lipper L, McCarthy N, Zilberman D, Asfaw S, Branca G (eds) Climate smart agriculture. Natural resource management and policy, vol 52, Springer, Cham

52. Khatri A, Regmi PP, Chanana N et al (2020) Potential of climate-smart agriculture in reducing women farmers' drudgery in high climatic risk areas. Clim Change 158:29–42. https://doi.org/10.1007/s10584-018-2350-8
53. Therond O, Duru M, Roger-Estrade J et al (2017) A new analytical framework of farming system and agriculture model diversities. a review. Agron Sustain Dev 37:21. https://doi.org/10.1007/s13593-017-0429-7
54. Singh R, Singh GS (2017) Traditional agriculture: a climate-smart approach for sustainable food production. Energ Ecol Environ 2:296–316. https://doi.org/10.1007/s40974-017-0074-7

Remote Sensing and Soil Quality

Graham Hay and Paul Krause

Abstract Soil health is an environmental factor that impacts a range of important issues including food production, water retention and soil organic carbon storage. Agriculture relies on healthy soil for crop growth and animal grazing. Water retention reduces the risks of desertification and of flooding as the capacity to retain water reduces the rate of surface water flow. In addition, soil organic carbon represents the largest terrestrial carbon stock and is second only to the oceans. Yet, soil health is threatened by intensive farming practices and changes of land use such as deforestation. Thus, it is important to manage soil health to maintain food security, avoid desertification and maintain or ideally increase soil organic carbon storage. A useful tool to inform this management function would be a machine learning model that can predict soil health given land cover and parameters of the abiotic context, such as terrain elevation and historical weather data. The first step in developing such a model is to be able to identify the land cover for a chosen area. Satellites provide multi-spectral images that include the visual bands. Land cover databases provide the ground truth labels for a supervised learning approach to train an image semantic segmentation model. This chapter describes how Sentinel-2 satellite image data was combined with data from the UK Centre for Ecology and Hydrology Land Cover Map 2015, to train a convolutional neural network for land cover classification for the South of England.

1 Introduction

Given the importance of soil health, there is a need to manage it to avoid soil degradation and environmental damage, and to maintain food supply and soil organic carbon storage. A useful tool to inform this management function would be a machine

G. Hay · P. Krause (✉)
Department of Computer Science, Surrey University, Guildford, UK
e-mail: p.krause@surrey.ac.uk

P. Krause and F. Xhafa (eds.), *IoT-based Intelligent Modelling for Environmental and Ecological Engineering*, Lecture Notes on Data Engineering and Communications Technologies 67, https://doi.org/10.1007/978-3-030-71172-6_7

learning model that can predict soil health, given land cover and parameters of the abiotic context, including digital elevation model (DEM) data and historical weather data.

Assessing soil health directly is a manual process of measuring below-ground properties with instruments, which is not a practical method of monitoring soil health at scale. Neither is it a suitable method of generating data in sufficient quantity to train a machine learning model. As an alternative, the Normalized Difference Vegetation Index (NDVI) is proposed as a proxy for soil health. NDVI is commonly derived from satellite imaging data. If satellite data covering an appropriate area, at a high enough resolution, can be identified, then the per pixel NDVI would provide enough data to be used as the target of the machine learning model. Furthermore, this model can be tested against NDVI measurements in the field, using a multi-spectral camera mounted on an unmanned aerial vehicle or drone.

Training a predictive machine learning model would need the following data at many points in time, over a number of seasons:

- raster digital elevation data aligned with the satellite image
- historical weather data for a period preceding the date of the satellite image
- per pixel land cover.

Other inputs such as physical soil parameters and human inputs, like fertilisers and pesticides, would be desirable but sources of this data are not known. The target of the prediction would be the per pixel NDVI, calculated from the spectral bands of the corresponding satellite image.

Developing a predictive model from remote sensing data alone is a challenging undertaking and may not be achievable without at least some linking in with terrestrial IoT data sources. This chapter will report on our progress in this direction in order to provide some pointers to results we may expect to extend in the near future the capabilities covered so far in this book.

1.1 Structure of this Chapter

The next section provides an overview of the importance of soil health. Section 3 then describes the ambition of using a combination of satellite imagery and NDVI (Normalized Difference Vegetation Index) data to enable soil health to be monitored routinely and remotely. Section 4 provides an overview of the current state of the art for soil health monitoring, and the challenges that need to be addressed to achieve our objective. Sections 5 and 6 provide an overview of our work so far on using supervised learning for land cover classification; a first step in achieving our goal. Section7 describes the details of our modelling approach, and Sects. 8 and 9 describe the test and evaluation of our model, and associated results. Section 10 then evaluates the work so far against our project objectives. We end the chapter with a discussion of what we can conclude from the work so far, and an outline of how we plan to take the work forward.

2 Soil Health

Soil health has been broadly defined as the capacity of a living soil to function, within natural or managed ecosystem boundaries, to sustain plant and animal productivity, maintain or enhance water and air quality, and promote plant and animal health [4].

But what is soil? Soil is a mixture of minerals and organic matter. The minerals generally comprise sand, silt and clay. The organic matter comprises plant and animal detritus, microbial biomass and stable organic matter (humus). Soil particles and organic matter bind together to form aggregates and this process is strongly influenced by plant roots and fungi. So, soil organic matter contributes to soil aggregate formation. The pores between soil aggregates contain air and water. The proportion of these constituents and the stability of soil aggregates vary under the influence of weather, vegetation and human management in the form of tillage and cropping systems. Soil aggregates are an essential component of soil health with a proportion of about 50% solid aggregate, 25% air and 25% water being considered optimal for plant growth [1].

The region of interaction between plant roots, soil and microbes is known as the rhizosphere. This interaction facilitates nutrient cycling in which fixed nitrogen, phosphorus and water are taken up by plants in complex processes. In the carbon cycle, photosynthetic plants take up atmospheric carbon dioxide and release oxygen back into the atmosphere, while decomposing soil organic matter releases carbon dioxide. The interaction between soil, plant and microbes is essential to the nutrient cycling processes [1]. Furthermore, it has been shown that the biodiversity of soil biota enhances the productivity, stability and sustainability of ecosystems [3].

Soil organic matter contains carbon, known as soil organic carbon. In 2012, it was estimated [28] that the carbon stored globally in biomass was 560 Giga tonnes (Gt), while soil organic carbon was 2,344 Gt. Only the oceans store more carbon with 38,400 Gt. Thus, it is important to maintain soil health to retain or even increase the quantity of soil organic carbon rather than let it be released into the atmosphere through soil degradation.

Degradation of soil health can arise from intensive farming practices and changes of land use.

Intensive conventional agriculture is designed to increase output, typically by extensive use of synthetic inputs such as chemical fertilizers and pesticides, and increased tillage. Pesticides kill not only the pests but other soil organisms, reducing biodiversity, and can leach into and contaminate ground water. Excessive use of chemical fertilizer can lead to acidification of soil and also contamination of ground water. Use of heavy machinery can result in soil compaction, reducing porosity and water retention and aeration. This leads to reduced soil organic matter and resistance to erosion from runoff. However, crop rotations and cover crops with reduced or no tillage can maintain soil health and reduce soil erosion [2].

Land use change can take the form of conversion of forest or natural grassland to pasture or cropland which "changes vegetation and disturbs soils, leading to loss of soil carbon and other nutrients, changes in soil properties and changes to above and below-ground biodiversity" [26]. This often results in a degradation of soil health. When this occurs in drylands, it is known as desertification [16].

The need to monitor soil health on a regular basis is of paramount importance. As we will see in the next chapter, enhanced soil health in agricultural areas has the potential to significantly improve the efficiency of agriculture (by reducing external inputs whilst maintaining yields). This means the environmental costs both upstream and downstream of an agricultural unit can be significantly reduced. In addition, we also see a growth in the carbon sequestration rates that if enacted globally will have a major beneficial effect on the anthropomorphic carbon budget.

3 Satellite Imagery and NDVI

Satellite imagery is a form of passive remote sensing that records multi-spectral electromagnetic radiation digital images of the earth's surface from an orbiting satellite. After passing through the atmosphere, electromagnetic energy from the sun is reflected from the surface of the earth and after passing through the atmosphere again, is received by the satellite's sensors. Multiple sensors record energy received in different spectral bands, such as the visual red, green and blue bands together with near-infrared (NIR) and short-wave infrared (SWIR). Some satellites also have sensors that record thermal infrared energy that is emitted from the surface rather than reflected.

Satellite imagery can be characterised by four types of data resolution: spectral, spatial, radiometric and temporal. The spectral resolution describes the number and range of the electromagnetic wave bands that are captured. Spatial resolution describes the size of the ground features that can be resolved and so corresponds to the width of the raster grid cells in the image. This is a function of the spectral band, since longer wavelengths contain less energy and so sensors need a wider field of view to collect enough energy. Sensors that detect a very large number of narrow spectral bands are known as hyperspectral sensors. Radiometric resolution describes the smallest difference in energy intensity that can be distinguished. The raw intensity level is known as the digital number (DN) and higher resolution requires a greater number of bits to encode. The temporal resolution describes the time interval between image collections of the same area of the earth's surface. This is primarily determined by the revisit period of the satellite, which is the time before the satellite passes over the same area, typically a number of days [13].

Researchers in the field of remote sensing have found that indices derived using combinations of reflectance values of different spectral bands yield useful indicators of land surface properties. One of the most widely used of these is the Normalized Difference Vegetation Index (NDVI) which gives a measure of the amount of healthy vegetation present [13]. The formula for the index is:

$$NDVI = \frac{NIR - Red}{NIR + Red}$$

This index is based on the fact that healthy, photosynthetically active vegetation strongly absorbs visible red light and strongly reflects near-infrared light. Therefore, an NDVI value close to one indicates the presence of healthy vegetation. Mathematically, the index can range from minus one to plus one, but in practice varies from slightly negative values to plus one.

4 Soil Health Monitoring

Soil health monitoring has been going on for many years. It has been and still currently is performed by manually taking soil samples from the ground. For example, the European Commission Joint Research Centre published sampling instructions for surveyors in 2017 [9] which describes the procedure for manual collection of soil samples. Similarly, a soil monitoring network in Belgium is described, in which soil samples are collected manually with an auger [14]. Manual collection and analysis of soil samples is labour intensive and expensive. It is also too time consuming to be capable of generating the large volumes of data that are often needed for training modern machine learning models such as neural networks.

The development of wireless sensor networks and the Internet of Things (IoT) has been taken up in agricultural research and farming, to implement local soil monitoring networks that collect measurements of soil properties automatically [5]. While this is a great step forward, the types of sensor that are currently commercially available are limited to measuring soil moisture, EC, salinity and temperature.

On the other hand, remote sensing multi-spectral satellite imagery is collected automatically with global coverage. If the data is collected at a high enough resolution, it should provide sufficient volume for training a machine learning model. We have seen that that, currently, soil health monitoring is labour intensive and expensive, making the case for a machine learning approach that is scalable and inexpensive. Object based satellite image classification methods are not guaranteed to produce the desired classes and may require manual intervention to do so. On the other hand, convolutional neural networks represent a state-of-the-art supervised learning approach that requires no manual intervention. This is the approach we will explore in the remainder of this chapter.

5 Supervised Learning for Land Cover Classification

Our ultimate aim of soil quality assessment using remote sensing is still a distance away. Our current hypothesis is that it will not be possible to use NDVI data alone without also having an assessment of the land cover classification; in order to interpret NDVI data, we do need to know at the very least whether it is being collected from forest, grassland, arable land or heathland, for example. We will review our progress in this direction in this section.

5.1 Data Collection and Preparation

Supervised learning requires training data examples with the corresponding label data. The Sentinel-2 program has been identified as a source of satellite images and the UK Centre for Ecology and Hydrology Land Cover Map 2015 data has been identified as a source of label data. The data was downloaded, and the necessary preparation and integration was performed to create training and test datasets.

The European Space Agency Sentinel-2 mission consists of two polar orbiting satellites, S2A and S2B, each carrying a passive multispectral instrument (MSI) that records high resolution images over a swath width of 290 km. The spectral resolution of the bands that are of interest is summarized in Table 1. These have a spatial resolution of 10 m, a radiometric resolution of 12-bits and a temporal resolution of 5 days [6].

The captured images are made available as data products, consisting of ortho-image tiles for all bands, covering a certain region of the earth's surface and a processing level. Sentinel-2 offers two processing levels. Level-1C provides Top-Of-Atmosphere reflectances in cartographic geometry, and Level-2A provides Bottom-Of-Atmosphere reflectances derived from the Level-1C product by applying an atmospheric correction algorithm. The Level-2A product also contains a true colour image (TCI) file, which is an RGB image that has been synthesized purely to aid human visualization [8].

Sentinel-2 data products can be downloaded from the Copernicus Open Access Hub [6], either manually or via an API. Two Level-2A data products were downloaded covering an area of the south of England, including Surrey, on different dates: 22/06/2020 and 25/06/2020. Data products can be explored, viewed and manipulated using the SNAP tool provided by the European Space Agency [7].

The data products include data quality reports and all checks were passed on both dates. However, an important factor in satellite imagery is cloud cover. For the 25/06/2020 data product, the cloud cover was virtually zero, but for the 22/06/2020 data product, the cloud cover was almost 27%.

Table 1 Sentinel-2 spectral bands at 10 m resolution

Band number	Band description	S2A		S2B	
		Central wavelength (nm)	Bandwidth (nm)	Central wavelength (nm)	Bandwidth (nm)
2	Blue	492.4	66	492.1	66
3	Green	559.8	36	559.0	36
4	Red	664.6	31	664.9	31
8	Near infrared	832.8	106	832.9	106

5.2 Data Preparation

Since it is virtually cloudless, the data product for 25/06/2020 was selected for the training data. A very low cloud cover region of the data product for 22/06/2020 was selected to test performance on a dataset from a different date.

The data product contains an image tile per band, at 10 m resolution in JPEG2000 format, of size 10,980 × 10,980, with a single 16-bit channel. For training the image segmentation model, the tile area was sub-divided into a set of 256 × 256 images with three 16-bit channels, containing the red, green and blue bands, in GeoTiff format. GeoTiff is a format, commonly used in Geographic Information System (GIS) software tools, containing meta-data specifying the location and co-ordinate reference system. Sentinel-2 uses the Universal Transverse Mercator (UTM) projection and the WGS84 datum. The location of each sub-image was calculated and the meta-data added to the GeoTiff file using the same co-ordinate reference system. Of the 884 sub-images so created, 10% were set aside to form a holdout test set.

Although the Level-2A product includes a true colour image file, after some initial exploration, this was not used. The image is created for the purpose of human visualisation and involves manipulation of the channel intensities that might alter the colour balance to improve the appearance. Training should be done on the original reflectance data, or if any transformations are to be applied to the data, then they should be applied to the original data.

The data product for 22/06/2020 was used for testing only. A low cloud region was selected and sub-divided into 256 × 256 images in exactly the manner as described above, yielding 28 image files.

However, it became apparent that the channel intensity distributions for the two data products were not the same. If the number of pixels with the same intensity is plotted as a histogram, the distribution of intensities can be visualised. A distribution containing many pixels with high intensity values is very bright. A distribution with pixel values spread across a wide range has high contrast. In this case, the test dataset of 22/06/2020 has a distribution that starts at a slightly lower intensity, peaks at a higher intensity value, and has slightly more contrast with a longer tail than the distribution for the training data of 25/06/2020 (see Fig. 1). The longer tail is almost certainly due to the presence of some remaining clouds, which adds a proportion of white. The difference in contrast is a little unexpected, given that the pixel values represent reflectance values. It would appear to be systematic and may arise from the pre-processing of the data product prior to release, such as the Level-2A processing. It was therefore decided to apply a transformation in creation of the training and test datasets.

The intensity distribution of an image can be modified by applying an intensity transformation. Adding to the pixel values will shift the distribution upwards, increasing the brightness. Multiplying the pixel values by a factor greater than one will stretch the distribution, increasing the contrast. Such operations are often used to improve the visual appearance of an image. More sophisticated transformations exist, such as histogram equalisation, which is still simple to implement, but has a

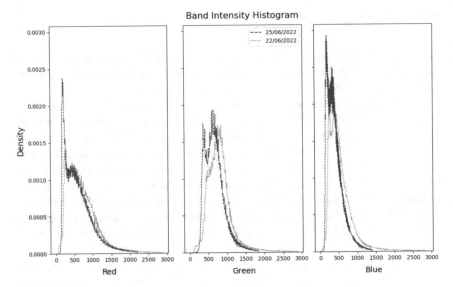

Fig. 1 Band intensity histograms

more pronounced effect on the distribution [10]. However, it was decided to use a simple contrast stretch, to keep the distribution as close to the original distribution as possible. The stretch factor is the ratio such that the mean of the pixel intensities taken over the region of interest in the source data product, across all three bands (red, green, blue) results in a value of 65,535 (the maximum value for a 16-bit channel) divided by 4. This single stretch factor is used for all three bands in the data product. Applying the contrast stretch to each of the two data products brings the distributions into closer alignment (see Fig. 2).

6 Land Cover Map 2015

The UK Centre for Ecology and Hydrology make a number of environmental datasets freely available for educational and academic use, including the Land Cover Map 2015 [30]. The data set was released in 2017 and covers nearly the whole of the United Kingdom including the south of England. It is available as raster data at 25 m resolution and also as a vector dataset. The vector dataset consists of over 6 million non-overlapping feature polygons at very high resolution [24]. Each feature is labelled with one of 21 target land cover classes as listed in Table 2.

Initial inspection of a few sample images was promising, showing that the labels matched well with the images. Figure 3 shows an image (top) and with the labels colour coded as in Table 2 and overlaid at 50% transparency (bottom). However, it was found during testing of the model predictions, that the land cover labels are not entirely accurate. Some shapes in the vector database clearly contain more than one

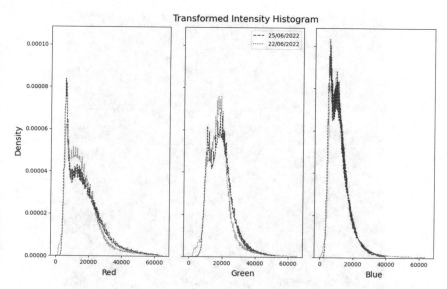

Fig. 2 Transformed band intensity histograms

Table 2 Land cover map 2015 target classes

Class Number	LCM 2015 Target Class	Standard colour mapping
1	Broadleaved woodland	
2	Coniferous woodland	
3	Arable and Horticulture	
4	Improved Grassland	
5	Neutral Grassland	
6	Calcareous Grassland	
7	Acid Grassland	
8	Fen, Marsh and Swamp	
9	Heather	
10	Heather grassland	
11	Bog	
12	Inland Rock	
13	Saltwater	
14	Freshwater	
15	Supra-littoral Rock	
16	Supra-littoral Sediment	
17	Littoral Rock	
18	Littoral sediment	
19	Saltmarsh	
20	Urban	
21	Suburban	

G. Hay and P. Krause

Fig. 3 Top: satellite image; Bottom: satellite image with labels overlaid

class. Some shapes appear to be visually the same, or at least very similar, but have been classified with different labels. Even in Fig. 3, some ambiguities can be seen. For example, there is a square field on the lower left (annotated with white outline) that is labelled as 'Arable and horticulture', but is visually indistinguishable from other fields that are marked as 'Improved grassland'. This is discussed further in the next section.

6.1 Data Preparation

The Land Cover Map 2015 is the source of ground truth data for training the image segmentation model. For each satellite sub-image created for training or testing, the ground truth label image must also be created for the same location at the same resolution. To do this, the sub-window of the Land Cover Map 2015 vector database must be rasterized to the required resolution. However, there are two issues that must be addressed first.

First, the Land Cover Map 2015 database contains 6,737,558 polygons, which makes it too unwieldy and slow to use directly. To address this issue, a sub-set of the Land Cover Map 2015 was created, covering a region that contains the Sentinel-2 image.

Second, the Land Cover Map 2015 uses a different co-ordinate reference system (CRS) from Sentinel-2 data products. Sentinel-2 uses UTM region 30 N and the WGS84 datum. This corresponds to the EPSG:32,630 co-ordinate reference system. The Land Cover Map 2015 uses the British National Grid system, which corresponds to the EPSG:27,700 co-ordinate reference system. For the location co-ordinates to align correctly, the co-ordinates must be converted to the same reference system. It was decided to standardize on the EPSG:32,630 CRS, and to reproject the sub-set of the Land Cover Map 2015 into this CRS.

Both of these operations were performed using the QGIS package [21]. The Land Cover Map 2015 subset was created by displaying and marking the region of interest, which selected 451,207 polygons that are contained within or touch this region. The Edit menu 'Copy Features' item was used, followed by the Edit menu 'Paste Feature As' then 'New Vector Layer'. This created a new vector database which was saved in GeoPackage format. The reprojection was performed by selecting the sub-set layer created in the previous operation and using the Vector menu 'Data Management Tools' item, then 'Reproject Layer'. A target CRS of EPSG:32,630 was specified and the reprojected layer was saved in GeoPackage format.

The satellite sub-image files and the ground truth label files were generated with Python code, using the rasterio [15] and GDAL [17] libraries.

7 Modelling

We used U-net for building our land cover classification model. U-net is a fully convolutional neural network with four contracting stages, a bottleneck, and four expansive stages [23]. Each stage comprises several layers, each of which can be implemented using built-in Keras layer types [12]. The network can be constructed using the Keras functional API [11] which allows layers to be connected to form a directed acyclic graph.

Each of the stages in the contracting path has five layers:

- 2D convolution with a 3 × 3 kernel
- 2D convolution with a 3 × 3 kernel
- Batch Normalization (optional)
- 2D Max Pooling with a 2 × 2 pool size
- Dropout.

The 2D convolution layers use a 'rectified linear unit' (ReLU) activation function which allows efficient training using back propagation. The weights are initialised using a normal distribution with a standard deviation given by $\sqrt{2/n_l}$, where n_l is the number of inputs from the previous layer to a neuron in layer 1 [23]. In this implementation, a padding of 'same' is used, which retains the same image size by padding the edges of the image with zero values. This keeps the architecture clean and simple.

The 2D max pooling layer down samples the image size by a factor of two, by taking the maximum value in each 2 × 2 pool. This causes the network to learn features at increasing scale.

The batch normalization layer is not present in the original u-net specification [23]. Batch normalization applies a transformation that subtracts the mean and divides by the standard deviation over a batch. This is intended to make the network less sensitive to fluctuations in pixel intensities. The network was tested with and without batch normalization.

Dropout layers are mentioned in the u-net specification 'at the end of the contracting path' but are not included in the architecture diagram [23]. Dropout is a regularization technique designed to reduce overfitting by introducing noise to the hidden units [27]. Inputs are randomly set to zero at the specified rate and all other inputs are scaled by 1/(1 − rate) so that the sum of the inputs remains the same.

The bottleneck stage simply has two 2D convolutions as described above. At this stage, the image size has reduced to 16 × 16, but the number of channels has reached 1024.

Each of the stages in the expansive path has five layers:

- 2D transpose convolution with a 2 × 2 kernel
- Concatenation
- Dropout
- 2D convolution with a 3 × 3 kernel
- 2D convolution with a 3 × 3 kernel.

The transpose convolution layer up-samples the image by a factor of two using weights that are learned during training. No activation function is applied.

The concatenation layer concatenates the up sampled channels with the normalized output of the convolutional layers from the corresponding contractive stage of the same image size.

The dropout layer provides regularization.

The 2D convolution layers combine the predictions with the higher resolution location information.

Finally, classification is performed by a 2D convolution with a 1×1 kernel to produce 22 channels with 'softmax' activation. Each channel represents the probability that the pixel belongs to each one of the 22 classes.

For training when there are two or more label classes and the ground truth labels are specified as integer numbers, the correct loss function is 'sparse_categorical_crossentropy'. The optimizer used for training is the 'Adam' optimizer. The metric that is computed and logged in training is accuracy.

The model is saved to file at the end of an epoch, if it is the best model so far in terms of the monitored metric. The metric used is the 'validation accuracy'.

All in all, the u-net, as defined here, has a total of 31,035,030 learnable parameters. The network takes many hours to train on a standard workstation with no GPU or TPU. For this reason, training of the neural network was performed using the Google CoLab platform, which provides GPU or TPU hardware. The training time on CoLab was around 1 h.

7.1 Data Augmentation and Refinement

The training data used comprises 755 images of size 256×256, giving 49,479,680 labelled pixels. Since, at this stage, the model is intended for use with other Sentinel-2 images of the same resolution, it was decided not to apply spatial or scaling transformations. However, it was observed that the pixel value distributions were not the same in different Sentinel-2 images, which were taken on different days. A correction for this was applied in the data set preparation stage, but to help the model to generalize across different Sentinel-2 images a shift in pixel intensity value was applied at random to each image.

The Keras library supports the use of Python generators to supply a batch of images. Thus, at each step of the training a batch is generated with a random selection of images. Each time an image is used in a batch, a different random shift is applied to the pixel values. The same shift is applied to the three colour channels. Since no spatial transformations are applied to the images, the label images do not need to be transformed.

7.2 Test Design

Since this is a supervised learning task, with labels available for the satellite images, test data was readily derived. Two test data sets were created, one for each of the two Sentinel-2 images from different dates.

The satellite image for 25/06/2020 is cloudless and was used for training. This was subdivided into smaller images of size 256×256 and corresponding label images were generated. These were separated into a training set and a test set, containing 90% and 10% of the sub-images respectively.

To assess how well the model generalizes to images from other dates, the satellite image for 22/06/2020 was used purely for testing. A low cloud cover region of this image was selected and similarly subdivided into smaller images with label images.

Testing was carried out by running the image segmentation model in inference mode. The pixel level classifications for each test sub-image were output as a GeoTiff image file, identical in structure to the label image files. The predictions could then be compared to the labels.

8 Test and Evaluation

To assess model performance, a number of metrics were computed. These allow the results of experiments with hyper parameters and data augmentation to be compared, and also enable comparison with results reported by other authors.

The results were evaluated to identify the best model and to gain insight into the performance of the model. The results were also compared to work reported by another author.

8.1 Test Metrics

Since this is a multi-class classification task, the model can be assessed using a confusion matrix and metrics derived from it. Two levels of test metrics were generated. A confusion matrix and associated metrics were generated for each of the test sub-images individually, primarily intended for investigative purposes. For model assessment, an overall confusion matrix and associated metrics were generated for the whole test set.

The metrics for the overall assessment were as follows:

- precision
- recall
- f1-score
- Intersection over Union (IU)

Precision is the proportion of pixel predictions for a class, that are correctly classified, while recall is the proportion of pixels that truly belong to a class that were classified correctly. If $n_{j,k}$ is the number of pixels with true class j and predicted class k, and the number of class labels is L, then, for class i:

$$\text{precision}_i = \frac{n_{i,i}}{\sum_{j=1}^{L} n_{j,i}}$$

$$\text{recall}_i = \frac{n_{i,i}}{\sum_{k=1}^{L} n_{i,k}}$$

The f1-score is the harmonic mean of the precision and recall, so for class i:

$$f1_i = \frac{2 \cdot \text{precision}_i \cdot \text{recall}_i}{\text{precision}_i + \text{recall}_i}$$

The intersection over union for class i, is given by:

$$IU_i = \frac{n_{i,i}}{\sum_{k=1}^{L} n_{i,k} + \sum_{j=1}^{L} n_{j,i} - n_{i,i}}$$

In general, intersection over union is lower than both precision and recall.

In addition to the per class metrics, the frequency weighted average of the metrics, weighted by the frequency of the true class, was calculated. This better reflects the overall performance, since classes with smaller support has has less impact.

The metrics for individual sub-images were generated using the 'scikit learn' Python library [25] standard confusion matrix functions, which do not include the intersection over union metric.

9 Results

During initial exploration, it was observed that two sets of classes were visually and semantically similar and so it was decided to merge these. In the Python training batch generation function, and also in the Python test script, the 'Coniferous woodland' class was mapped to the 'Broadleaved woodland' class, and the 'Urban' class was mapped to the 'Suburban' class'.

A number of experiments or tests were run. First the model was trained, and then the model was tested on both test datasets. The test data is used to independently test the performance of the model on unseen data, and so is not used in the training process. To monitor and assess the progress of training, a proportion of the training data is randomly split off into a validation data set. Training proceeds by epochs, consisting of a number of batches of a specified size, in this case 8 training images per batch. The number of batches is the size of the training set divided by the batch size. At the end of each epoch, the model logs the training loss, training accuracy, validation loss and validation accuracy.

In each experiment, the models were tested against the unseen test datasets for 25/06/2020 and 22/06/2020. The overall metrics were recorded and also the metrics for one sample test image which is of particular interest. This is the image containing Bookham in Surrey, and the fields to the south, as shown in Fig. 2. Overall pixel classification weighted average metrics are shown in Table 3. The single image pixel classification weighted average metrics for Bookham are shown in Table 4.

The first experiment was to establish a baseline, with no data augmentation and no batch normalization. As can be seen in the first line of Table 3, the metrics for the test images collected on 22/06/2020 are significantly lower than the metrics for the images collected on 25/06/2020, which are from the same satellite image as was used for training. This means that this model does not generalize so well to satellite images from other dates.

Further experiments were run to try to improve the ability of the model to generalize. Since batch normalization is intended to make the model less sensitive to variations in brightness, an experiment was run with no data augmentation, but with batch normalization layers added to the network. For the same reason, experiments were run with data augmentation, but no batch normalization. The batch generation function sampled a random shift each time an image file was loaded, and this was added to the red, green and blue channels. Two levels of shift were run, one sampled from a uniform distribution in the range [-2500, 2500] and another in the range [-5000, 5000]. See Tables 3 and 4.

From both Tables 3 and 4, it can be seen that adding batch normalization improved the metrics for both test dataset dates. However, the improvement for the test dataset of 22/06/2020 was not as great as when using the intensity shift data augmentation. Both levels of intensity shift produced very similar test results. The shift of U [-2500,2500] produced slightly better metrics and might also be preferred on the basis that the simplest model is best.

10 Evaluation

The objective of this project was to train a semantic image segmentation model for pixel-level land cover classification of satellite images, using the visual bands only, with no post processing steps that require human intervention. This has been done with some success, achieving a frequency weighted f1 score of 80% on unseen sub-images from the same satellite image as the training data and 78% on sub-images from a different date.

The confusion matrix for the model with the shift of U [−2500,2500] and the test data for 25/06/2020 is shown in Fig. 4. For this dataset, the main pixel land cover labels are 'Broadleaved woodland' (19%), 'Arable and horticulture' (33%), 'Improved grassland' (33%), 'Suburban' (13%) and other classes (2%). The confusion matrix for the same model and the test data for 22/06/2020 is shown in Fig. 5.

Table 3 Overall pixel classification weighted average metrics

Experiment	Validation Accuracy	Test data: 25/06/2020				Test data: 22/06/2020			
		Precision	Recall	f1	IU	Precision	Recall	f1	IU
Baseline	82	80	80	80	67	78	64	70	52
Batch normalization	82	82	82	82	70	76	70	73	56
Intensity shift U [−2500,2500]	81	80	80	80	67	79	77	78	64
Intensity shift U [−5000,5000]	82	80	81	80	68	80	74	77	63

Table 4 Single image pixel classification weighted average metrics

Experiment	Validation Accuracy	Bookham: 25/06/2020				Bookham: 22/06/2020			
		Precision	Recall	f1	IU	Precision	Recall	f1	IU
Baseline	82	87	87	87		81	72	72	
Batch normalization	82	89	89	89		85	83	83	
Intensity shift U [−2500,2500]	81	87	87	87		86	86	86	
Intensity shift U [−5000,5000]	82	87	87	87		86	86	86	

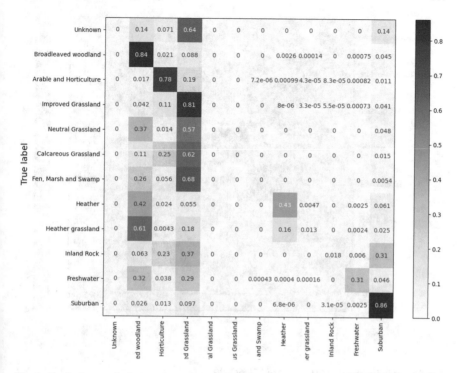

Fig. 4 Confusion matrix for test data of 25/06/2020

For this dataset, the main pixel land cover labels are 'Broadleaved woodland' (35%), 'Arable and horticulture' (12%), 'Improved grassland' (39%), 'Suburban' (13%) and other classes (1%). The matrix has been normalized over the true class, so the figures on the diagonal correspond to the recall for each class.

For both datasets, the main confusion is between 'Improved grassland and 'Arable and horticulture', with the confusion being somewhat more pronounced for the 22/06/2020 dataset. There is also a tendency to misclassify 'Broadleaved woodland' as either 'Arable and horticulture' or 'Improved grassland' and also to misclassify 'Suburban' as either 'Arable and horticulture' or 'Improved grassland'.

To try to gain a better understanding of the source of this confusion, a closer inspection of the datasets was made. For this purpose, sub-images were displayed with the land cover labels colour coded as in Table 2 and overlaid at 50% transparency. As noted in section earlier, it can be seen that the land cover labels are not entirely accurate. Some examples are shown in Figs. 6 and 7. Close inspection by the reader will reveal further inconsistencies.

Figure 6 shows several regions, annotated with white outlines, that appear to be inaccurately labelled. At the top left, there is a region labelled as 'Broadleaved woodland' that is clearly not woodland. At the top middle, there is a region labelled as 'Improved grassland' that appears to be predominantly woodland. At the middle left, there are two regions labelled as 'Arable and horticulture' that appear to be

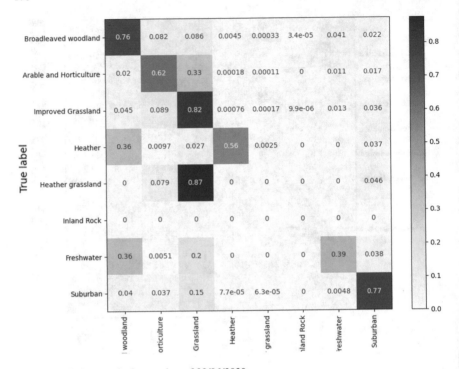

Fig. 5 Confusion matrix for test data of 22/06/2020

grassland. At the bottom middle, there is a region labelled as 'Freshwater' that is apparently not water.

Figure 7 shows more regions, annotated with white outlines, that are apparently misclassified between 'Improved grassland' and 'Arable and Horticulture'. At the bottom right, there is a region labelled as 'Improved grassland' that is visually indistinguishable from surrounding regions that are labelled as 'Arable and horticulture'. Near the centre, there is a region labelled as 'Arable and horticulture' that is apparently grassland.

Since the land cover data is derived from satellite images collected in 2015, it is to be expected that land use changes have occurred. Farmers have rotated crops and changed fields from pasture to arable, and vice versa. The Land Cover Map 2015 dataset is itself generated using machine learning methods and so will have its own level of accuracy, but no figures for this appear to be published.

It is very difficult to quantify the impact of these inaccuracies, but as stated earlier, it will have a detrimental effect on both training and testing. A detailed analysis of labelling inaccuracies requires visual inspection of the images and making a judgement as to which class label should be assigned to the region. To review and adjust the image labels for the entire region of interest would be a very labour intensive and time-consuming task. Ultimately, what is needed is a set of training images with

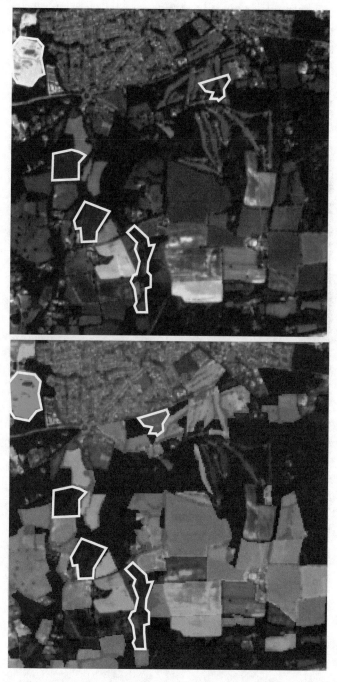

Fig. 6 Top: sub-image from dataset for 25/06/2020; Bottom: sub-image with labels overlaid

Fig. 7 Top: sub-image from dataset for 25/06/2020; Bottom: sub-image with labels overlaid

matching pixel labels in sufficient quantity for training a machine learning model. In this respect, u-net has been shown to be effective with relatively small datasets [23].

For comparison, in an evaluation of convolutional neural networks for segmentation of Sentinel-2 imagery [29], the authors assessed the accuracy of several network architectures, including u-net. For this study, the near-infrared band was used as well as the visual bands from images at 10 m resolution of a region of the Netherlands. Another difference in this study was that the u-net architecture used six contractive and expansive stages instead of four, giving it greater capacity. In addition, the training data covered a greater temporal range. In other respects, the set up was broadly similar, with the models being trained end-to-end. The image size was varied, but the optimal size was found to be 244 pixels, similar to the 256 pixels used here. For u-net, the authors reported a weighted f1 score of 85%, compared to the 78% or 80% reported here. The difference in performance is likely to be due to:

- the use of the near-infrared band, increasing the spectral information content in the data
- the difference in volume and temporal range of the training data
- the detrimental effect of the land cover label inaccuracies.

11 Conclusions and Next Steps

This project is part of a larger, more ambitious program to assess soil health from parameters of the abiotic context, such as terrain elevation and historical weather data, together with the land cover. For arable land, the land cover classes would need to be to the level of vegetation type or crop type. This project is an initial step towards the larger goal, in which the feasibility of performing land cover classification from satellite imagery was investigated.

The objective of this project was to train a semantic image segmentation model for pixel-level land cover classification of high-resolution satellite images, using the visual bands only. This was done successfully using supervised learning of a u-net convolutional neural network. The model achieved a frequency weighted f1 score of 80% on unseen sub-images from the same satellite image as the training data and 78% on sub-images from a different date.

The following conclusions are drawn, based on the experience of this project.

Performance of the model could be improved by more accurate land cover label data. The Land Cover Map 2015 data has been found to have some inaccuracies and inconsistencies. These are likely to be due to changes in land use since the data was collected, but may also arise from the modelling used to generate the dataset. A more up to date Land Cover Map may help but would inevitably become out of date.

Performance of the model could be improved by using a greater temporal range of satellite training data. In this project, data augmentation was employed to help the model generalize to other satellite image dates. A better approach could be to use images from a large number of collection dates that would encompass a wide range of atmospheric conditions. This would also address the issue of seasonality.

The issue of cloud cover was avoided by selecting images with very little or no cloud cover. This could restrict the amount and range of images that could be used for training. Handling of cloud cover within images should be implemented to exclude just the regions of cloud.

Performance of the model could be improved by using spectral bands other than the visual bands. In particular, the near-infrared band contains a lot of information relating to vegetation and is provided at 10 m resolution. A model, based on visual bands only, will not be able to discriminate vegetation type from satellite imagery at 10 m resolution.

It was assumed that the Land Cover Map 2015 label data was, and remains, accurate. It was also assumed that the convolutional neural network requires large amounts of training data and that combining the Land Cover Map 2015 data with as large a region of the satellite image as possible would provide this. In fact, the UK Centre for Ecology and Hydrology has just released new Land Cover Maps for 2017, 2018 and 2019 [31]. To validate and adjust the image labels for the entire region of interest would be prohibitive in time and effort. Therefore, efforts should be made to quantify the impact of any inaccuracies, using the most up-to-date data. The u-net architecture, used here, is known to be effective on relatively small training datasets, by making use of image augmentation. Based on this, an alternative strategy could be to manually select, by visual inspection, a smaller, more accurate set of images and labels that are representative of the classes of interest. This would be done over a range of image collection dates. The set of training data could then be extended and refined iteratively, based on the confusion matrix generated in testing. This would still be a manual process, but it is to be hoped, significantly less work than validating the entire region of interest.

The next step would be to build a land cover classification model that discriminates vegetation type. This would need to take into account the seasonality of crop lifecycles. It would need to make use of other spectral bands in addition to the visual bands. A suitable source of training data would need to be identified.

When a predictive model of soil health has been built, it will need to be validated. Since the proposal is to use NDVI as a proxy for soil health, a means of measuring NDVI in the field will be required. While field surveys of NDVI are offered as a commercial service, a low-cost alternative would be preferable. One alternative that we have performed some initial experiments on is to mount a Raspberry Pi and NOIR camera on a drone. The Raspberry Pi 3B [22] is a low-cost device that supports an optional high quality digital camera module. Modern digital cameras are sensitive to near-infrared (NIR) and, in standard configuration, a filter is used to block the unwanted NIR signal. The Raspberry Pi Camera V2 has this filter, but the Pi NOIR Camera V2 [19] has the filter removed. Researchers at Public Lab [20] have combined the NOIR camera with a red filter to record images containing the spectral information required to calculate the NDVI. Our initial experiments on such a set up were promising.

Developing a predictive model of soil health is an ambitious project. Nevertheless, the results of the work so far are encouraging, indicating that building a predictive model is feasible. Such a predictive model could yield high impact results.

References

1. Al-Kaisi M, Lal R, Olson KR, Lowery B (2017) Fundamentals and functions of soil environ-ment. In: Lowery B (ed) Soil health and intensification of agroecosytems. Academic Press, London, pp 1–23

2. Arriaga FJ, Guzman J, Lowery B (2017) Conventional agricultural production systems and soil functions. In: Lowery B (ed) Soil health and intensification of agroecosytems. Academic Press, London, pp 109–125

3. Bender SF, Wagg C, van der Heijden MG (2016) An underground revolution: biodiversity and soil ecological engineering for agricultural sustainability. Trends Ecol Evol 31(6):440–452

4. Doran JW (2002) Soil health and global sustainability: translating science into practice. Agr Ecosyst Environ 88(2):119–127

5. Elijah O, Rahman TA, Orikumhi I, Leow CY, Hindia MN (2018) An overview of internet of things (IoT) and data analytics in agriculture: benefits and challenges. IEEE Internet Things J 5(5):3758–3773

6. European Space Agency (2020) Copernicus Open Access Hub. https://scihub.copernicus.eu/dhus/#/home. Accessed July 2020

7. European Space Agency (2020) Science toolbox exploitation program. https://step.esa.int/main/download/. Accessed June 2020

8. European Space Agency (2020) Sentinel-2 level-2A alogrithm overview. https://earth.esa.int/web/sentinel/technical-guides/sentinel-2-msi/level-2a/algorithm. Accessed July 2020

9. Fernández-Ugalde O, Orgiazzi A, Jones A, Lugato E, Panagos P (2017) European soil data centre - LUCAS 2018 - SOIL COMPONENT: sampling instructions for surveyors. https://esdac.jrc.ec.europa.eu/content/lucas-2018-soil-component-sampling-instructions-surveyors. Accessed 6 Aug 2020

10. Gonzalez R, Woods R (2008) Intensity transformation and spatial filtering. In: Digital image processing, 3rd edn. Pearson Education, Upper Saddle River, New Jersey, pp 104–192

11. Google (2020) Keras - the functional API. https://keras.io/guides/functional_api/. Accessed Aug 2020

12. Google (2020) Keras API reference. https://keras.io/api/. Accessed Aug 2020

13. Khorram S, Koch FH, van der Wiele CF, Nelson SA (2012) Remote sensing. (J. N. Pelton, ed.) Springer, New York

14. Kruger I, Chartin C, van Wesemael B, Malchair S (2017) Integrating biological indicators in a soil monitoring network (SMN) to improve soil quality diagnosis – a case study in Southern Belgium (Wallonia). Biotechnol Agron Soc Environ 21(3):219–230

15. Mapbox (2020) Rasterio - Introduction. https://rasterio.readthedocs.io/en/latest/intro.html#. Accessed August 2020

16. Mirzabaev AJ-O (2019). Desertification. In: Climate Change and Land: an IPCC special report on climate change, desertification, land degradation, sustainable land management, food security, and greenhouse gas fluxes in terrestrial ecosystems. https://www.ipcc.ch/srccl/

17. Open Source Geospatial Foundation (2020) GDAL documentation. https://gdal.org. Accessed Aug 2020

18. Phiri D, Simwanda M, Nyirenda V, Murayama Y, Ranagalage M (2020) Decision tree algo-rithms for developing rulesets for object-based land cover classification. ISPRS Int J Geo-Infor 9(5):329.

19. Pi NOIR Camera V2 (2020) https://www.raspberrypi.org/products/pi-noir-camera-v2/. Accessed Aug 2020

20. Public Lab (2020) Near-Infrared Camera. https://publiclab.org/wiki/near-infrared-camera. Accessed Aug 2020

21. QGIS (n.d.).https://www.qgis.org/en/site/. Accessed July 2020

22. Raspberry Pi 3 Model B+ (2020) https://www.raspberrypi.org/products/raspberry-pi-3-model-b-plus/. Accessed Aug 2020

23. Ronneberger O, Fischer P, Brox T (2015). U-Net: convolutional networks for biomedical image segmentation. In: International Conference on medical image computing and computer-assisted intervention, pp 234–241
24. Rowland C, Morton R, Carrasco L, McShane G, O'Neil A., Wood C (2017) Land Cover Map 2015 (vector, GB).
25. scikit learn (2020) https://scikit-learn.org/Accessed Aug 2020
26. Smith P, House JI, Bustamante M (2016) Global change pressures on soils from land use and management. Glob Change Biol 22(3):1008–1028
27. Srivastava N, Hinton G, Krizhevsky A, Sutskever I (2014) Dropout: a simple way to prevent neural networks from overfitting. J Mach Learn Res 15(1):1929–1958
28. Stockmann U, Adams MA, Crawford JW, Field DJ (2013) The knowns, known unknowns and unknowns of sequestration of soil organic carbon. Agr Ecosyst Environ 164:80–89
29. Syrris V, Hasenohr P, Delipetrev B, Kotsev A, Kempeneers P, Soille P (2019) Evaluation of the potential of convolutional neural networks and random forests for multi-class segmentation of sentinel-2 imagery. Remote Sens 11(8):907
30. UK Centre for Ecology and Hydrology (2015) Land Cover Map 2015. https://www.ceh.ac.uk/services/land-cover-map-2015. Accessed June 2020
31. UK Centre for Ecology and Hydrology (2020) LCM2019, LCM2018 and LCM2017. https://www.ceh.ac.uk/services/lcm2019-lcm2018-and-lcm2017. Accessed Aug 2020

Machine Learning Modelling-Powered IoT Systems for Smart Applications

Seifeddine Messaoud, Olfa Ben Ahmed, Abbas Bradai, and Mohamed Atri

Abstract With the rapid development of Internet of Things (IoT) and the use of smart devices and social networks in our daily lives, applications-based on IoT are growing exponentially in many fields such industries, business and daily life activities. The IoT technology brings a lot of promise for humanity by improving life quality and comfort and by strengthening human bonds, among others. In the next few years, billions of connected devices will be spread across smart homes, vehicles, cities, and industries. Such connected devices, with restricted resources, will interchange with users and the surrounding environment. In this context, Machine Learning (ML), Which is able to provide embedded intelligence in the IoT devices and networks, can be leveraged to decode the meaning and behavior behind the device's data, implement accurate predictions, and make decisions for several tasks. In this chapter, we present an overview of research works about ML-base IoT systems in different areas of applications. First, we present a deep overview of IoT's technology. Then, we highlight the most fundamental concepts of ML categories and algorithms. After that, we shed light on the ML-based IoT critical challenges and provide some potential future research directions. Eventually, we present an IoT-based ML technique scenario for smart irrigation in Agriculture 4.0.

S. Messaoud (✉)
Laboratory of Electronics and Microelectronics, University of Monastir, Monastir, Tunisia
e-mail: seifeddine.messaoud@fsm.rnu.tn

O. Ben Ahmed · A. Bradai
XLIM Institute, University of Poitiers, Bât SP2MI, 11 Bd Marie et Pierre Curie, 86962
Chasseneuil Cedex, France
e-mail: olfa.ben.ahmed@univ-poitiers.fr

A. Bradai
e-mail: abbas.bradai@univ-poitiers.fr

M. Atri
College of Computer Science, King Khalid University, Abha, Saudi Arabia
e-mail: matri@kku.edu.sa

© Springer Nature Switzerland AG 2021
P. Krause and F. Xhafa (eds.), *IoT-based Intelligent Modelling for Environmental and Ecological Engineering*, Lecture Notes on Data Engineering and Communications Technologies 67, https://doi.org/10.1007/978-3-030-71172-6_8

List of Acronyms

ADAS	Advanced Driver Assistant Systems
ADM	Automating Design Methodology
AI	Artificial Intelligent
ANN	Artificial Neural Network
DSS	Decision Support Systems
GMM	Gaussian Mixture Model
IIoT	Industrial Internet of Things
IoMT	Internet of Medical Things
IoT	Internet of Things
IT	Information Technology
KNN	K Nearest Neighbor
LAN	Local Area Network
M2M	Machine to Machine
ML	Machine Learning
NFC	Near Field Communication
NNC	Neural Network Classifier
QoS	Quality of Service
RFID	Radio Frequency Identification
RL	Reinforcement Learning
SARSA	State Action Reward State Action
SS	Semi-Supervised
SSL	Semi-Supervised Learning
SVM	Support Vector Machine

1 Introduction

In recent years, there is a significant growth in interest in IoT worldwide. It will not be an exaggeration to consider IoT as the most researched areas in the last decade [1]. IoT can be defined as a global network infrastructure composed of various connected devices that rely on communication, sensory, information processing technologies, and networking [2]. Yet, IoT technology offers numerous advantages over conventional networking solutions, such as reliability, accuracy, lower costs, and ease deployment that enable their use in a wide range of diverse fields and applications [3].

The increasing IoT's number will enhance network coverage but on the other hand, it will also increase the collected data size as well as computational complexity at the centralized base station [4]. Meanwhile, IoT's collaborative nature brings several advantages, including self-organization, flexibility, rapid deployment, and the processing capacity [5]. However, it also comes with several challenges like application design, communication protocols, heterogeneity, network coverage, energy conservation, QoS, security and privacy to name a few [6]. However, an IoT's technology must address these challenges to realize numerous envisioned applications and meet their requirements [7]. Therefore, new methods and techniques are needed to overcome such challenges [8].

Artificial Intelligence (AI) is a modern science for discovering patterns and making predictions from data based on statistics, data mining, pattern recognition, and predictive analytics [9, 10]. ML, a sub-field of AI, is a process of development, analysis and implementation leading to establish a systematic process [11]. It provides machines' capabilities to find solutions to complicated problems, by exploiting big data collections [12]. This offers an opportunity to analyze and highlight the correlations that exist between two or more given situations, and to predict their different implications [13, 14].

In this chapter, we provide a thorough overview on IoT technology, principal ML categories, and their important role in IoT-based ML applications. After that, we propose smart irrigation scenario based on IoT and ML techniques. In this context, our chapter consists of six sections organized as follows: In the Sect. 2, we introduce IoT's technology and challenges. Section 3 highlights the main ML categories. In addition, the use case-based IoT application will be surveyed in this section. In Sect. 4, we provide a discussion about the critical issues of ML tools-based IoT systems and the promising future solutions to solve them. Section 5, introduces the proposed smart irrigation system-based IoT scenario for Agriculture 4.0. Finally, we conclude the chapter in Sect. 6.

2 Internet of Things (IoT): An Overview and Challenges

2.1 Overview of IoT Technologies

The IoT is an evolving technology delivering positive economic and societal impact. Indeed, IoT is characterized by a powerful analytic capabilities with an enormous potential to improve the quality of human life. In the next few years, several real life items will be combined with internet connections such as industrial components, durable goods, consumer products, cars, sensors, and other daily used objects. Yet, by 2025 the economical impact of IoT will reach $11 trillion and the number of connected IoT devices will be about 75 billion devices [1].

The definition of IoT as "the combination of computers and network concepts aiming to control and monitor devices" had been around longer. However, recent technological advancements allow actually researcher to define IoT system based on more than one architecture concept. Hence, different architectures have been proposed, where the most basic one consists of three layers [2, 3]: the perception layer, the network layer, and the application layer. As presented in Fig. 1, the perception layer represents the system's physical layer, which consists of connected sensors for sensing and collecting environmental information. It aims at sensing physical

Fig. 1 Three layers
architecture

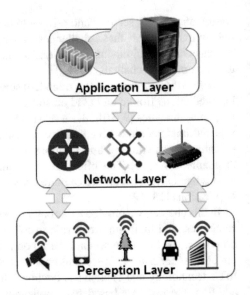

parameters and identifying environmental objects or tasks. Then, the network layer
is in charge of linking to other network devices, smart things, and servers. The output features of the last mentioned layer will be exploited later for processing and
transmitting data. Finally, the application layer aims to deliver application-specific
services to the final user. It is in charge of defining various use case applications-based
IoTs systems, for example: smart city, smart home and smart industry.

The aforementioned basic layers-based IoT concept defines the main idea of the
IoT architecture, but it remain insufficient for research on IoT which often focuses
on finer aspects. Hence, a five-layers IoT architecture was proposed and it comprise
additional layers: processing and business layers [15]. Basically, the five layers are
perception, transport, processing, application, and business layers, as denoted in
Fig. 2. It is to note that the main idea of the perception and application layers is the
same as the first architecture.

The transport layer is in charge of transferring the sensed data from the perception
layer to the processing layer via the communications network technologies such as
wireless, 3G/4G/5G, LAN, Bluetooth, RFID, and NFC technologies. Where, the
processing layer is defined also as the middleware layer, its tasks comprise storing,
analyzing, and processing large datasets that transferred from the transport layer.
However, the processing layer provides several services set to the lower layers and
exploit wide technologies such as Cloud Computing, Edge Computing, Databases,
and Big Data processing modules. Then, the business layer is in charge of piloting
the whole IoT system, including business and profit models, applications, and users'
privacy. This classification of architecture is based on protocols presented in [16].

From another point of view, data processing can be done in a large centralized
manner based on cloud computers. In this context, an architecture-based cloud-
centric gives the global orchestration task to the cloud while applications are above

Fig. 2 Five layers
architecture

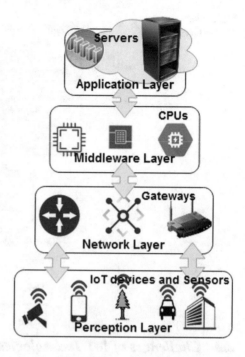

it, and the smart network things are below it [17]. However, the priority factor is given
to cloud computing regards its considerable resilience and scalability. In addition, it
provides many services like core infrastructure, software, platform, and storage.

Due to the data heterogeneity and the QoS needed by users, there is a shift across
other IoT architectures, fog/edge computing, where the physical sensors and gateways make part of the data processing and analytics. A fog architecture presents a
layered approach as shown in Fig. 3, which introduces monitoring, pre-processing,
storage, and security layers between the physical and transport layers [18].

The monitoring layer is in charge of monitoring resources, power, services, and
responses. The pre-processing layer comprises filtering and processing tasks, and
analyzes sensed data. The temporary storage layer stores functionalities as well as
distribution and data replication. Finally, the security layer is in charge of performing
encryption/decryption and ensuring the privacy and integrity of data. While, the pre-
processing and monitoring is done at the edge network before sending data to the
cloud.

Regarding communications techniques, different communication models with different characteristics could be used when implementing IoT systems. Four common
communication models are used in the IoT architecture which are device-to-device,
device-to-cloud, device-to-gateway, and back-end data-sharing. These models highlight the flexibility in how IoT devices can connect and deliver value to the end-user
[19, 20].

Fig. 3 Fog
architecture-based IoT

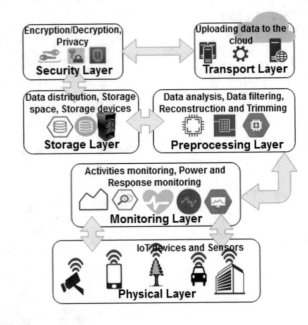

2.2 Challenges of IoT Technologies

In the coming years, IoT trends becoming reality, it will change thinking from trend evolution to the implications and issues. This development will make a hyper-connected world which is an era to the general-purpose nature of the IoT itself, which does not place inveterate restrictions on the applications and services that can benefit from the technology. However, several issues and problems, as well as security, privacy, interoperability, and reliability, will be real challenges for the IoT's concept which need to be resolved to move forward and take profit.

2.2.1 Security

The security foresight is not a novel context in Information Technology IT, in which attributes of many IoT applications present individual security issues. Addressing these issues and guaranteeing IoT services security should be a major priority. Users require confidence in IoT applications and related data and security from vulnerabilities, mostly when this technology becomes more pervasive and incorporated into our everyday lives. Hence, in the case when IoT device is poorly secured, it can be susceptible to possible cyberattacks entry and therefore exposes the data of users to steal [4].

The interconnected IoT devices nature means that every poorly ensured system connected online can impact the safety and flexibility of the Internet globally. This issue is expanded by other considerations like the mass-scale deployment of homogeneous IoT devices, the capacity of some devices to automatically link to other devices, and the likelihood of fielding these devices in unsecure surrounding. Hence, IoT devices' developers and users have to ensure that they do not disclose users and the internet itself to possible damage. Appropriately, a collaborative technique to the security issues will be required to create an ensure and powerful IoT's safety solutions suited to the large complexity issues [4].

2.2.2 Privacy

The entire IoT prospect relies on strategies and techniques that consider individual privacy choices towards a vision broad of anticipations. The streaming data and user's specificity provided by IoT devices can open unthinkable and unique IoT user's value. Therefore, worry about privacy and harming possibility, might hold back full IoT adoption. This means that privacy rights and respect for user privacy anticipations are complementary to maintain confidentiality on the Internet, connected devices, and related services [4].

Moreover, the privacy discussion issue is redefined by IoT trends. However, many IoT implementations can dramatically change the ways how data are gathered, analyzed, exploited, and preserved. Meantime, IoT dilates solicitude about the chance for increased monitoring and pursuit, hardness in the ability to opt-out data collection, and the aggregating power of IoT data streams to paint detailed digital users' portraits. While these are important issues, they are not indomitable. In order to perceive the occasion, strategies will require to be sophisticated to esteem personal privacy choices towards a broad spectrum of expectations, while still fostering innovation in new technology and services [5].

2.2.3 Interoperability

A fragmented IoT technology environment in such an application will rein industry and user's values. While full interoperability across services and products is not ever possible or vital, purchaser may be resistant to buy IoT products and services if there is a flexibility deficiency in integration, concern about restricting vendors, and high possession complexity [6].

Additionally, poorly configured and prepared IoT devices can have negative results for the whole network resources. Suitable standards, reference models, and best practices could be helpful to reduce the device proliferation that may be operating in disruptive ways. Using open, public and widely available standards as the technical building blocks for IoT devices and services (such as the Internet Protocol) will support major user interest, invention, and economic opportunity [7].

2.2.4   Reliability

Reliability, at a primary level, is worried about the study of failures. More specifically, it is troubled with how failures are caused, how they can be treated, and how they can be prevented. The basic aim of having boosted reliability is to raise the IoT service success rate, by virtue of its capability of information delivery. It becomes a critical side. Hence, a chain of checksum is desired to be implemented over the hardware and the software part of the IoT framework. A shortcoming due to system failure or threats from intrusion always holds the framework reliability as one of the major challenges [8].

3   ML Categories-Based IoT Applications

The learning activity is fundamental for epidermal existence in order to comprehend and perceive several parameters such as a voice, a person, an object, etc. One mostly differentiates learning which depends on information, memorizing and learning by generalization in which we commonly create a model for learning examples to meet new examples and scenarios. For the machines, it is easy to handle a spacious data amount, but hard to create an optimal model which can successfully recognize new objects in a new test sample [9]. In this section, we briefly highlight the four ML categories mainly Supervised Learning, Unsupervised Learning, Semi-supervised Learning, and Reinforcement Learning [10]. After that we shed light on the ML techniques-based IoT applications.

3.1   Machine Learning Categories

3.1.1   Supervised Learning

Supervised Learning is an automatic tool that learns a prediction model from a collected data in order to resolve a problem related to a real world task or situation. The main aim is to classify up-coming data (test phase) using the created classification model (on the training phase using training data). In Supervised Learning the desired output labels of the training data are known from the beginning. Therefore, human interference (Supervisor) is important in the learning phase whose goal is to build a conscious model based on previously known parameters [11]. Figure 4 depicts the supervised Learning scheme [10].

Generally, the aim of building a ruling model is to predict the value of an output variable using the values taken by the other input variables for each new data or observation. In the case of a binary or a multi-class classification problem, the explained variable is a nominal variable, each modality of which corresponds to one of the possible classes. Whereas, in the regression problem, the explained variable is

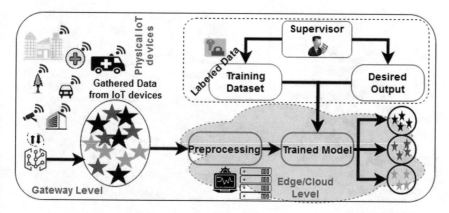

Fig. 4 Supervised Learning Concept-based IoT scenario

a quantitative variable. Similarly, in a structured prediction problem, the explained variable takes values in a structured data domain [21]. Supervised Learning algorithms include K Nearest Neighbor (KNN), Support Vector Machine (SVM), Naive Bayes, Decision Trees, Neural Network Classifier (NNC), Linear Regression, Nonlinear regression, Logistic Regression, and Neuro-Fuzzy Learning. For interested readers, the work in [10] provides a clarified review on ML algorithms.

3.1.2 Unsupervised Learning

Unsupervised Learning is a ML techniques type which mostly handles with diverse data. It splits the data based on heterogeneity into subgroups (clusters). The data considered to be most identical are correlated within a similar group. On the other hand, the dissimilar data are associated with other groups. Usually, the main is to authorize an extraction learning from data. In this learning class, the data are not labeled, so that the learning algorithm finds alone common points among the data inputs. Since unlabeled data are more plentiful than labeled data, ML techniques that facilitate unsupervised Learning are mostly useful [12].

Unsupervised Learning is in charge of illustrating the data distribution and relationships between variables without discriminating between observed variables and variables to be predicted. In other words, the clustering of individuals in classes occurs without any prior learn of those groups or samples that make it up. This process, as shown in Fig. 5 [10], called clustering which goal is to organize a data collection, so that the elements of a cluster are more identical to each other [13]. However, many algorithms are used in the context of unsupervised Learning as well as, K-means Clustering, Fuzzy C-means Clustering, and Hidden Markov Model. For more details about unsupervised Learning strategy and algorithms, we refer the reader to the work of [10].

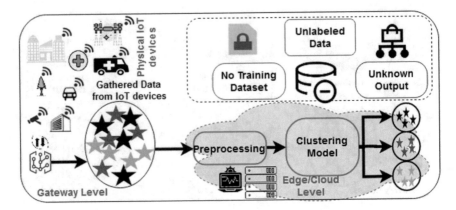

Fig. 5 Unsupervised Learning Concept-based IoT scenario

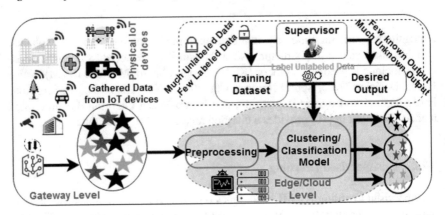

Fig. 6 Semi-Supervised Learning Concept-based IoT scenario

3.1.3 Semi-Supervised Learning

Semi-Supervised Learning (SSL) is an ML tool that learns from both pre-classified and unclassified samples as illustrated in Fig. 6 [10]. Indeed, SSL techniques are used among traditional supervised Learning schemes in which all input samples are pre-labeled and through unsupervised learning methods where no class is assigned at all [14]. Officially, SSL is about creating a classy model from a limited and partially labeled data set and then allocating labels the input data. SSL is in charge of generalizing the process of maintaining input sample subsets that are identical with class labels [22].

The SSL defines several contexts as well as the Semi-Supervised (SS) classification and the SS clustering [23]. Where the former is an extension of the supervised classification problem. The training data consist of both labeled and unlabeled instances. It is generally considered that there are more unlabeled data than the labeled data. The aim of this classification is to train a classifier from unlabeled and labeled data, so that it is better than the supervised classifier which is trained on only labeled data. While the latter explains an unsupervised clustering extension. The training data comprises a non-labeled instances, as well as cluster-supervised information. The goal of such scheme is to obtain an optimal clustering than the one of the only unlabeled data. Hence, several algorithms are used as powerful tools is SSL as well as Gaussian Mixture Model (GMM), Manifold Regularization, and Semi-Supervised Support Vector Machine (S3VM) [10].

3.1.4   Reinforcement Learning

Reinforcement Learning (RL) indicates a ML class which learns from consecutive expertise what's necessary to be done in order to find out the best solution. More exactly, RL is in charge of learning via interaction with the environment and observing the actions, outcome [24]. It gives the ability for machines to automatically locate the ideal conduct in a specific context, in order to maximize its performance. For that, a modest return of the results is necessary to learn how the machines must act.

As depicted in the Fig. 7 [10], the RL aim is based on an iterated interaction, in the form of the execution at each instant n of an action a_n. Since the current state is s_n which leads to the new state that is s'_n and provides the reward r_n. On the basis of this interaction, a policy is gradually improved. However, in practice, most RL algorithms do not work directly on the policy but go through the iterative approximation of the value function [25]. Therefore, the main task of RL to learn that how to associate actions with situations in order to maximize a reward quantitatively. A brief description of RL process is given as follows [10]:

- The agent observes an input state
- An action is determined by a decision-making function (policy)

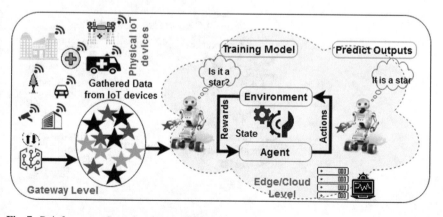

Fig. 7 Reinforcement Learning Concept-based IoT scenario

- The action is performed
- The agent receives a result according to its environment
- Information about the result given for this state or action is recorded.

In this context, many algorithms are used in this category as well as Q-learning, State Action Reward State Action (SARSA, and Deep Q Network [10].

3.2 ML Tools-Based IoT Applications

ML ML has earned a considerable amount of worldwide popularity, which is possible through the *Internet of Things (IoT)*. However, ML is considered to be one of the most suitable computational tools to provide intelligence in the IoT systems-based applications and their challenges. Various domains adopt IoT as a significant area, whereas others have removed pilot projects to map the potential of IoT in large projects.

In this context, IoT-based applications has a lot of potential for social, environmental, and economic impact. Medical and Healthcare, Industrial Processing, Agriculture and Breeding, smart Mobility, Smart Grid, Smart Homes/Buildings, and Public Safety and Environment Monitoring are some of the IoT use case areas. All of these areas are related to us in one way or another. During recent years, their existence and usability have attained a visionary scale and have become of paramount importance. It may not be incorrect to state that the future of the Internet is purely based on the IoT concept, which drives us into the future practically.

Various IoT use case and applications have been depicted in Fig. 8. This section highlights some IoT-based applications and ML techniques.

Fig. 8 IoT Applications

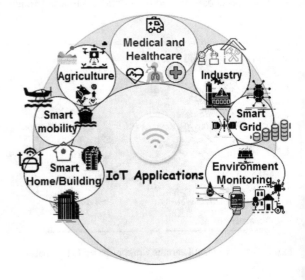

3.2.1 Medical and Healthcare

Oftentimes it is denominated as the Internet of Medical Things (IoMT). It represents the IoT-based application that connects healthcare services to the IT system through on line various computer networks and servers. Medical devices are equipped with in-built Wi-fi systems that further allow machine-to-machine communication based on the IoMT concept.

In this context, authors in [26] proposed utilizing generated data for future reference and usage by uploading the data onto the cloud server. The study suggested a platform based on cloud computations for managing the data. Authors in [27], proposed the usage of RFID based applications for body-centric systems for gathering information on human behavior in compliance with power and sanitary regulations. In [28], the authors suggested networked sensors, either worn or embedded. The framework of these devices was capable of gathering rich information that indicated the physical and mental health of an individual. Authors in [29] highlighted that IoT based smart rehabilitation systems were better techniques to mitigate problems associated with aging populations and a shortage of health professionals. Furthermore, the study proposed an Automating Design Methodology (ADM) for smart rehabilitation systems in IoT. Authors in [30] highlighted the potential benefits of using M-IoT in non-invasive glucose level sensing and potential M-IoT architecture based on diabetes management has been proposed in the study. The observations from the framework were sent on mobile for information updates. Authors proposed, in [31], an intelligent home-based platform, the iHome Health-IoT system that had an open-platform based on intelligent medicine box (iMedBox). Authors in [32] highlighted the significant importance of IoT in the field of healthcare. Proactive healthcare analytics for cardiac disease prevention has been the Usage of smartphone for an exemplary case has been depicted in the study. However, the proposed technique was only able to detect the threat of heart attack and no alarm or notification framework was proposed in the study. Authors in [33] proposed an IoT-based framework to control blood sugar, then Markov model is used to determine the appropriate insulin dose. Authors in [34] proposed an approach based on IoT and fuzzy inference system to recognize coronavirus disease cases. Authors in [35] employed Naive Bayes and SVN to predict kidney disease. In [36], authors proposed several decision tree techniques as well as NB tree, J48, LADTree, BFTree, LMT, and Random Forest to predict the thyroid disease.

3.2.2 Industry

The concept of IoT has also blossomed in the industry and is the main contributor to Industry 4.0. However, the equipment and industrial requirements are so intense that the functional capabilities of the IoT are either molded or tailored to meet the needs of the industry.

In this context, authors in [37] proposed the use of a standardized IoT architecture as suggested by IETF. A data-centric scheduling algorithm, Traffic-Aware Scheduling

Algorithm (TASA) has also been proposed in this work. Authors in [38] proposed a framework based on a super frame structure for slotted MAC. The framework designed by IWSN designers achieved the objective by minimizing the MAC access latency. In [39] studied a deep knowledge on various public logistics services. A Supply Hub in Industrial Park (SHIP) application was proposed in the study to achieve the objective of information sharing and real-time visibility. This suggested tool had an efficiency of sharing physical assets and services along with effectiveness. Authors in [40] proposed an IoT infrastructure for collaborative warehouse. The study integrated a bottom-up approach with numerous mechanisms like Decision Support Systems (DSS), self-organizing, and negotiation protocols among agents. The work in [41] evaluated various available resources and methodologies for the Industrial Market perspective. Evolution in terms of technology has been observed. Context-Aware computing theories, evaluation framework, and communication mediums have been discussed.

On the other hand, *Industrial Internet of Things (IIoT)* will be the most talked about in the world, in which the world economic landscapes will be changed accordingly. In this context, resource allocation is denoted as bottleneck issues that researchers seek to find out solutions. However, authors in [42] proposed Mini batch gradient descent (MBGD)-based network slicing paradigm for resource allocation in *Industry 4.0* . Authors in [43] proposed a deep federated learning-based network slicing for slice resource allocation in IIoT IindustryIoT. A multi-game theory-based network slicing was proposed in [44] for slice resource allocation.

3.2.3   Agriculture

Climate-smart agriculture (CSA) is defined as an approach to transform and reorient agricultural development according to climate change. There have been enormous changes in the techniques and methodologies of performing agricultural activities. The new farmer has now moved from concept farming to modernized concepts. Researchers working in this field have developed theories and practices that incorporate smart devices to assess the parameters that contribute to the growth and irrigation of the plant, and according to observations, agricultural activities are carried out.

In this context, the authors in [45] highlighted an up-to-date framework for major technology drivers in IoT. The study also envisioned the help farmers could get from IoT applications to obtain information to deliver crops directly to customers within a small area. Authors in [46] proposed the Intelligent Agriculture Management and Information Management (IAMIS) framework to identify the intact characteristics of agricultural activities. In addition, the information stored in the proposed framework was accessible to anyone who further wished to perform the same process via personal computers, tablets, or mobile phones. The authors in [47] centered on modern agriculture and management perspectives with the help of WSN. The proposed study introduced a micro-irrigation system for agriculture based on hardware and network architecture along with software operations that control the micro-irrigation system. Authors in [48] proposed a connected dairy based on decision tree tool to monitor the cows health and detect their health conditions. Authors in [49] proposed a smart beehive in which decision tree technique was exploited to predict hive state, with the aim is to protect hive health, based on sensed vital sign like CO_2 and O_2. The K-means technique was used with the GPS technology to classify and distinguish soil classes [50]. The ann scheme was employed by [51] to classify eggs as fertile on infertile. Many works and studies are introduced in [52–54] and references therein.

3.2.4 Smart Mobility

Nowadays, smart and green cities represent the dynamic change in the world that made smart mobility the methodology that enables smooth, efficient and flexible mobility across different modes in these cities. However, Vehicular Ad-hoc Network (VANETs) is the most talked subjects. Hence, it is a new shift paradigm to a more flexible, multi-modal, and smart transport system. It is indeed the the Internet of Vehicles (IoV) pillar that faces to improve road safety either by reducing accidents, giving new solutions towards optimized and smart transportation modes [55].

In this context, many challenges related to traffic problems and moving from one to another place, which were and still are the research focus of many researchers. Authors in[56] proposed a new communications specifically to automotive applications focusing on electric mobility. Ethernet has been widely used for various experiments. The most talked-about is the Advanced Driver Assistant Systems (adas), which meets the studied navigation, positioning, multimedia, and communication systems. Authors in [57] highlighted the most prominent impact of smart vehicles based on Smart City, IoT, and Eco-Conscious Cruise Control for public transportation, which aims to use the resources available in the framework that provided driving recommendations in an environmentally friendly and efficient way. Authors in [58] highlighted the traffic congestion challenge needs to be addressed which impeded the entire network. The study experimented with the concept of a deep constrained Boltzmann machine and recurrent neural network architecture that was able to predict the traffic congestion development based on GPS data obtained from cars. A parallel computing environment framework based on GPU methodological was proposed in the study. Authors in[59] proposed Bayesian Network Seasonal to predict Short-term traffic in large traffic network. Authors in [60] proposed CNN-based framework to detect blind spot for a smart connected car. In [61], authors used fuzzy c-means to predict short-term traffic in large road network. Authors in [62] employed in their study the Markov Random Field to detect parking space. Several other work can be found in [63, 64] and references therein.

3.2.5 Smart Grid

Smart Grid is another technological advancement proof, that represents an electricity distribution network that promotes the information flow between suppliers and consumers, using digital technology, in order to adjust the electricity flow, after performing observations that are made with the help of sensors. in real-time and allow for more efficient management. However, the grid directs the power according to the calculated requirements that have been established before.

Much research works have been provided in this context. However, authors in [65] proposed Game theory to formulate scheduled energy consumption where strategies are based on daily schedules of household applications and their energy loads. Authors in [66] proposed a M2M communication model called the Cognitive M2M. The framework is intended for an energy efficiency spectrum discovery scheme. A coordination-based energy-saving spectrum discovery scheme was also proposed, for the smart grid, to provide a significant reduction in energy consumption. Authors in [67] proposed an SVM-based approach to monitor voltage sensitivity and power system. The authors in [68] provided a Decision Tree combined with SVM, which can achieve top-down classification to detect and locate real-time energy theft at each level in the transmission and distribution process of the power system. Authors in [69] proposed an ann to examine energy consumption data in order to report energy fraud. Authors in [70] deep convolutional neural networks to analyze data and detect electricity theft. Authors in [71] employed many ML technique as well as Random forest, decision tree, neural networks, and SVM to detect intrusion based on gathered data from synchrophasor devices.

4 IoT-Based ML Open Issues

4.1 Lightweight IoT-Based ML Approaches

The widespread deployment of IoTs particularly in a smart city surroundings generates a huge data amount. Current ML techniques cannot orchestrate a dynamic and huge amount data in a real-time environment, so a lot of data is lost without information extracting [72]. The great unlabeled data amount can be combined with a small set of disaggregated data for better convergence of ML diagrams. In this context, lightweight ML techniques can be sophisticated that are proper for treatment the large IoT data [73]. The notion of data analytics can be applied in this context as sensor location, type and data can aid in developing lightweight models.

4.2 Distributed IoT-Based ML Tools

The ML applications for IoT have to overcome with huge datasets. The real industrial datasets for ML-based applications can be thousands of GBs [74]. In such a scenario, the ML-based models which are normally complex and power intensive cannot be run on a single machine. The overall workload can be divided using distributed ML with worker machines, but it also opens certain issues to be met [75]. Bandwidth is one of the crucial challenges to be faced by powerful worker machines. The worker machines have to frequently exchange data between them at a high transfer rate but such high bandwidth is usually not available which creates a bottleneck [76]. The machines should also need to coincide to perform sequential tasks [77]. In a realistic scenario, all worker machines are not exactly of identical processing power which slows down the learning and optimization process.

4.3 Federated IoT-Based ML Tools

IoT ML implementations have to deal with traditional centralized learning networks that face increasing issues in terms of privacy, preservation and scalability communications costs. In such a scenario, it is not possible to run complex ML technologies with heterogeneous aggregated data in a centralized manner and satisfy QoS for users [78]. Federated Learning Networks has been proposed as a promising model to support ML [79]. In disparity to centralized data storage and processing in centralized learning, Federated Learning exploits edge servers to store data and perform training distributively [80]. Incidentally, the edge devices in federated Learning can maintain training data locally, which preserves privacy and reduces communications overhead. However, because typical training within federated Learning relies on the contributions of advanced devices, the training process can be disrupted if some high-end machines upload incorrect or bogus training results [81].

4.4 IoT-Based ML at the Edge

The vast amount of connected devices has transformed the entire network community in a new era called the IoTs [82]. The IoT concept has simplified society in one way, but delay-sensitive and context-informed applications have put specific challenges on the performance of lightweight IoT devices [83]. To meet the demand for real-time data computing, edge computing has provided as a promising key solution by executing data computing requests for IoT devices through some neighboring devices [84]. Traditional ML may be confused with terminal-generated data since it is more complex to identify real data from the complex and noisy environment [85]. Deep learning can play its role in edge devices to improve learning and also to maintain privacy of saved data during intermediate data transfer.

4.5 IoT-Based ML Data Munging

Data collected from IoT devices, which are bulky, inconsistent, inconsistent, and full of typos, cannot be used as input for complex ML applications [86]. To beat this issue and obtain data trends by making it incorporated, the IoT data collected must go through the cleansing process [87], called as data munging . This process includes transformation, data cleaning or cleaning, data exploration, metadata enrichment exploitation, entering missing values, removing unnecessary or invalid data that is not required to obtain basic trends of data, and then data validation [88]. This traditional method has several limitations, especially when there are huge amounts of data daily from IoT Industry 4.0 and smart cities. For them, need for accurate and trustworthy real-time analytics remains critical, in order to deal promptly with sudden issues and problems that have occurred.

5 Proposed IoT-Based ML Scenario for Smart Agriculture 4.0: Smart Irrigation System

5.1 Overview of the Agriculture 4.0

With latest disruptive technologies in Industry 4.0, agriculture reached a major technological revolution known as Agriculture 4.0. In fact, Industry 4.0 approach (or IIoT) allowed creating an environment in which all elements are linked together continuously and effortlessly. In this context, all devices (such as cyber-physical systems) and functionalities are addressed as services that continuously communicate with each other, thus achieving a high level of coordination [89]. In this way, the ability to coordinate activities is essential for improved supply chain management, where optimization normally requires the contemplation of many elements in constant competition with each other [90]. In recent years, this industrial innovations has supported revolutions in agriculture sector. The first revolution started with Agriculture 1.0 that relies on animal power, then the fuel engine revolution defined the Agriculture 2.0. Recently, Agriculture 3.0 is defined with the guidance systems and precision farming, once military GPS signals were made publicly available [91]. Hence, Agriculture 4.0 farm activities collected by sensors are connected to the cloud.

The evolution of Agriculture 4.0 is happening in parallel with comparable developments in the industrial sector, based on future manufacturing ideas. Agriculture 4.0 , stands for the common internal and external interaction of agricultural operations, providing digital information across all farm sectors and operations. As in the industrial sector, Agriculture revolution 4.0 represents a great opportunity to look at the disparity and uncertainties involved in the Agri-food production chain [92]. In this vein, farms will become smarter, more efficient, and more environmentally sustainable, due to the combination and integration of technologies such as data and services integration in the network infrastructure, information and communication systems, and ML techniques [93].

5.2 Proposed Smart Irrigation System

IoT technology has revolutionized every human life area by making everything connected and smart. IoT refers to a network of things that make up a self-configuring network. The development of a system based on smart farming technology and IoT is a day-to-day transformation from classical to smart agriculture, especially in the *Agriculture 4.0* era. Thus, by improving it and making it cost-effective and reducing waste.

Nowadays, farmers need to manually turn on/off the water pump to irrigate their lands. However, to prevent plant damage due to the water supply failure, they need to check manually at regular intervals whether the soil needs irrigation or not. In this context, we propose a IoT-based smart irrigation system as denoted in Fig. 9. The aim of this scenario is to provide smart farming system based on IoT and ML techniques and help them to monitor environment effectively, which will allow them to increase total yield and quality of the product.

The proposed system is about five connected sensors implemented in the soil. The Humidity sensor to sens soil humidity value (the amount of water vapor in the air), the Temperature sensor to sense temperature value, the Moisture sensor to estimate volumetric water content in the soil, the Light intensity sensor, and finally the Connected Relay sensor to switch on/off water pump motors. All these sensors collect and transfer sensed data to the cloud, where the Linear regression based-approach [10] analyses sensed data and predict the required water units. After that, the cloud system sends an information signal, after controlling the actual motor's state, to switch on/off the relay of the water pump water to start or finalize land irrigation.

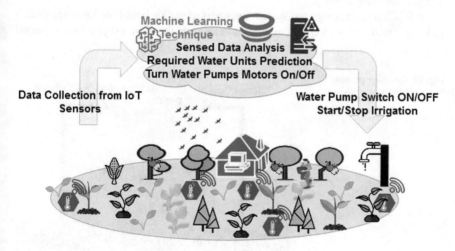

Fig. 9 Smart Irrigation System based on IoT and ML tool

The proposed smart irrigation system based on IoT and ML consists of an offline phase then an online phase. However, in the former, the smart irrigation-based approach collects data only from sensors and store them, to train after the proposed scheme. This training phase aims at tuning the linear regression model weight parameters. The second phase is denoted as the test or online phase, in which the proposed model acts as a brain for this system. However, in this phase, Humidity, Temperature, Light Intensity, and the Connected Relay sensors are connected all to the cloud-based system and collect and transfer their data. In this case, The proposed system based on the trained model will predict immediately the required amount of water units and switch on/off the connected water pumps relays.

5.3 Results and Discussions

The training process used to train our proposed framework based ML approach is depicted in Fig. 10. The Mean Square Error (MSE) [42] is used as an optimization technique, where the model to be trained was the linear regression technique. The input dataset [94] consist of 46 samples captured at different interval of time. Each sample consists of Humidity (H), Temperature (T), Light (L) intensity and Moisture (M) values.

The figures below depict the numerical results of the proposed IoT-based smart irrigation system. The deployed results are in terms of train phase (curve fitting) and many test scenarios. However, the accuracy of our proposed scheme is about 99% and the Fig. 11 is the best prove of that. However, the linear regression weight parameters are tuned on 0.276, 1.135, −1.982, for the Humidity, Light intensity, and Temperature respectively.

At this stage, our proposed system for smart irrigation based on linear regression technique will be tested at different input values and the output signal for connected

Fig. 10 Training process

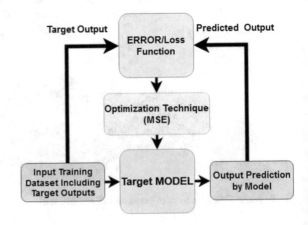

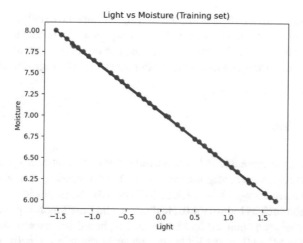

Fig. 11 Training phase result of the linear regression model

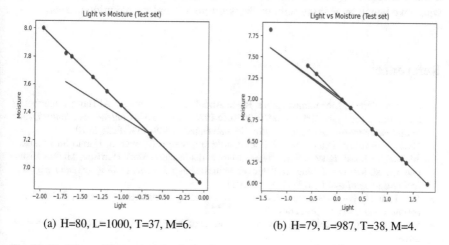

(a) H=80, L=1000, T=37, M=6.   (b) H=79, L=987, T=38, M=4.

Fig. 12 Test phase of the smart irrigation system

relays will be treated accordingly. However, in Fig. 12a with input parameters (H=80, L=1000, T=37, M=6) the relay water pump is turned off and the system predicts 4.7 water unit is required but the moisture at this time is much high (M=6), that is way relay is turned off. This, saving water and protecting plants. Figure 12b with input parameters (H=79, L=987, T=38, M=4), the predicted required water units is about 5.35 which is greater than the sensed Moisture (4) that is way the relay water pump is turned on.

Soil moisture measurement is important in agriculture 4.0 to help farmers manage their irrigation systems more efficiently. Not only are farmers able to generally use less water to grow a crop, but they are also able to increase yields and crop quality through better soil moisture management during critical plant growth stages.

6   Conclusion

In this chapter, we presented IoT' architecture types and challenges. Then, we summarized ML tools in four categories and detailed their learning strategies. In Sect. 3, we surveyed research works that tackled IoT and ML challenges in various domains and application such as Industry 4.0, Healthcare, and Smart mobility. Next, we highlighted and analyzed some critical issues in ML-based IoT as well as Distributed ML, Federated ML, ML at the Edge, and ML data Munging. Finally, we proposed a smart irrigation system based on IoT scenario and ML tool for Agriculture 4.0, where we began by shining light on Agriculture 4.0 and its evolution process.

References

1. Shadi Al-Sarawi, Mohammed Anbar, Rosni Abdullah, and Ahmad B Al Hawari. Internet of things market analysis forecasts, 2020–2030. In *2020 Fourth World Conference on Smart Trends in Systems, Security and Sustainability (WorldS4)*, pages 449–453. IEEE, 2020
2. Ovidiu Vermesan, Peter Friess, Patrick Guillemin, Sergio Gusmeroli, Harald Sundmaeker, Alessandro Bassi, Ignacio Soler Jubert, Margaretha Mazura, Mark Harrison, Markus Eisenhauer, et al. Internet of things strategic research roadmap. *Internet of things-global technological and societal trends*, 1(2011):9–52, 2011
3. Da Li Xu, He Wu, Li Shancang (2014) Internet of things in industries: A survey. IEEE Transactions on industrial informatics 10(4):2233–2243
4. Spyros G Tzafestas. Ethics and law in the internet of things world. *Smart cities*, 1(1):98–120, 2018
5. Mohemed Almorsy, John Grundy, and Amani S Ibrahim. Collaboration-based cloud computing security management framework. In *2011 IEEE 4th International Conference on Cloud Computing*, pages 364–371. IEEE, 2011
6. Rose Karen, Eldridge Scott, Chapin Lyman (2015) The internet of things: An overview. The Internet Society (ISOC) 80:1–50
7. James Manyika. *The Internet of Things: Mapping the value beyond the hype.* McKinsey Global Institute, 2015
8. Atzori Luigi, Iera Antonio, Morabito Giacomo (2017) Understanding the internet of things: definition, potentials, and societal role of a fast evolving paradigm. Ad Hoc Networks 56:122–140
9. Luo Xiong, Liu Ji, Zhang Dandan, Chang Xiaohui (2016) A large-scale web qos prediction scheme for the industrial internet of things based on a kernel machine learning algorithm. Computer Networks 101:81–89
10. Seifeddine Messaoud, Abbas Bradai, Syed Hashim Raza Bukhari, Pham Tran Anh Qung, Olfa Ben Ahmed, and Mohamed Atri. A survey on machine learning in internet of things: Algorithms, strategies, and applications. *Internet of Things*, page 100314, 2020

11. Trent D Buskirk, Antje Kirchner, Adam Eck, and Curtis S Signorino. An introduction to machine learning methods for survey researchers. *Survey Practice*, 11(1):2718, 2018
12. Jennifer G Dy and Carla E Brodley. Feature selection for unsupervised learning. *Journal of machine learning research*, 5(Aug):845–889, 2004
13. Tsai Cheng-Fa, Tsai Chun-Wei, Han-Chang Wu, Yang Tzer (2004) Acodf: a novel data clustering approach for data mining in large databases. Journal of Systems and Software 73(1):133–145
14. Xiaojin Jerry Zhu. Semi-supervised learning literature survey. Technical report, University of Wisconsin-Madison Department of Computer Sciences, 2005
15. Said Omar, Masud Mehedi (2013) Towards internet of things: Survey and future vision. International Journal of Computer Networks 5(1):1–17
16. Miao Wu, Ting-Jie Lu, Fei-Yang Ling, Jing Sun, and Hui-Ying Du. Research on the architecture of internet of things. In *2010 3rd International Conference on Advanced Computer Theory and Engineering (ICACTE)*, volume 5, pages V5–484. IEEE, 2010
17. Gubbi Jayavardhana, Buyya Rajkumar, Marusic Slaven, Palaniswami Marimuthu (2013) Internet of things (iot): A vision, architectural elements, and future directions. Future generation computer systems 29(7):1645–1660
18. Flavio Bonomi, Rodolfo Milito, Preethi Natarajan, and Jiang Zhu. Fog computing: A platform for internet of things and analytics. In *Big data and internet of things: A roadmap for smart environments*, pages 169–186. Springer, 2014
19. Lapide Larry (2004) Rfid: What's in it for the forecaster. Journal of Business Forecasting Methods and Systems 23(2):16–19
20. Klaus Doppler, Mika Rinne, Carl Wijting, Cássio B Ribeiro, and Klaus Hugl. Device-to-device communication as an underlay to lte-advanced networks. *IEEE communications magazine*, 47(12):42–49, 2009
21. Ethem Alpaydin. *Introduction to machine learning*. MIT press, 2020
22. Xiaojin Zhu and Andrew B Goldberg. Introduction to semi-supervised learning. *Synthesis lectures on artificial intelligence and machine learning*, 3(1):1–130, 2009
23. Kulis Brian, Basu Sugato, Dhillon Inderjit, Mooney Raymond (2009) Semi-supervised graph clustering: a kernel approach. Machine learning 74(1):1–22
24. Leslie Pack Kaelbling, Michael L Littman, and Andrew W Moore. Reinforcement learning: A survey. *Journal of artificial intelligence research*, 4:237–285, 1996
25. Karl Cobbe, Oleg Klimov, Chris Hesse, Taehoon Kim, and John Schulman. Quantifying generalization in reinforcement learning. In *International Conference on Machine Learning*, pages 1282–1289. PMLR, 2019
26. Charalampos Doukas and Ilias Maglogiannis. Bringing iot and cloud computing towards pervasive healthcare. In *2012 Sixth International Conference on Innovative Mobile and Internet Services in Ubiquitous Computing*, pages 922–926. IEEE, 2012
27. Amendola Sara, Lodato Rossella, Manzari Sabina, Occhiuzzi Cecilia, Marrocco Gaetano (2014) Rfid technology for iot-based personal healthcare in smart spaces. IEEE Internet of things journal 1(2):144–152
28. Moeen Hassanalieragh, Alex Page, Tolga Soyata, Gaurav Sharma, Mehmet Aktas, Gonzalo Mateos, Burak Kantarci, and Silvana Andreescu. Health monitoring and management using internet-of-things (iot) sensing with cloud-based processing: Opportunities and challenges. In *2015 IEEE International Conference on Services Computing*, pages 285–292. IEEE, 2015
29. Yuan Jie Fan, Yue Hong Yin, Li Da Xu, Yan Zeng, and Fan Wu. Iot-based smart rehabilitation system. *IEEE transactions on industrial informatics*, 10(2):1568–1577, 2014
30. Robert SH Istepanian, Sijung Hu, Nada Y Philip, and Ala Sungoor. The potential of internet of m-health things "m-iot" for non-invasive glucose level sensing. In *2011 Annual International Conference of the IEEE Engineering in Medicine and Biology Society*, pages 5264–5266. IEEE, 2011
31. Yang Geng, Xie Li, Mäntysalo Matti, Zhou Xiaolin, Pang Zhibo, Da Li Xu, Kao-Walter Sharon, Chen Qiang, Zheng Li-Rong (2014) A health-iot platform based on the integration of intelligent packaging, unobtrusive bio-sensor, and intelligent medicine box. IEEE transactions on industrial informatics 10(4):2180–2191

32. Arijit Ukil, Soma Bandyoapdhyay, Chetanya Puri, and Arpan Pal. Iot healthcare analytics: The importance of anomaly detection. In *2016 IEEE 30th international conference on advanced information networking and applications (AINA)*, pages 994–997. IEEE, 2016

33. Mwaffaq Otoom, Hussam Alshraideh, Hisham M Almasaeid, Diego López-de Ipiña, and José Bravo. Real-time statistical modeling of blood sugar. *Journal of medical systems*, 39(10):123, 2015

34. Syyada Abeer Fatima, Naveed Hussain, Asma Balouch, Iqra Rustam, Muhammad Saleem, and Muhammad Asif. Iot enabled smart monitoring of coronavirus empowered with fuzzy inference system. *International Journal of Advance Research, Ideas and Innovations in Technology*, 6(1), 2020

35. Vijayarani S, Dhayanand S et al (2015) Data mining classification algorithms for kidney disease prediction. International Journal on Cybernetics & Informatics (IJCI) 4(4):13–25

36. Turanoglu-Bekar Ebru, Ulutagay Gozde, Kantarcı-Savas Suzan (2016) Classification of thyroid disease by using data mining models: a comparison of decision tree algorithms. Oxford Journal of Intelligent Decision and Data Sciences 2:13–28

37. Maria Rita Palattella, Nicola Accettura, Luigi Alfredo Grieco, Gennaro Boggia, Mischa Dohler, and Thomas Engel. On optimal scheduling in duty-cycled industrial iot applications using ieee802. 15.4 e tsch. *IEEE Sensors Journal*, 13(10):3655–3666, 2013

38. Yan Hairong, Zhang Yan, Pang Zhibo, Da Li Xu (2014) Superframe planning and access latency of slotted mac for industrial wsn in iot environment. IEEE Transactions on Industrial Informatics 10(2):1242–1251

39. Xuan Qiu, Hao Luo, Gangyan Xu, Runyang Zhong, and George Q Huang. Physical assets and service sharing for iot-enabled supply hub in industrial park (ship). *International Journal of Production Economics*, 159:4–15, 2015

40. Paul J Reaidy, Angappa Gunasekaran, and Alain Spalanzani. Bottom-up approach based on internet of things for order fulfillment in a collaborative warehousing environment. *International Journal of Production Economics*, 159:29–40, 2015

41. Charith Perera, Chi Harold Liu, Srimal Jayawardena, and Min Chen. A survey on internet of things from industrial market perspective. *IEEE Access*, 2:1660–1679, 2014

42. Seifeddine Messaoud, Abbas Bradai, and Emmanuel Moulay. Online gmm clustering and mini-batch gradient descent based optimization for industrial iot 4.0. *IEEE Transactions on Industrial Informatics*, 16(2):1427–1435, 2019

43. Messaoud Seifeddine, Bradai Abbas (2020) Olfa Ben Ahmed, Pham Quang, M Atri, and M Shamim Hossain. Deep federated q-learning-based network slicing for industrial iot, IEEE Transactions on Industrial Informatics

44. Dawaliby Samir, Bradai Abbas, Pousset Yannis (2019) Distributed network slicing in large scale iot based on coalitional multi-game theory. IEEE Transactions on Network and Service Management 16(4):1567–1580

45. Bandyopadhyay Debasis, Sen Jaydip (2011) Internet of things: Applications and challenges in technology and standardization. Wireless personal communications 58(1):49–69

46. Duan Yan-e. Design of intelligent agriculture management information system based on iot. In *2011 Fourth International Conference on Intelligent Computation Technology and Automation*, volume 1, pages 1045–1049. IEEE, 2011

47. Sanbo Li. Application of the internet of things technology in precision agriculture irrigation systems. In *2012 International Conference on Computer Science and Service System*, pages 1009–1013. IEEE, 2012

48. Anitha Ilapakurti and Chandrasekar Vuppalapati. Building an iot framework for connected dairy. In *2015 IEEE First International Conference on Big Data Computing Service and Applications*, pages 275–285. IEEE, 2015

49. Fiona Edwards-Murphy, Michele Magno, Pádraig M Whelan, John O'Halloran, and Emanuel M Popovici. b+ wsn: Smart beehive with preliminary decision tree analysis for agriculture and honey bee health monitoring. *Computers and Electronics in Agriculture*, 124:211–219, 2016

50. Verheyen Kris, Adriaens Dries, Hermy Martin, Deckers Seppe (2001) High-resolution continuous soil classification using morphological soil profile descriptions. Geoderma 101(3–4):31–48

51. Das K, Evans MD (1992) Detecting fertility of hatching eggs using machine vision ii: Neural network classifiers. Transactions of the ASAE 35(6):2035–2041
52. Kaloxylos Alexandros, Eigenmann Robert, Teye Frederick, Politopoulou Zoi, Wolfert Sjaak, Shrank Claudia, Dillinger Markus, Lampropoulou Ioanna, Antoniou Eleni, Pesonen Liisa et al (2012) Farm management systems and the future internet era. Computers and electronics in agriculture 89:130–144
53. Yifan Bo and Haiyan Wang. The application of cloud computing and the internet of things in agriculture and forestry. In 2011 International Joint Conference on Service Sciences, pages 168–172. IEEE, 2011
54. Yibo Chen, Jean-Pierre Chanet, and Kun Mean Hou. Rpl routing protocol a case study: Precision agriculture. In First China-France Workshop on Future Computing Technology (CF-WoFUCT 2012), 2012
55. Saif Al-Sultan, Moath M Al-Doori, Ali H Al-Bayatti, and Hussien Zedan. A comprehensive survey on vehicular ad hoc network. Journal of network and computer applications, 37:380–392, 2014
56. Peter Hank, Steffen Müller, Ovidiu Vermesan, and Jeroen Van Den Keybus. Automotive ethernet: in-vehicle networking and smart mobility. In 2013 Design, Automation & Test in Europe Conference & Exhibition (DATE), pages 1735–1739. IEEE, 2013
57. Dimosthenis Kyriazis, Theodora Varvarigou, Daniel White, Andrea Rossi, and Joshua Cooper. Sustainable smart city iot applications: Heat and electricity management & eco-conscious cruise control for public transportation. In 2013 IEEE 14th International Symposium on" A World of Wireless, Mobile and Multimedia Networks"(WoWMoM), pages 1–5. IEEE, 2013
58. Ma Xiaolei, Haiyang Yu, Wang Yunpeng, Wang Yinhai (2015) Large-scale transportation network congestion evolution prediction using deep learning theory. PloS one 10(3):e0119044
59. Gaetano Fusco, Chiara Colombaroni, Luciano Comelli, and Natalia Isaenko. Short-term traffic predictions on large urban traffic networks: Applications of network-based machine learning models and dynamic traffic assignment models. In 2015 International Conference on Models and Technologies for Intelligent Transportation Systems (MT-ITS), pages 93–101. IEEE, 2015
60. Donghwoon Kwon, Suwoo Park, SunHee Baek, Ritesh K Malaiya, Geumchae Yoon, and Jeong-Tak Ryu. A study on development of the blind spot detection system for the iot-based smart connected car. In 2018 IEEE International Conference on Consumer Electronics (ICCE), pages 1–4. IEEE, 2018
61. Hitoshi Kanoh, Takeshi Furukawa, Souichi Tsukahara, Kenta Hara, Hirotaka Nishi, and Hisashi Kurokawa. Short-term traffic prediction using fuzzy c-means and cellular automata in a wide-area road network. In Proceedings. 2005 IEEE Intelligent Transportation Systems, 2005., pages 381–385. IEEE, 2005
62. Qi Wu, Chingchun Huang, Shih-yu Wang, Wei-chen Chiu, and Tsuhan Chen. Robust parking space detection considering inter-space correlation. In 2007 IEEE International Conference on Multimedia and Expo, pages 659–662. IEEE, 2007
63. Sahil Garg, Kuljeet Kaur, Syed Hassan Ahmed, Abbas Bradai, Georges Kaddoum, and Mohammed Atiquzzaman. Mobqos: Mobility-aware and qos-driven sdn framework for autonomous vehicles. IEEE Wireless Communications, 26(4):12–20, 2019
64. LaFrance Adrienne (2015) Self-driving cars could save 300,000 lives per decade in america. The Atlantic 29:
65. Amir-Hamed Mohsenian-Rad, Vincent WS Wong, Juri Jatskevich, Robert Schober, and Alberto Leon-Garcia. Autonomous demand-side management based on game-theoretic energy consumption scheduling for the future smart grid. IEEE transactions on Smart Grid, 1(3):320–331, 2010
66. Zhijing Qin, Grit Denker, Carlo Giannelli, Paolo Bellavista, and Nalini Venkatasubramanian. A software defined networking architecture for the internet-of-things. In 2014 IEEE network operations and management symposium (NOMS), pages 1–9. IEEE, 2014
67. George M Messinis, Alexandros E Rigas, and Nikos D Hatziargyriou. A hybrid method for non-technical loss detection in smart distribution grids. IEEE Transactions on Smart Grid, 10(6):6080–6091, 2019

68. Jindal A, Dua A, Kaur K, Singh M, Kumar N, Mishra S (2016) Decision tree and svm-based data analytics for theft detection in smart grid. IEEE Transactions on Industrial Informatics 12(3):1005–1016
69. Vitaly Ford, Ambareen Siraj, and William Eberle. Smart grid energy fraud detection using artificial neural networks. In *2014 IEEE Symposium on Computational Intelligence Applications in Smart Grid (CIASG)*, pages 1–6. IEEE, 2014
70. Zheng Zibin, Yang Yatao, Niu Xiangdong, Dai Hong-Ning, Zhou Yuren (2017) Wide and deep convolutional neural networks for electricity-theft detection to secure smart grids. IEEE Transactions on Industrial Informatics 14(4):1606–1615
71. K Vimalkumar and N Radhika. A big data framework for intrusion detection in smart grids using apache spark. In *2017 International Conference on Advances in Computing, Communications and Informatics (ICACCI)*, pages 198–204. IEEE, 2017
72. Erwin Adi, Adnan Anwar, Zubair Baig, and Sherali Zeadally. Machine learning and data analytics for the iot. *Neural Computing and Applications*, pages 1–29, 2020
73. Furqan Alam, Rashid Mehmood, Iyad Katib, Nasser N Albogami, and Aiiad Albeshri. Data fusion and iot for smart ubiquitous environments: A survey. *IEEE Access*, 5:9533–9554, 2017
74. Yongrui Qin, Quan Z Sheng, Nickolas JG Falkner, Schahram Dustdar, Hua Wang, and Athanasios V Vasilakos. When things matter: A survey on data-centric internet of things. *Journal of Network and Computer Applications*, 64:137–153, 2016
75. Mohammad Saeid Mahdavinejad, Mohammadreza Rezvan, Mohammadamin Barekatain, Peyman Adibi, Payam Barnaghi, and Amit P Sheth. Machine learning for internet of things data analysis: A survey. *Digital Communications and Networks*, 4(3):161–175, 2018
76. Jiang Tigang, Fang Hua, Wang Honggang (2018) Blockchain-based internet of vehicles: Distributed network architecture and performance analysis. IEEE Internet of Things Journal 6(3):4640–4649
77. Renjie Gu, Shuo Yang, and Fan Wu. Distributed machine learning on mobile devices: A survey. arXiv preprint arXiv:1909.08329, 2019
78. Kato Nei, Mao Bomin, Tang Fengxiao, Kawamoto Yuichi, Liu Jiajia (2020) Ten challenges in advancing machine learning technologies toward 6g. IEEE Wireless Communications
79. S. Messaoud, A. Bradai, O. Ben Ahmed, P. Quang, M. Atri, and M. S. Hossain. Deep federated q-learning-based network slicing for industrial iot. *IEEE Transactions on Industrial Informatics*, pages 1, 2020
80. Viraj Kulkarni, Milind Kulkarni, and Aniruddha Pant. Survey of personalization techniques for federated learning. arXiv preprint arXiv:2003.08673, 2020
81. Latif U Khan, Walid Saad, Zhu Han, and Choong Seon Hong. Dispersed federated learning: Vision, taxonomy, and future directions. arXiv preprint arXiv:2008.05189, 2020
82. Francis Griffiths and Melanie Ooi. The fourth industrial revolution-industry 4.0 and iot [trends in future i&m]. *IEEE Instrumentation & Measurement Magazine*, 21(6):29–43, 2018
83. Daiwat A Vyas, Dvijesh Bhatt, and Dhaval Jha. Iot: trends, challenges and future scope. *IJCSC*, 7(1):186–197, 2015
84. Mahmut Taha Yazici, Shadi Basurra, and Mohamed Medhat Gaber. Edge machine learning: Enabling smart internet of things applications. *Big data and cognitive computing*, 2(3):26, 2018
85. Devki Nandan Jha, Khaled Alwasel, Areeb Alshoshan, Xianghua Huang, Ranesh Kumar Naha, Sudheer Kumar Battula, Saurabh Garg, Deepak Puthal, Philip James, Albert Y Zomaya, et al. Iotsim-edge: A simulation framework for modeling the behaviour of iot and edge computing environments. arXiv preprint arXiv:1910.03026, 2019
86. Ge Mouzhi, Bangui Hind, Buhnova Barbora (2018) Big data for internet of things: A survey. Future generation computer systems 87:601–614
87. Salvador García, Sergio Ramírez-Gallego, Julián Luengo, José Manuel Benítez, and Francisco Herrera. Big data preprocessing: methods and prospects. *Big Data Analytics*, 1(1):9, 2016
88. Milenkovic Milan (2020) Internet of Things: Concepts and System Design. Springer
89. Jan Schlechtendahl, Matthias Keinert, Felix Kretschmer, Armin Lechler, and Alexander Verl. Making existing production systems industry 4.0-ready. *Production Engineering*, 9(1):143–148, 2015

90. Wiendahl Hans-Hermann (2011) Auftragsmanagement der industriellen Produktion: Grundlagen, Konfiguration. Springer-Verlag, Einführung
91. Marucci Alvaro, Colantoni Andrea, Zambon Ilaria, Egidi Gianluca (2017) Precision farming in hilly areas: The use of network rtk in gnss technology. Agriculture 7(7):60
92. Burak Ozdogan, Anil Gacar, and Huseyin Aktas. Digital agriculture practices in the context of agriculture 4.0. *Journal of Economics Finance and Accounting*, 4(2):186–193, 2017
93. Strozzi Fernanda, Colicchia Claudia, Creazza Alessandro, Noè Carlo (2017) Literature review on the 'smart factory' oncept using bibliometric tools. International Journal of Production Research 55(22):6572–6591
94. https://github.com/Chinukapoor/Smart-Agriculture-using-IoT-and-Machine-Learning

Seifeddine Messaoud received his M.S degree (2017) in electronic and microelectronics from Monastir University, Tunisia. He is currently completing his Ph.D. degree at the Electronic and Micro-Electronic Laboratory (LR99ES30), Sciences Faculty of Monastir. He is working in collaboration with XLIM Laboratory, Poitiers, France. His current research interest is machine learning, wireless network, IoT, network slicing, Image processing, IPs and SoC.

Olfa Ben Ahmed is an associate professor in the university of Poitiers and the XLIM research center. She obtained her PhD degree in Computer Science from the University of Bordeaux and the LaBRI (France) in January 2015. She served as Research and Teaching Assistant at the University of Bordeaux, ENSEIRB MATMECA and the university of Limoges. She was a post-doc researcher at the data science department of the EURECOM research center in Nice. Her topics of research include artificial intelligence, computer vision and image processing.

Abbas Bradai received the Ph.D. degree from the University of Bordeaux, France, in 2012. He is currently as an Associate Professor with the University of Poitiers and has been a member with the XLIM Research Institute (CNRS UMR7252) since September 2015. His research interests include multimedia communications over wired and wireless networks, IoT, software defined network, and virtualization. He is/was involved in many French and European Projects (FP7 and H2020), such as ENVISION and VITAL.

Mohamed Atri is Professor in the Department of Computer Engineering at College of Computer Science, King Khalid University, Saudi Arabia. He received his Ph.D. Degree in Microelectronics from the University of Monastir, Tunisia in 2001. He has obtained the HDR degree from the University of Monastir, Tunisia in 2011. He is currently a member of the Laboratory of Electronics & Microelectronics, Faculty of Science of Monastir. His research includes Circuit and System Design, Image processing, Network Communication, IPs and SoC.

Enabling IoT Wireless Technologies in Sustainable Livestock Farming Toward Agriculture 4.0

Eleni Symeonaki, Konstantinos G. Arvanitis, Dimitrios Loukatos, and Dimitrios Piromalis

Abstract As part of the latterly introduced approach of Agriculture 4.0, practices involving ubiquitous computing advancements and conceptual innovations of "smart" agricultural production tend to be adopted. This fact is considered to be critical in addressing the challenge of securing adequate food supplies for the constantly increasing world population, taking also into regard the imperative necessity of exploiting natural resources according to policies related to sustainable growth. Moreover it has already been recognized that enabling the Internet of Things (IoT) wireless technologies into livestock farming systems for the benefit of sustainable growth is of high significance. To this end, potential solutions should be provided for developing responsive and adaptive IoT integrated systems which will deliver a wide variety of qualitative low-cost services in accordance with the objectives of modern sustainable livestock farming. This work presents and critically analyzes the existence, functionality and interoperability of various approaches in this area, as well as their maturity to be integrated toward the concept of Agriculture 4.0. In addition to this, some key challenges are identified regarding the management, process and exchange of the large amounts of heterogeneous sensory raw data that are acquired remotely in precision livestock farming environments.

E. Symeonaki (✉) · K. G. Arvanitis · D. Loukatos
Agricultural University of Athens, Iera Odos, 75, Athens 11855, Greece
e-mail: esimeon@uniwa.gr

K. G. Arvanitis
e-mail: karvan@aua.gr

D. Loukatos
e-mail: dlouka@aua.gr

E. Symeonaki · D. Piromalis
University of West Attica, Thivon 250 & P. Ralli, Egaleo 12244, Greece
e-mail: piromali@uniwa.gr

© Springer Nature Switzerland AG 2021
P. Krause and F. Xhafa (eds.), *IoT-based Intelligent Modelling for Environmental and Ecological Engineering*, Lecture Notes on Data Engineering and Communications Technologies 67, https://doi.org/10.1007/978-3-030-71172-6_9

List of Acronyms with Explanation

AI	Artificial Intelligence
AR	Augmented Reality
ECA	Event-Condition-Action
EID	Electronic Identification
EPRS	European Parliamentary Research Service
FAO	Food and Agriculture Organization
GNSS	Global Navigation Satellite System
GPS	Global Positioning System
GRAP	Grup de Recerca en AgròTICa i Agricultura de Precisió (Research Group in AgroICT and Precision Agriculture)
ICT	Information and Computer Technology
IoT	Internet of Things
ISPA	International Society for Precision Agriculture
MEC	Mobile Edge Computing
ML	Machine Learning
PA	Precision Agriculture
PLF	Precision Livestock Farming
UN	United Nations
VR	Virtual Reality
WEF	World Economic Forum
WOW	Walk-over-Weighing
WSAN	Wireless Sensor and Actuator Network

1 Introduction

Taking into account recent United Nations (UN) projections, the world population is about to reach 8.5 billion in 2030, and further increase to 9.7 billion by 2050 and 11.2 billion by 2100 [1]. To address the challenge of providing adequate supplies of safe and healthful food at reasonable costs, so as to satisfy the increasing nutritional needs of the constantly growing world population, agricultural production has to be raised up to 60% during the 21st century, according to estimates of the Food and Agriculture Organization (FAO) of the United Nations [2]. What is more, due to the issues which arise from the global climate change as well as the escalating pressures on ecosystems, and biodiversity, this aim has to be accomplished taking into regard the national and international policies related to sustainable growth [3]. To this end, in order for agriculture to be physically, socially and politically sustainable, it is essential to incorporate practices and technologies that enable farming systems to [4]:

(a) efficiently increase long-term productivity without failing to maintain the high quality and commercial competitiveness of agricultural products,

(b) protect the environment and conserve natural resources,

(c) enhance the living standard for the farming community and the society in overall.

Precision Agriculture (PA), as a contemporary approach which refers to the overall management of farming systems [5] based on quality and quantity coefficients [6], assists significantly in controlling environmental and economic risks, by accomplishing more suitable and targeted usage of finite resources [6, 7]. Despite the fact that the concept of PA first emerged in the early '90s referring to the undertaking of proper actions at the right intensity at the right place and time [7], it has not been attributed with a distinct definition concerning the kind of technologies which are included in it. As a matter of fact, the Research Group in AgroICT and Precision Agriculture (GRAP) of the University of Lleida (Universitat de Lleida) in Spain assembled a non-exhaustive list of 27 various definitions of PA as found in the literature, making additionally discrete reference to the definition officially adopted by the International Society for Precision Agriculture (ISPA) in July 2019 [8]. Precisely, according to the ISPA official definition *"Precision Agriculture is a management strategy that gathers, processes and analyzes temporal, spatial and individual data and combines it with other information to support management decisions according to estimated variability for improved resource use efficiency, productivity, quality, profitability and sustainability of agricultural production"* [9]. As deriving from this definition, credible, cost-effective, and user-friendly solutions are considered to be a significant factor for the benefit of sustainable growth. Concisely, the aims of PA in terms of sustainability are, according to the European Parliamentary Research Service (EPRS), as follows [10]:

- Optimization of the available resources usage as well as of the reliance on natural processes with emphasis on long-term highly sustainable, efficient and profitable production.
- Reduction in the usage of inputs with the highest negative impacts for the environment as well as for food safety.
- Improvement of the social aspects quality and the working conditions for all relevant stakeholders (e.g. farmers, veterinarians, nutritionists, etc.)

Although the PA approach originally addressed to arable farming, its practices are nowadays adopted in all types of biological production systems, among which livestock production [11]. Livestock Farming is indeed one of the core branches of the agricultural sector, dealing with the breeding and care of animals (including but not limited to sheep, swine, beef and dairy cattle, poultry, horses) as well as pasture and turf management [2, 10, 11].

In particular, integrating the PA concept to livestock production is attributed with the term Precision Livestock Farming (PLF). The aim of PLF is to manage animals, even within large herds, individually as a biological unit and continuously monitor in real-time, parameters which are related to the health, the welfare, the breeding as well as the production and reproduction of each animal [12, 13]. Additionally PLF focuses on the environmental impact of the related processes (including pasture and turf management) and integrates external data, such as weather forecasts, leading to evidence-based decisions on livestock production [13]. The PLF approach incorporates various collateral fields, such as animal science, bio-engineering, computing, and socio-economics, in order to realize systems that can provide advanced monitoring and managing techniques for the benefit of the animals, the implicated stakeholders, the community and the environment. In accordance with the objectives of PA for sustainable growth, PLF systems, as a constantly developing technological trend incorporating modern Information and Computer Technology (ICT) for the real-time monitoring and management of animals, can be integrated with ubiquitous computing advancements and conceptual innovations of smart technologies in the context of Industry 4.0, focusing on increasing the levels of production automation.

The Industry 4.0 concept, which is based on a strategic framework initially introduced by the German National Academy of Science and Engineering (acatech) in 2013 [14] and was the theme of the World Economic Forum (WEF) annual meeting in 2016 [15], is progressively affecting the production potentials of all industry sectors [16, 17], through the establishment of a digitalized environment where physical and virtual objects can interconnect and interact autonomously [17, 18]. The implementation of smart technologies within agriculture (which is considered to be the most important sector of primary industry), as part of the Industry 4.0 approach, has latterly set forth to be attributed with the term "Agriculture 4.0" [19, 20]. Among the key technologies involved in Agriculture 4.0, according to the guidelines for the operative implementation of Industry 4.0 in the agricultural sector, the Internet of Things (IoT) wireless technologies appear to be particularly significant, taking into consideration their capabilities and effects for the benefit of sustainability [21].

More precisely, the IoT introduces a concept according which, physical objects are benefited with sensory and computational support for communicating and interacting with each other, based on standard protocols via a highly distributed public network, such as the Internet [22]. What is more, the IoT comprises a set of various technologies combining, among others, wired or wireless sensory devices, information systems and enhanced machinery. Consequently, integrating the IoT into PLF systems is regarded to turn into a major asset for sustainable livestock management [19] as it may lead to advanced automation in such a way that the overall production process performance will be optimized by accounting for variability and uncertainties within livestock farming systems [23, 24]. Indeed, enabling IoT wireless technologies in PLF systems through a wide range of potential applications, farmers can be granted with the ability to remotely monitor conditions and receive real-time information about pasture lands and animals, so as to make the right decisions for achieving production increase, cost reduction as well as savings in inputs. In particular, automating production reduces the cost of the human labor involved in daily tasks, such as feeding, milking, and cleaning while adapting production inputs site-specifically within a pasture land and individually for each animal, allows better use of resources to ensure environmental protection, while securing the quantity and quality of the food supply. Furthermore, IoT wireless technologies can provide the means of monitoring and managing the food production chain since important features of animal welfare and quality of life (indicating that animals are positively stimulated in their environment, both physically and cognitively) can be insomuch enhanced as farmers enable them to be, veterinarians define them, and consumers request them. This is in line with the recent concept of "one welfare" which acknowledges the direct and indirect interconnection among animal and human well-being as well as environmentally friendly livestock production systems.

Taking into consideration that the research community has already recognized the high significance of enabling IoT wireless technologies in PLF systems for the benefit of sustainable growth, this work presents and critically analyzes the existence, functionality and interoperability of various approaches in this area, as well as their maturity to be integrated toward the concept of Agriculture 4.0. Additionally to this, some key challenges are identified regarding the management, process and exchange of the large amounts of heterogeneous sensory raw data that are acquired remotely in PLF environments.

2 Context of IoT Wireless Technologies in PLF

To date, several architectures have been developed for being applied in PLF production systems, depending on various application scenarios. A generic architectural diagram describing the data management and control process in a typical PLF system [25] is depicted in Fig. 1. According to this structure, the input of data which is acquired automatically via sensors is then subjected to data fusion and analysis along with any manually acquired and external data input (e.g. environmental data acquired via weather stations) so as to control the system based on automated real-time decision making. Furthermore the generated information is transferred to external control agents such as professional consultants (e.g. veterinarians, nutritionists, etc.) in order to assist the system in the making of more informed management decisions for real-time actions as well as for future predictions. The key concept of this structure is that the monitoring outcomes, after being integrated with the other incoming information, are fed back into the data fusion and analysis stage, enabling the control system to be optimized with machine learning algorithms and therefore adjust to various requirements addressing to each individual animal of a specific farm as well as to different livestock farming systems. Considering the fact that the system generates an extremely large amount of heterogeneous data which constantly increases with the time, all data are relayed to the cloud for being analyzed, communicated and stored. Hosting these operations in the cloud provides a highly flexible, secure and cost efficient solution for any probable horizontal or vertical scaling required due to the large volumes of data.

When integrating the IoT approach into typical PLF systems, wireless technologies are considered to be the most significant feature for enabling smart real-time

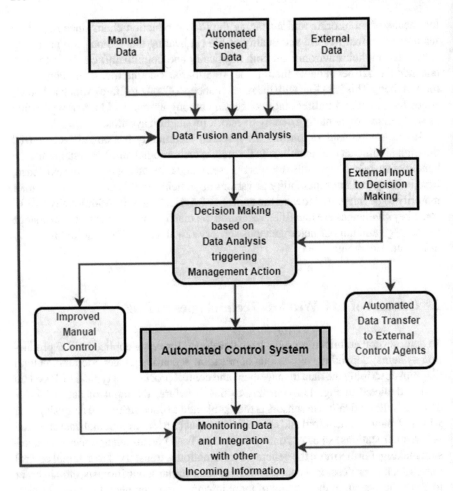

Fig. 1 Generic architectural diagram of data management and control process in a typical PLF system

interactions amongst the animals and multiple devices with or without human mediation [3, 24, 26–28]. A schematic diagram of such a smart PLF system integrating the IoT approach is depicted in Fig. 2.

Considering this approach, raw data which are acquired remotely by a Wireless Sensor and Actuator Network (WSAN) deployed in a livestock facility (a farm or a field), are relayed via gateways to a cloud control center [29], along with any manually acquired and external data input, for being analyzed diagnosed, and managed with Artificial Intelligence (AI) and data fusion techniques [30, 31] in order to obtain higher reliability and lower detection error probability. The cloud responds back to the facility equipment with control actions as well as to the end users providing them with monitoring information and services. The versatility of a PLF system integrated

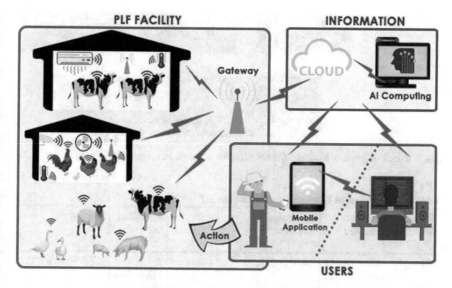

Fig. 2 Schematic diagram of a smart PLF system integrating the IoT approach

with the IoT wireless technologies lies in its ability to employ the cloud for performing self-discovery and self-configuration as well as for offering on-demand services to livestock facilities placed in distant rural regions.

At this point it should be noted that although the proposed system can be hosted to a public, a private as well as to a hybrid cloud, it is preferable all cloud services (IaaS, PaaS and SaaS) to be offered for usage by anyone via a public or a hybrid cloud, since the IoT promotes the idea of "anything/anyone—anytime—anyplace" [3]. Nevertheless due to the fact that most farming enterprises are of small to medium size, hosting the proposed system in a public cloud is considered to be the most suitable solution due to the fact that hardware infrastructures and computing power of the cloud service provider are used for the cloud services.

In a more detailed reference to the WSAN integrated in the livestock facility, this consists of groups of self-powered nodes deployed in a mesh network topology with communication range adequate for operating over distance. These nodes incorporate sensors for remotely acquiring real-time data about various physical parameters concerning the animals and the facility in general, as well as actuators for interacting with the facility equipment enabling the proper physical actions. Regarding the wireless nodes hardware design architecture, several solutions, specially addressing to the harsh operating environment of rural regions have been introduced. The block diagram of a typical WSAN node for livestock management is illustrated in Fig. 3.

One of the main challenges in the development of IoT integrated PLF systems is the choice of frequency bands to be used by the WSANs for the transmission process. Although to date there are several communication protocols available offering a wide range of frequency bands, as shown in Table 1, there is no standard frequency to be used by IoT WSANs as characteristics differ for each frequency band and available

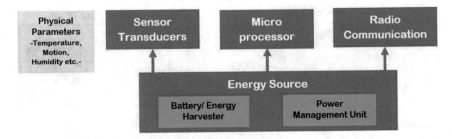

Fig. 3 Block diagram of a typical WSAN node for livestock management

Table 1 Comparison of Wireless Protocols for the IoT

Wireless technology	Wireless standard	Operating frequencies	Max. Coverage range	Max. Data rate	Security
WiFi	IEEE 802.11 a/b/g/n/ac/ad	2.4/5/60 GHz	100 m	2 Mbps–6.93 Gbps	WEP/WPA/WPA2
Bluetooth	IEEE 802.15.1	2400/2483.5 MHz	100 m	1–3 Mbps	56/128 bit
BLE	IEEE 802.15.1	2400/2483.5 MHz	100 m	1Mbps	128 bit AES
6LowPAN	IEEE 802.15.4	908.42/2483.5 MHz	100 m	250 Kbps	128 bit
Zigbee	IEEE 802.15.4	2400/2483.5 MHz	10 m	250 Kbps	128 bit
GPRS	3GPP	GSM 850/1900 MHz	10 km	171 Kbps	GEA2/GEA3/GEA4
LTE	3GPP	700/2600 MHz	28 km	1 Gbps	SNOW 3G
Sigfox	Sigfox	908.42 MHz	30-50 km	10–1000 bps	N/A
LoRaWAN	LoRa Alliance	868/915 MHz	15 km	50–100 Kbps	128 bit
DASH7	ISO/IEC 18000-7	433.92/868/915 MHz	10 km	28–200 kbps	128 bit AES

transmission frequencies are regulated by laws which differ worldwide. Comparison results deriving from literature [32–34], suggest that, among the several protocols available for the IoT (i.e. WiFi, Bluetooth, 6LowPAN, SigFox, Zigbee, etc.), the LoRaWAN as well as the DASH7 are far more appropriate for deployment in distant rural areas, as they offer the optimum combination in terms of communication range, data rate and power consumption [35].

Whichever the requirements or the specifications of the PLF system are, the integration of IoT wireless technologies may generate a unique framework for livestock management depending on the types of the devices, the available services and the complete setup of the system. In this way, the PLF system will be able to maintain the animal welfare standards whilst also manage efficiently the available resources so as to ensure maximum livestock production of optimum quality while the farm operates in environmentally and economically sustainable conditions.

3 Applications of IoT Wireless Technologies in PLF

The deployment of WSANs into IoT integrated PLF facilities should meet several operation requirements that may have to be applied and processed in parallel as forth below:

- Monitoring and/or control of biological and physical parameters of the animals such as temperature, heart rate, etc. as well as animal behavior and location.
- Forecast and control of imminent animal diseases such as parturient paresis (milk fever), listeriosis, tetanus, etc.
- Monitoring and/or control of ambient conditions such as temperature, humidity, gas emissions, etc. as well as meteorological data.
- Security surveillance and control of livestock facilities.

3.1 Animal Monitoring—Forecast and Control of Imminent Animal Diseases

Animal monitoring is a significant requirement for PLF management, aiming for smart and effective farming as well as for preventing or detecting imminent diseases.

In this context wireless sensing technologies are essential to the development of IoT integrated livestock monitoring systems. For this there are several types of biometric devices and sensors available for automatically identifying individual animals and locating their position as well as for acquiring data related to animal biological parameters, detecting imminent diseases and other problem indications which may affect the livestock production.

3.1.1 Electronic Identification (EID)

Given that a primary objective of the PLF approach is to measure and manage variability at the level of each animal individually, the ability of automated identification is undeniably essential within PLF systems. EID tags operating at ultrahigh frequencies seem to be the current trend in animal electronic identification as they can increase the reading distance and speed as well as the number of tags that can be read at the same time [36]. The EID tags are most usually attached to the ear of the individual animal it is designated to identify (Fig. 4), as it is less likely to injure the animal compared to other methods available [37].

Fig. 4 Indicative EID tag
and EID-reader

Furthermore each EID tag can incorporate internal memory for digitally storing information related to the animal, such as veterinary treatment, overcoming in this way data capture issues due to imminent unavailability of network access (e.g. over Wi-Fi or cellular network). However as local data storage limits the integration of data involving different animals (e.g. data from an entire herd) [38], the stored data can be uploaded into remote cloud servers at any time the network is accessible.

3.1.2 Health, Behavior and Location Monitoring

According to animal sciences, there are various acknowledged relationships between the state of health and metabolism of an animal and basic physical or biological factors such as weight, body temperature, heart and breathing rate, etc. [39, 40]. Additionally, an animal's behavior can be an evident indicator of its physiological state such as the stage of its activity level which may be linked to its reproductive state. Moreover a diseased animal may indicate physiological and behavioral anomalies as for instance in the case of "Listeriosis" the symptoms of which include uncoordinated movement of the animal's limbs, stiffness of its neck as well as tendency of moving in circles or leaning against a fence or a wall [41]. In this sense, health and behavior monitoring assists significantly in the early detection and prevention of imminent diseases as well as in the management of livestock production.

As presented in Sect. 2 of the present chapter, a smart PLF system integrating the IoT approach (Fig. 2) employs several groups of self-powered nodes which incorporate sensors (such as wearable sensing and biometric devices) for remotely acquiring real-time data about the animals' physiology and behavior. These data are relayed to the cloud for being analyzed, communicated and stored. The cloud responds back triggering up to farmers and veterinarians real-time information about any probable physiological and behavioral anomalies so as to undertake proper actions. Additionally, given the existence of specific conditions meeting pre-established criteria, the

Table 2 Types of sensing and biometric devices according to monitored animal physical and biological parameters

Type of sensing and biometric device	Physical and biological parameters	Monitoring
Thermometer	Body temperature	Health
Pulsometer	Heart rate	Health
Respiration monitor	Breathing rate	Health
Walk-over-Weighing (WOW)	Weight	Health
Accelerometer/Gyroscope	Position and inclination of the animal's head (Assists to estimate if the animal is moving, eating, ruminating or just resting)	Health/Behavior
Global Navigation Satellite System (GNSS)	Anomalous behavior patterns and possible areas with more ambient stress specially inside barns	Location/Behavior
Pedometer	Number of movements over a given period	Behavior/Health

corresponding actuators are triggered in the PLF facility via Event-Condition-Action (ECA) rules so as to enable proper operations.

There are various types of wearable sensing and biometric devices [42–45], most commonly attached to a collar around the animal's neck or leg (Fig. 5), which are responsible for acquiring data related to the animal's physical and biological parameters indicating probable symptoms (e.g., fevers, stress, etc.) that affect its health conditions and livestock production as in Table 2.

3.2 Ambient Conditions and Meteorological Data Monitoring

Monitoring ambient conditions such as temperature, humidity, gas emissions, etc. into indoor facilities, such as barns, is essential for the welfare of animals while the surveillance of meteorological data is significant for effective pasture and turf management.

PLF systems incorporate sensing devices so as to measure the ambient conditions for animals living indoors (e.g. barns) since these may affect their welfare and health [46, 47]. These sensing devices are classified to sensors for estimating the index of temperature (air thermometer) and humidity (hygrometer) which is related to animal stress, as well as to gas sensors for measuring potentially dangerous levels of gas emissions (such as methane, hydrogen sulphide, ammonia, formaldehyde) in the air of indoor facilities.

Furthermore advanced PLF systems tend to involve agro-meteorological stations for monitoring atmosphere and soil parameters concerning pasture and turf fields [48–50] so as to come to the optimum decisions related to their management (e.g.

Fig. 5 Indicative wearable
sensing and biometric device
(collar)

irrigation). The agro-meteorological stations most commonly incorporate sensors
for measuring air temperature (air thermometer), relative air humidity (hygrometer),
wind speed and direction (anemometer), solar radiation (pyranometer), rainfall and
artificial irrigation (pluviometer), as well as soil temperature and moisture.

3.3 Security Surveillance and Control of Livestock Facilities

3.3.1 Predator Alert

Domestic animals (e.g. sheep) often suffer attacks from feral carnivores (e.g. wolves),
directing negative impact on the animal welfare as well as financial cost on the
livestock facility. For this remote camera "traps" have been developed to alert farmers
to the presence of predators on or near their livestock facility. These "traps" are
actually digital cameras with night vision (so that they can work at night) that are
activated by movement sensors. The images which are captured by the camera "traps"
are automatically processed to identify the species of the predator and alert the
farmer for its presence. An additional action allowing the farmer to manage the
predator attack is the automated activation of intensive lighting and sound alarms so
as to intimidate the predator and force it to get away. Research indicates that such
solutions are generally reliable [51].

3.3.2 Automated Sorting Gates

Automatic sorting (also known as shedding or drafting) gates allow animals to be
automatically partitioned as they move down a passageway [52]. Actually, gates
driven pneumatically, direct the animals to exit a crate individually or into small

Fig. 6 Indicative sorting gate

groups. The sorting gates can be interlinked to an EID system permitting the selection of specific animals based on their EID tag or alternatively, to an automated weighing system selecting the animas based on their weight (Fig. 6).

3.3.3 Pasture Access Control

Dynamic control of pasture access is facilitated with several technological solutions. The simplest one proposes the application of timed remotely controlled gate handles which release an electric fence gate in order to allow animals access the paddock or the feeding area at a specific time without the need for human intervention. A more sophisticated approach proposes the application of a fully robotic fence which can move across the paddock, given the precondition that the ground is sufficiently flat to accommodate the movement of the robots (Fig. 7).

The virtual fencing approach (also known as geo-fencing) proposes a radically different solution in controlling the access of animals to pasture since rather than relying on physical barriers, virtual fences rely on the animal learning how to access the paddock through reinforcement [53]. In greater detail, when the animal approaches a virtual boundary, which is delimited with a signal cable laid on the ground, a sensor attached to the collar of the animal detects the proximity and if needed activates a warning sound signal. Should the animal stand back the sound signal is deactivated, but if it chooses to proceed it runs into an electric fence and receives a stimulus while the unit attached on the collar relays the related data to the

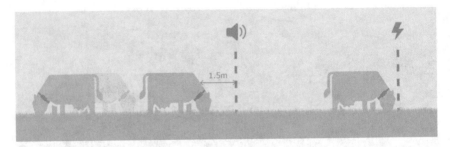

Fig. 7 Virtual fencing operation for pasture access control

farmer via a mobile application. In this way the animals can learn the boundaries of their pasture through trial and error [54].

4 Discussion

Sustainable livestock farming requires a holistic approach to farm planning and management based on knowledge for decision making that could reduce, if not eliminate, a great number of the existing constraints to agricultural productivity while conserving the natural resources and protect the environment. The ability of livestock farming to successfully participate in the transformative change in agriculture and food systems may well depend on its ability to move from the traditional industrial paradigm, designed to increase productivity, to a knowledge-based paradigm leading to long-term economic, environmental, and social sustainability. It is considered that IoT wireless technologies can make a breakthrough in livestock management by increasing the level of intelligence in monitoring, control and decision-making along with the synergies between physical and virtual objects according to the guidelines for operative implementation of Agriculture 4.0. Nevertheless there are some significant challenges that need to be addressed regarding the further integration of the IoT wireless technologies to sustainable livestock farming.

At first, the IoT integrated PLF systems involve a great number of devices which highly differentiate in processors, memory, programming languages, communication protocols and interfaces resulting inevitably to device as well as data heterogeneity issues that ultimately affect the scalability of the systems. Although there is great research interest to the standardization of IoT wireless technologies in livestock farming, there is still need to establish a complete and evident architectural framework as well as protocols and standards to interconnect the various heterogeneous devices and services.

Furthermore, security and privacy vulnerabilities consist a critical barrier to the large scale deployment of IoT wireless technologies in livestock farming. Since the integration of the rapidly increasing interconnected things provides the ability to remotely control their operations, several security issues arise concerning the interactions between the physical and virtual objects. Indeed, the IoT wireless applications operate in highly vulnerable and unprotected environments where sensitive data might be compromised to eventual hacking attacks. The necessity of protecting these objects and ensuring their long lifespan enhances the necessity of providing cloud-based security solutions with resource efficient, thing-to-cloud interactions that will be adequate for the complex dynamic networks in livestock farming.

Finally the strong need of real-time data processing imposes constant and high speed network connectivity for the support of unobstructed streaming. In this context, edge computing, as an evolved version of the cloud computing technology, may enable new IoT applications and services which meet the aforementioned constraints of farming environments, by applying new sophisticated data management as well as services analytics. Furthermore, the association of Mobile Edge Computing (MEC) and 5G emerging technologies for supporting mobility in distant rural areas (i.e. drone control, AR/VR, etc.) seem to be an interesting subject of scientific research for capturing future growth opportunities in sustainable livestock farming.

Concisely, generalizing the scope of the interconnected things beyond the sensing services, it is not hard to imagine the transformation of the entire livestock farming sector into a smart web of context-sensitive remotely identified, sensed, and controlled virtual objects that will be shared and reused in different contexts, projecting the vision of Agriculture 4.0 for the global sustainable growth.

5 Conclusion

Enabling IoT wireless technologies to PA is already having a great impact in many areas of agricultural production. This work presented the influences that IoT wireless technologies brought to PLF toward the guidelines for the operative implementation of Agriculture 4.0 as well as for the benefit of sustainable growth. In this context IoT integrated PLF systems ensure optimized production of high quality agricultural products, in a way that safeguards the welfare and health of all farmed species, protects the natural environment as well as improves the economic and social conditions of farmers and local communities in general.

The research in literature came up with findings which indicate that the IoT wireless technologies provide advanced livestock management enabling smart ubiquitous and autonomous interconnection with and among physical and virtual objects, leading in this way to the vision of Agriculture 4.0. Enabling the IoT wireless technologies in PLF systems increases the level of intelligence in monitoring, control and decision-making. Thereby the efficiency and quality of livestock production is increased while the utilization of natural resources is improved and the climate dependency is reduced assisting in environment and ecosystem protection, and promoting sustainable development.

As, given the interest shown, more affordable and feasible solutions which address the challenges of IoT wireless technologies implementation in PLF are expected to be widely available, transforming the entire sector into a smart web of interconnected objects that will be remotely identified, sensed, and controlled. Eventually, livestock farms are about to be transformed into self-adaptive autonomous systems involving smart interconnected objects that will be able to make decisions and operate without the need of human intervention.

References

1. United Nations (2019) Population. https://www.un.org/en/sections/issues-depth/population/index.html. Accessed 30 Apr 2020
2. FAO (2017) The future of food and agriculture—Trends and challenges. http://www.fao.org/3/a-i6583e.pdf. Accessed 30 Apr 2020
3. Symeonaki EG, Arvanitis KG, Piromalis DD (2019) Cloud computing for IoT applications in climate-smart agriculture: a review on the trends and challenges toward sustainability. In: Theodoridis A, Ragkos A, Salampasis M (Eds) Innovative approaches and applications for sustainable rural development. HAICTA 2017. Springer Earth System Sciences. Springer, Cham, pp 147–167
4. Gromis di Trana M, Bava F, Pisoni P (2019) A sustainable value generator in the Italian wine industry. British Food J 122(5):1321–1340
5. Crookston RK (2006) A top 10 list of developments and issues impacting crop management and ecology during the past 50 years. Crop Sci 46(5):2253–2262
6. Mulla D, Khosla R (2016) Historical evolution and recent advances in precision farming. In: Rattan L, Stewart BA (Eds) Soil-specific farming precision agriculture, 1st edn. CRC Press Taylor & Francis Group, Florida, pp 1–36
7. Leonard EC (2016) Precision agriculture. In: Wrigley C, Corke H, Seetharaman K, Faubion J (eds) Encyclopedia of food grains, 2nd edn. Elsevier, Oxford, pp 162–167
8. GRAP, University of Lleida (2020) Precision Agriculture definitions. http://www.grap.udl.cat/en/presentation/pa_definitions.html. Accessed 30 Apr 2020
9. ISPA (2019) Precision Ag Definition. https://www.ispag.org/about/definition. Accessed 30 Apr 2020
10. EPRS (2016) Precision Agriculture and the future of farming in Europe. http://www.europarl.europa.eu/RegData/etudes/STUD/2016/581892/EPRS_STU(2016)581892_EN.pdf. Accessed 30 Apr 2020
11. Berckmans D (2015) Smart farming for Europe: value creation through precision livestock farming. In: Halachmi I (ed) Precision livestock farming applications: making sense of sensors to support farm management. Wageningen Academic Publishers, Wageningen, pp 25–36
12. Rutter SM (2017) Advanced livestock management solutions. In: Ferguson DM, Lee C, Fisher A (eds) Herd and flock welfare: advances in sheep welfare. Woodhead Publishing, Sawston, pp 245–261
13. Berckmans D (2017) General introduction to precision livestock farming. Animal Front 7(1):6–11
14. Kagermann H, Wahlster W, Helbig J (2013) Recommendations for implementing the strategic initiative INDUSTRIE 4.0—Securing the Future of German Manufacturing Industry. acatech-National Academy of Science and Engineering, Munich. http://forschungsunion.de/pdf/industrie_4_0_final_report.pdf. Accessed 30 Apr 2020
15. WEF (2016) Mastering the fourth industrial revolution. http://www3.weforum.org/docs/WEF_AM16_Report.pdf. Accessed 30 Apr 2020
16. Pfeiffer S (2017) The vision of "Industrie 4.0" in the making—a case of future told, tamed, and traded. NanoEthics 11(1):107–121

17. Xu LD, Xu EL, Li L (2018) Industry 4.0: state of the art and future trends. Int J Prod Res 56(8):2941–2962
18. Rojko A (2017) Industry 4.0 concept: background and overview. Int J Interact Mobile Technol 11(5):77
19. Bonneau V, Copigneaux B, Probst L, Pedersen B (2017) Industry 4.0 in agriculture: focus on IoT aspects. European Commission. https://ec.europa.eu/growth/tools-databases/dem/mon itor/sites/default/files/DTM_Agriculture%204.0%20IoT%20v1.pdf. Accessed 30 Apr 2020
20. De Clercq M, Vats A, Biel A (2018) Agriculture 4.0: the future of farming technology. world government summit in collaboration with Oliver Wyman. https://www.worldgovernmentsum mit.org/api/publications/document?id=95df8ac4-e97c-6578-b2f8-ff0000a7ddb6. Accessed 30 Apr 2020
21. Rose DC, Chilvers J (2018) Agriculture 4.0: broadening responsible innovation in an era of smart farming. Front Sustain Food Syst 2:1–5
22. Talavera JM, Tobón LE, Gómez JA, Culman MA, Aranda JM, Parra DT, Quiroz LA, Hoyos A, Garreta LE (2017) Review of IoT applications in agro-industrial and environmental fields. Comput Electron Agric 142:283–297
23. Lakhwani K, Gianey H, Agarwal N, Gupta S (2019) Development of IoT for smart agriculture a review. In: Rathore V, Worring M, Mishra D, Joshi A, Maheshwari S (Eds) Emerging trends in expert applications and security. Advances in Intelligent Systems and Computing, vol 841. Springer, Singapore, pp 425–432
24. Fresco R, Ferrari G (2018) Enhancing precision agriculture by internet of things and cyber physical systems. Atti della Società Toscana di Scienze Naturali 125:53–60
25. Banhazi T, Marcus H (2018) Development of precision livestock farming technologies. In: Chen G (ed) Advances in agricultural machinery and technologies. CRC Press-Taylor & Francis Books, London, pp 179–194
26. An W, Chang Y (2017) A study on the livestock feed measuring sensor and supply management system implementation based on the IoT. J Korea Inst Inf Electron Commun Technol 10(5):442–454
27. Zhang J, Kong F, Zhai Z, Han S, Wu J, Zhu M (2016) Design and development of IoT monitoring equipment for open livestock environment. Int J Simul Syst Sci Technol 17(26):2–7
28. Internet of Food and Farm 2020 (2020) IOF2020 Reference architecture for interoperability, replicability and reuse. https://www.iof2020.eu/open-call/d3.3-iof2020-reference-architecture. pdf. Accessed 30 Apr 2020
29. Mahesh DS, Savitha S, Dinesh KA (2014) A cloud computing architecture with wireless sensor networks for agricultural applications. Int J Comput Netw Commun Secur 2(1):34–38
30. Jha K, Doshi A, Patel P, Shah M (2019) A comprehensive review on automation in agriculture using artificial intelligence. Artif Intell Agric 2:1–12
31. Alam F, Mehmood R, Katib I, Albogami NN, Albeshri A (2017) Data fusion and IoT for smart ubiquitous environments: a survey. IEEE Access 5:9533–9554
32. Piromalis D, Arvanitis K (2015) Radio frequency identification and wireless sensor networks application domains integration using DASH7 Mode 2 standard in agriculture. Int J Sustain Agric Manag Inf 1(2):178–189
33. Jawad HM, Nordin R, Gharghan SK, Jawad AM, Ismail M (2017) Energy-efficient wireless sensor networks for precision agriculture: a review. Sensors 17:1781
34. Ayoub W, Samhat AE, Nouvel F, Mroue M, Prevotet J (2018) Internet of mobile things: overview of LoRaWAN, DASH7, and NB-IoT in LPWANs standards and supported mobility. IEEE Commun Surv Tutor 21(2):1561–1581
35. Symeonaki E, Arvanitis K, Piromalis D (2020) A context-aware middleware cloud approach for integrating precision farming facilities into the IoT toward agriculture 4.0. Appl Sci 10(3):813
36. Cappai MG, Picciaua M, Nieddua G, Bittib MPL, Pinnaaa W (2014) Long term performance of RFID technology in the large scale identification of small ruminants through electronic ceramic boluses: implications for animal welfare and regulation compliance. Small Rumin Res 117:169–175

37. Umstatter C (2014) Precision sheep management-new approaches and future development. Adv Anim Biosci 5(1):112
38. Morgan-Davies C, Lambe N, Wishart H, Waterhouse T, Kenyon F, McBean D, McCracken D (2018) Impacts of using a precision livestock system targeted approach in mountain sheep flocks. Livestock Sci 208:67–76
39. Okada H, Itoh T, Suzuki K, Tsukamoto K (2009) Wireless sensor system for detection of avian influenza outbreak farms at an early stage. IEEE Sens 2009:1374–1377
40. Hennessy DA, Wolf CA (2018) Asymmetric information, externalities and incentives in animal disease prevention and control. J Agric Econ 69(1):226–242
41. Chaters GL, Johnson PCD, Cleaveland S, Crispell J, de Glanville WA, Doherty T, Salvador LCM (2019) Analysing livestock network data for infectious disease control: an argument for routine data collection in emerging economies. Philos Trans Royal Soc B 374(1776):20180264
42. Hussain SJ, Khan S, Hasan R, Hussain SA (2020) Design and implementation of animal activity monitoring system using TI sensor tag. In: Mallick P, Balas V, Bhoi A, Chae GS (Eds) Cognitive informatics and soft computing. Advances in Intelligent Systems and Computing, vol 1040. Springer, Singapore
43. Mohamad G, Gaber T (2019) Wireless sensor networks-based solutions for cattle health monitoring: a survey. In: The international conference on advanced intelligent systems and informatics (AISI2019), 26–28 October 2019, Cairo, Egypt
44. Vidic J, Manzano M, Chang CM, Jaffrezic-Renault N (2017) Advanced biosensors for detection of pathogens related to livestock and poultry. Vet Res 48(1):11
45. Harrop P (2016) Wearable technology for animals 2017-2027: technologies, markets. Forecasts, IDTechEx
46. Benjamin M, Yik S (2019) Precision livestock farming in swine welfare: a review for swine practitioners. Animals 9(4):133
47. Kaufman JD, Saxton AM, Ríus AG (2018) Short communication: relationships among temperature-humidity index with rectal, udder surface, and vaginal temperatures in lactating dairy cows experiencing heat stress. J Dairy Sci 101(7):6424–6429
48. Madushanki A, Halgamuge M, Wirasagoda H, Syed A (2019) Adoption of the Internet of Things (IoT) in agriculture and smart farming towards urban greening: a review. Int J Adv Comput Sci Appl 10:11–28
49. Popović T, Latinović N, Pešić A, Zečević Ž, Krstajić B, Djukanović S (2017) Architecting an Iot-enabled platform for precision agriculture and ecological monitoring: a case study. Comput Electron Agric 140:255–265
50. Van Evert FK, Fountas S, Jakovetic D, Crnojevic V, Travlos I, Kempenaar C (2017) Big data for weed control and crop protection. Weed Res 57:218–233
51. Woodford L, Robley A (2011) Assessing the effectiveness and reliability of a trap alert system for use in wild dog control. Arthur Rylah Institute for Environmental Research, Technical Report Series No. 218, Department of Sustainability and Environment, Victoria State Government, Australia
52. Kammel DW, Burgi K, Lewis J (2019) Design and management of proper handling systems for dairy cows. Vet Clin: Food Anim Pract 35(1):195–227
53. Umstatter C (2011) The evolution of virtual fences: a review. Comput Electron Agric 75:10–22
54. Umstatter C, Morgan-Davies J, Waterhouse T (2015) Cattle responses to a type of virtual fence. Rangeland Ecol. Manage. 68:100–107

Eleni Symeonaki is a Ph.D. candidate in Agricultural University of Athens at the Department of Natural Resources Management and Agricultural Engineering. She has studied Automation Engineering at BA level and she obtained her Master's degree in Industrial Automation. She is employed as Laboratory Teaching Staff in the Department of Industrial Design and Production Engineering at the University of West Attica (UNIWA) in Athens, Greece and member of the research lab of Electronic Automation, Telematics and Cyber-Physical Systems. Her research interests include amongst others, IoT and Cloud Computing Technologies for the benefit of Precision Agriculture Information Systems, Operations Management, HMI, M2M, Cyber-physical Systems. Her work has been published in several journals, book chapters and international scientific conference proceedings.

Dr. Konstantinos G. Arvanitis received the B.Sc. and Ph.D. degrees from the National Technical University of Athens, Department of Electrical and Computer Engineering, in 1986 and 1994, respectively. He currently serves as a Professor of "Automation in Agriculture" in Agricultural University of Athens, Department of Natural Resources Management and Agricultural Engineering, Laboratory of Farm Machinery, where he also served as Head/Deputy Head of the Section of Farm Structures and Farm Machinery. His research work has received significant international recognition (more than 3100 citations, h-index = 26, g-index = 52). His main research interests are: Electrification and Automation in Agriculture, Advanced Process Control, Wireless Sensor Networks, ICT and Artificial Intelligence Applications in Agriculture, Remote Sensing, Optimization Techniques, Energy Management and Control of Autonomous Micro-Grids, Internet of Things and Cloud Computing.

Dr. Dimitrios Loukatos received the diploma in Electrical and Computer Engineering and the Ph.D. degree in Telecommunications and Computing, both from the National Technical University of Athens (NTUA), Greece. He currently is staff member of the Agricultural Engineering Laboratory of the Agricultural University of Athens, Greece. He has worked as a research associate at the NTUA and the National Centre for Scientific Research 'Demokritos'. He also worked as a senior engineer for the Institute of Geodynamics of the National Observatory of Athens, enhancing their seismic sensor network. His research interests include hardware/software platforms evaluation and optimization, management and applications of wired and/or wireless (sensor) networks, artificial intelligence, human-computer interaction, robotics and applications. His most recent work is focused on the area of IoT and Physical Computing. His research work has been published in many national and international scientific conference proceedings, book chapters and journals.

Dr. Dimitrios D. Piromalis is Assistant Professor in the Industrial Design and Production Engineering Department at the University of West Attica (UNIWA) in Athens, Greece. His is the director of the research lab of Electronic Automation, Telematics and Cyber-Physical Systems. His research interests include the design and development of electronic embedded systems in application areas such as autonomous vehicles, internet of things, and cyber-physical systems. He is author of more than one hundred and twenty publications in international scientific research journals, conferences and book chapters with high impact and many citations. Also, he is author of many books covering engineering subjects. In addition, prof. Piromalis has been collaborated with top multinational semiconductors industries as Field Application Engineer and Technical Consultant for the last twenty-five years.

A Cloud-Based Decision Support System to Support Decisions in Sow Farms

Jordi Mateo, Dídac Florensa, Adela Pagès-Bernaus, Lluís M. Plà-Aragonès, Francesc Solsona, and Anders R. Kristensen

Abstract In the pig farming industrial sector, innovation is a crucial factor in maintaining competitiveness. On the research side, there exists a large body of models and decision analysis tools that too often do not reach the end-user. In this chapter, we propose a software-as-a-service based on a cloud-based Decision Support System architecture that should overcome the main adoption barriers spotted in the literature. The service proposed takes advantage of existing herd management models feed with historical farm data and economic parameters recorded by the most popular farm management software used by pig companies. The approach includes a sow farm model and offers a set of analytic tools to help farmers in making better strategic, tactical and operational decisions based on their own data. This chapter highlights the advantages of optimization and simulation models hosted in a cloud computing platform to deliver a service of knowledge discovering and data analytics

J. Mateo (✉) · D. Florensa · F. Solsona
Department of Computer Science and INSPIRES, University of Lleida, Jaume II 69, 25001 Lleida, Spain
e-mail: jordi.mateo@udl.cat

D. Florensa
e-mail: didac.florensa@gencat.cat

F. Solsona
e-mail: francesc@diei.udl.cat

A. Pagès-Bernaus
Department of Business Administration, University of Lleida, Jaume II 73,25001 Lleida, Spain
e-mail: adela.pages@udl.cat

A. Pagès-Bernaus · L. M. Plà-Aragonès
Agrotècnio Research Center, University of Lleida, Rovira Roure 191, 25198 Lleida, Spain
e-mail: lmpla@matematica.udl.cat

L. M. Plà-Aragonès
Department of Mathematics, University of Lleida, Jaume II 73, 25001 Lleida, Spain

A. R. Kristensen
Department of Veterinary and Animal Sciences, University of Copenhagen, Grønnegårdsvej 2, 1870 Frederiksberg C, Denmark
e-mail: ark@sund.ku.dk

© Springer Nature Switzerland AG 2021
P. Krause and F. Xhafa (eds.), *IoT-based Intelligent Modelling for Environmental and Ecological Engineering*, Lecture Notes on Data Engineering and Communications Technologies 67, https://doi.org/10.1007/978-3-030-71172-6_10

to sow farms. The success in adoption depends on the added value and usability through software integration with current management tools used by pig producers. Preliminary results show that the proposed service helps pig managers to make better supervision of sows and to obtain the competitive advantages of using complex mathematical models in a practical, flexible and transparent way.

1   Introduction

The development of decision support tools for the pig sector has deserved lots of research for several decades and stimulated by the general use of computers on farms like in other agricultural sectors [1]. Already in the late 1950s, descriptive and mathematical modelling was applied to solve agricultural problems as mentioned in [2]. Since then, and even more in the last decades, researchers have developed new models aiming at improving different production process within the pig farming sector focusing mainly on breeding farms. Initially, models were for single farms [3], and recently they account for the coordination of different farm units within pig companies or cooperatives under pig supply chain operation structures [4]. For instance, there are proposals on optimization models for diet composition [5], models to control the climate of the farm by monitoring temperature and humidity [6] or the sow replacement problem among others [7, 8]. Several commercial software companies had developed software applications for Personal Computers (PCs) to register, organize, monitor and visualize farms' activity. This amount of data had a great value and constituted the basis for the improvement of mathematical models that would help further in the decision making process. In this sense, researchers continued to develop advanced models aimed at improving the overall performance of the different production units in the pig farming sector [9, 10].

Despite the initial expectation of the impact for a new set of PC-based decision tools for the sector, the success was limited [11]. The lack of integration with existing herd management applications, the required computational power and maintenance tasks or required upgrades for multiple PCs were the main obstacles for on-farm use. The development of Internet and Communication Technologies (ICT) have mitigated in part past problems since current state-of-the-art commercial solvers, and Decision Support Systems (DSS) for farm management tend to be adapted to the new ICT paradigm (multi-site, cloud-based services) in substitution of stand-alone PC-applications [12]. The challenge is the addition of smart capabilities in terms of business intelligence rules and data analytics [13]. As has been observed by several authors [11, 14] there are several reasons why a mathematical farm model does not arrive at the farmer's hands. The main reasons can be summarized as:

- unclear focus of the project
- steep learning curve
- lack of interface with the current management information systems
- hard system maintenance.

In the same way and from the broader perspective of agricultural systems, Jones et al. [15] analyse the current state-of-the-art of the agricultural systems science. They emphasize the definition of the Use Case as a central pivot for the development of interactions between models and users, and claim for the difficulties for the final user in finding easily accessible and usable applications.

Nowadays, cloud computing opens new research opportunities to design a new wave of DSS capable to transfer qualified knowledge from academia to society more easily. For instance, cloud-based services are ideal for scaling, growing, and dealing with uncertain computing workload derived from many concurrent users or makes easier the maintenance and update of complex optimisation models.

According to the National Institute of Standards and Technology (NIST) [16], cloud computing is a model for enabling ubiquitous, convenient, on-demand network access to a shared pool of configurable computing resources (e.g. networks, servers, storage, applications, and services) that can be rapidly provisioned and released with minimal management effort or service provider interaction. In this context, computational resources can be offered to users as a service (XaaS). The most commons are infrastructure as a service (IaaS), platform as a service (PaaS) and software as a service (SaaS).

Clouds can be classified depending on their administration and the sensitivity to the data access. Public clouds are externally available to everyone, such as Amazon Elastic Compute Cloud (EC2). However, there are companies that do not trust on public clouds or in the regulation laws where the public cloud is deployed and so, they build their own private cloud. Finally, hybrid clouds are a mixture of public and private cloud serve ices. Companies can either maintain crucial data inside a private cloud and rely on public clouds for ensuring the quality of the service. In short, when starting to develop a cloud service, building an hybrid cloud may have advantages for farmers since their raw data can be kept safe inside a private cloud (data tier), and computation and analytics can be done either in the private or public cloud depending on resource saturation.

In this chapter, we present eSow, a cloud-based DSS for breeding farms developed as a SAAS and aiming at helping farmers in sow replacement and performance analysis. The design of the DSS avoids the adoption barriers spotted in the literature by developing an easy-to-use cloud-based service. The main advantage of the proposed tool is the reduction of the learning curve to the end-user, who providing his/her farm data will be able to retrieve reports, perform sow herd performance analysis and simulations in a direct and comprehensive way. Knowledge related to the calibration of the model, running of the model or submodels is encapsulated within the cloud platform and integrated seamlessly in the web application. In addition, the development of a SAAS hosted in a cloud avoids pig companies or cooperatives to have powerful computers with an environment dedicated to solve complex mathematical models.

With only a web browser connected to the Internet, decision-makers get access to almost unlimited computing power, no matter which device is used (hand-handled device, desktop computer or laptops).

Moreover, the elastic architecture employed in the deployment of this chapter contributes to pave the development of emergent decision support systems extending the functionality to other stages in the pig sector. Users can take advantage of the cloud computing features as processing and storage power to get a tool fitted and integrated with his/her current farm management software [1, 13].

The main contribution of the chapter is showing how putting complex mathematical models available as a service to the farmers, advisors or consultants may help and support sow herd management decisions. Furthermore, the data integration, management, analysis and organization are also important aspects considered in this chapter, leaving the database system open to being enriched with sensor data in future refinements.

The remainder of the chapter is organized as follows: Sect. 2 presents an overall vision of related work of sow replacement models and decision support systems. Section 3 presents our proposal for a decision support system in the replacement of sows in piglet production farms. It follows Sect. 4 which describes the cloud architecture that orchestrates the decision support system. In Sect. 5, the results from a real case study are evaluated and discussed. Finally, Sect. 6 outlines the main conclusions and future work.

2 Sow Herd Management and DSS Tools

2.1 Sow Herd Management Models

Livestock systems are complex and involve models at different levels (the animal and the herd) and types (e.g. biological models or feed models among others) in addition to the interaction among them. Being the piglet production one of the key parts of the whole pig supply chain (it is the start of the production cycle), the optimal management of sow farms are of paramount importance [17].

Currently, the replacement of sows is mostly based on the farmer's experience, which relies on technical performance indicators, such as litter size, the age or the parity. There can be other thumb rules based on the company policy such as culling sows after an abortion or with a litter size below some threshold. Nevertheless, the final choice is based mostly on technical index measures, and economic analysis does not back it.

The fact that economic measures are not part of the decision process is because it requires advanced calculations such as determining the ideal culling rate and associated herd structure (i.e. the distribution of sows over parities), litter size curve, feeding cost and prices. Farms are dynamic systems, and this adds complexity in analyzing whether a decision is economically sound or not. The goal of a farm producing piglets is to produce as many piglets as possible, and for that, the number of piglets weaned per sow per year is a technical index highly correlated with the economic performance of the farm. A stable farm performance over time requires a balanced structure of the herd since sow production varies also over time. However, herd dynamics can be disturbed affecting piglet production and jeopardizing the existence of other negative effects. To account for all these sources of variability, stochastic and dynamics models are imposed.

In the literature, there are several approaches to the sow replacement problem. For instance, in the 80s Dijkhuizen et al. [18] developed Porkchop as a spreadsheet based model. They introduced a new index called the Retention Pay-Off that indicated the total extra profit that retention of a sow is expected to yield over her replacement. Later, Huirne et al. [19] presented a stochastic dynamic programming model that maximized the present value of the expected annual net returns over a specified planning horizon of a sow herd. This model was refined later by introducing other attributes such as the piglet mortality rate or the number of undersized piglets at birth [20]. Kirchner et al. [21] proposed the use of decision trees based on the gain ratio criterion to see whether a sow should be replaced or not. In this case, sows were not ranked but classified. A multi-level hierarchical Markov process with decisions on multiple time scales was presented by Kristensen et al. [7, 22]. The model incorporated the biological performance of sows through a dynamic linear model, the dynamic dropout structure and a feed intake model. Later the model was also extended to include clinical observations [8]. Plà et al. [23] proposed an equivalent semi-Markov model to represent farm dynamics to optimize the expected rewards. Finally, Rodràguez et al. [24] presented a two-stage stochastic programming model to optimize the sow replacement and the scheduling of gilt purchases.

In the present study, the model proposed in [7] is used to obtain the optimal sow replacement policy that maximizes the benefits of the farms considering the current herd dynamics, predictions about the productive behaviour, casualties and also the derived cost of the activity. The model optimizes sow replacement from an economic point of view and not only accounting for litter size performance. A more in-depth presentation of the model is presented in Sect. 3.

2.2   *Review of Decision Support Systems*

Decision support systems can be defined as information systems that offer solutions (one or more alternatives) to help a user, which can be a single person or an organization, in the decision-making process [13]. We focus on DSS which provides smart (i.e. data-driven and prescriptive and usable solutions capable of managing the data, feeding the mathematical models with precise data and integrating into the system all available resources (see Fig. 1).

The first generation of DSSs focused on simplifying the tasks of storing data, tracking changes and provided custom reports based on descriptive statistics. Advanced smart DSSs have extended these capabilities and added the use of predictive or

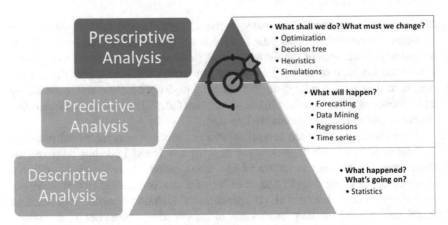

Fig. 1 Main features of decisions support systems

prescriptive methods, by for example sending system alerts, suggesting optimized alternatives or providing different scenarios to explore. Another important characteristic is the capability of being integrated into existing systems and getting access by devices that farmers are using in their daily tasks. Fig. 1 summarizes the main differences between the first generation of DSS focused on answering the questions: What happened? and What is going on? By using basic statistical methods and the new generation of DSS aimed at explaining: What will happen? What shall we do? or What must we change? By applying simulation or optimization models, among other techniques.

In order to identify the current state-of-the-art on DSS, we have surveyed what commercial applications are offering at the moment. The search is not exhaustive and has been limited to few illustrative examples of available farm management software in Spain offering at present cloud-based services. We have analyzed the following features:

- Storage: the capacity of the application to store data in an organized and hierarchical way.
- Traceability: the capacity of the application to track data changes.
- Data format: the ability to combine data from disparate and heterogeneous sources into meaningful and valuable information.
- Reporting: the delivery of (usually aggregated data) information in the form of descriptive analysis.
- Analytics: the ability to explore and interpret data, discovering patterns and meaningful information.
- Integration: the capacity to cooperate with the other agents in the current context.
- Alerts: whether the system can alert the users of special situations, either to anticipate or to mitigate the detected problems quickly or for taking advantage of some situations. This type of messages is usually the output of predictive statistical models.

Table 1 Comparison of different commercial cloud-based applications in agriculture

	Agrivi	Agroptima	FarmLogic	CloudFarms
	[26]	[27]	[28]	[25]
Storage	✓	✓	✓	✓
Traceability	✓	✓	✓	✓
Data format	✓	✓	✓	✓
Reporting	✓	✓	✓	✓
Analytics	✓	✗	✓	✓
Integration	✗	✗	✗	✓
Alerts	✓	✗	✗	✗
Suggestions	✗	✗	✗	✗
Scenarios	✗	✗	✗	✗

- Suggestions: whether the system can generate and deliver proposals for improvements. This type of messages is usually the output of prescriptive mathematical models.
- Scenario analysis: the capacity of the application to explore different unreal situations or scenarios, reproducing what would happen if this situation becomes true. This is usually the result of an underlying model of the system employing mathematical modelling or a simulation model.

Table 1 shows the features that are offered by four commercial farm management applications in agriculture, one of them for pig production [25]. As it can be observed, current DSSs do not offer optimization or scenario analysis capabilities. The work presented here aims at filling this gap.

3 The eSow DSS Overview

The overall structure of the eSow DSS and the flow of information is represented in Fig. 2. To start with eSow, it requires a set of parameters, either public such as prices or private related to the farm performance or management policies. The input data, however, is based on data already available through current farm management software and can be imported automatically by the farmer. With the evolution of sensors and smart farming, shortly some of the current models can likely be updated with more precise information [29, 30]. This data serves in a first stage to estimate several parameters of the system like the prolificity curve, the dropout structure or

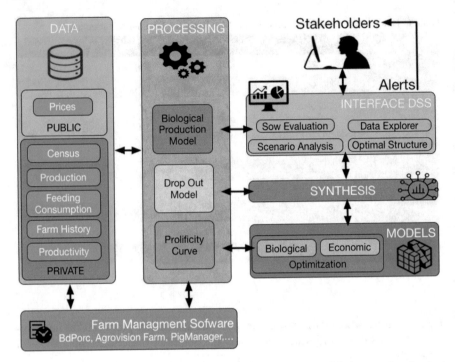

Fig. 2 Representation of the different parts of the data structure, models integrated and output delivered by the current eSow DSS

the biological production model. This processing stage is named calibration and it is required only time to time in case estimations should be updated.

3.1 Model Management Subsystem

The eSow DSS integrates the multi-level hierarchical Markov process presented in [7] in a modular way and with a different architecture of that proposed in [31]. The DSS is designed in a way that a step-by-step process is performed by the user leading to a smooth learning curve.

After the model is run, several outputs are generated. To support several decision making tasks, the application offers the main results: the value of each sow in the farm, the comparison between the current herd structure and its ideal composition and main KPIs of the farm at steady state. The sow valuation allows the user to rank them and see which sows are below expectations. Therefore, sows worse ranked are the candidates for culling. This information has a clear operational focus. There are other results providing information about the current state of the farm and what is the expected value in the ideal (optimal) situation for the given set of farm parameters.

This information is of great interest to managers that look at the farm evolution with a tactical/strategical vision. Also, in the strategic domain, a manager may wonder what would be the effects of changing some of the current practices by performing a sensitivity analysis.

These features are achieved after automatizing and encapsulating two important phases: the calibration of the farm parameters and the optimization processes.

- **Calibration of the farm parameters**: The calculations involved at the calibration stage are time-consuming and require dedicated computing resources. It requires the proper farm data importation and transformation to feed the R scripts and the execution of different models coded in R. The output of a calculation is the input for the next step inside the calibration process. This stage is only performed the first time that a farm is registered.
- **Optimization**: In this stage, the optimizer is executed, and the optimal policy and the sow rank list for sow replacement is obtained. Two types of outputs are be generated: several KPIs related to the ideal steady state of the farm or the economic index of the sows present in the farm. From the latter the ranking table are displayed showing current reproductive state of each sow. Note, that all the data required by the optimizer is heterogeneous and belongs to different sources like sow farm inventory, piglet selling price or feed consumption, thus in this step, it is also required data manipulation scripts to unify the information.

The numerical data required and generated by eSow depends on the number of farms and sows managed by a pig company. It is possible to estimate or approximate a lower bound for these data volume per model run $|DSS|$ in Megabytes (**MB**) by using the function suggested in Eq. (1) where $f \in F$ represents the number of farms, and $s_f \in f$ the number of sows in each farm f and p the number of parities per sow. Furthermore, the constants, $60 * 4$ represent the amount of floating-point numbers (1 single float precision is equivalent to 4 bytes), and the amount of integer values $20 * 2$ (1 integer is equivalent to 2 bytes)

$$|DSS| = \frac{\sum_{f \in F} s_f * p * (60 * 4 + 20 * 2)}{1024 * 1024} \tag{1}$$

3.2 Data Base Management Subsystem

The main features of the eSow DSS can be summarized as:

- **Data integration**: eSow can be fed with data of any farm management software that a Spanish farmer is already using or even if the farmer does not have any digital data source, the farmer can use default values and start a digital transformation using eSow.
- **Data explorer:** After the data processing, the user can obtain an overview of the current status of their sow farm and expected economic performance.

- **Recommendations**: One of the novel features of the eSow DSS presented in this chapter is the capability of making data-driven recommendations. These recommendations are based on the execution of a complex mathematical model that is able to optimize sow replacement and provide a ranking of sows according to their economic value.
- **Scenario Analysis**: Another important contribution of this work, is the capability of simulating and analyzing different scenarios. Farmers can evaluate what is going to happen if they intend to make changes on the farm. For example, they may wish to evaluate the impact and the consequences in terms of farm structure of modifying their prolificity curve by making some tactical decisions. Generally speaking, this feature helps the farmer to anticipate and mitigate the undesired effects of some decisions beforehand.

3.3 Interface Subsystem and Workflow

Figure 3 illustrates the work-flow when using the eSow DSS, separating the steps corresponding to the first use (i.e. involving calibration) from the rest. It also depicts the common actions that the farmer is allowed to do inside the eSow DSS.

The main part of the eSow DSS is the **Dashboard**, the central part shown in Fig. 3. Dashboard is whereby farmers can interact and manage the data, launch operations, check the results, consult the recommendations, configure their sessions, alerts and notifications and more. This area is private and contains all the information about the farm.

First of all, the farmer needs to declare a new farm and create a new session inside the eSow DSS to start using it. When a new farm is registered, a usable step-by-step form allows the user to introduce the farm data and perform the first calibration and optimization of the farm. This form allows the user to introduce historical data files, feeding parameters, cost parameters, production parameters, census parameters and also productivity parameters.

One important result derived out of the optimal replacement policy is the herd structure of the farm, i.e. the distribution of sows over parities. The optimal replacement policy and associated herd structure summarizes the information resulting from the optimisation model solved using specific farm parameters, either technical and economic ones. Hence, the farmer can compare the ideal herd structure with the current structure and ultimately detect unbalanced situations leading to performance troubles. Optimal policy is narrowly depending on the prolificity curve of the farm [32], estimated from the farm data and is displayed for informative purposes.

The main information delivered by the tool is the recommendation about sow candidates to be replaced. This information is displayed as an interactive ranking of sows. This ranking can be sorted using different criteria, being the most useful the economic criterion. Therefore the farmer has a tool to support critical decisions, like the replacement of sows, which are based not only in the farmer experience and

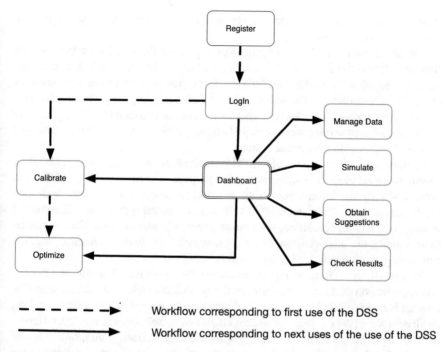

Fig. 3 Workflow within the DSS. Dashed lines depict the steps required the first time the eSow is used. Continuous lines, depict all the operations the farmer can realize with the eSow DSS

technical performance (e.g. litter size) but are also based on mathematical models and economic indicators.

The eSow has been conceived as a cloud-based DSS to avoid the barrier of software installation and maintenance inside farmers' computers. Moreover, a cloud-based service can perform software updates and upgrades seamlessly and delivering a generic product that can be used independently of the device used to get access, the farm management software and the level of knowledge in data science.

4 Cloud Architecture

The cloud architecture provides a platform to offer knowledge as a service to the pig farmers. This architecture orchestrates the execution of the models, the presentation of the results and the administration of the data. The architecture is composed of three layers to split the functionality: the presentation layer, the business logic, and the data layer. This separation makes this architecture flexible and hybrid; capable of being adapted to either endogenous or exogenous mathematical models, to integrate novel algorithms and even to incorporate more heterogeneous data sources, such as non-relational databases.

The architecture is built under the platform OpenNebula [33]. The main reason for choosing OpenNebula was to take advantage of the Stormy server. This is a private cloud deployed with the OpenNebula platform that belongs to the University of Lleida [34]. Nevertheless, the architecture is capable of being deployed or integrated into any commercial cloud, such as Microsoft Azure or Amazon Cloud. All the parts of the cloud-based service were deployed on this platform using virtual instances of Centos7 images as the operating system.

Figure 4 shows the layers skeleton. The architecture was designed to be flexible, elastic, scalable, easy to maintain and multi-purpose. The main function of the data layer is storing the information using different heterogeneous sources, such as a file server and a relational database. The business logic represents the mathematical models, the data manipulation, the reports generation and the analytical computations. Finally, the main purpose of the presentation layer is the interaction with the end-user, gathering and displaying all the information.

First of all, the data layer represents all the data tools that deal with the storage and organization of the data managed by the system. Nowadays, the data integration concept is considered a keystone in the digital transformation of any sector. Big data, small data, smart data or even not structured data have their space in this layer. Hence, the data layer proposed in this work helps farmers to collect, group, organize and store the data involved in the daily sow farms activity. In this work, the data layer is implemented using two different servers, a relational database, and a file server to store historical and log files.

Secondly, the presentation layer is the visible part of the application representing the tool that the farmers (end-users) see and interact with it. The presentation layer is a web application accessible from any modern browser and any device connected to the Internet. This is a soft layer that does not require too many computational resources. This layer is executed to obtain the reports and the analytics on whatever farmer' device just having an active internet connection. The main purposes of this layer are allowing farmers to register to the eSow DSS, upload and store their data, obtain the analytics, get access to an interactive dashboard and support their tactical and strategical decisions. To sum up, the user interface of the presentation layer makes the interaction between the mathematical models, the data, the analytics and the farmers as easy, transparent and usable as much as possible.

Finally, the core layer making everything to work is the business logic. It contains all the rationale of the application and is the communication bridge between the client, the database, and the different computing nodes. The business logic receives all the data from the presentation layer, transforms and manipulates this data and ensures that this data is stored in the data layer. On the one hand, the Backend server can be considered the brain that orchestrates all the skeleton. On the other hand, the computing nodes can be understood as the heart and the lungs, responsible for solving the hard computational tasks. The server was implemented using novel technologies such as Spring framework, flask framework and others. All communication between the server and client is done through REST API calls.

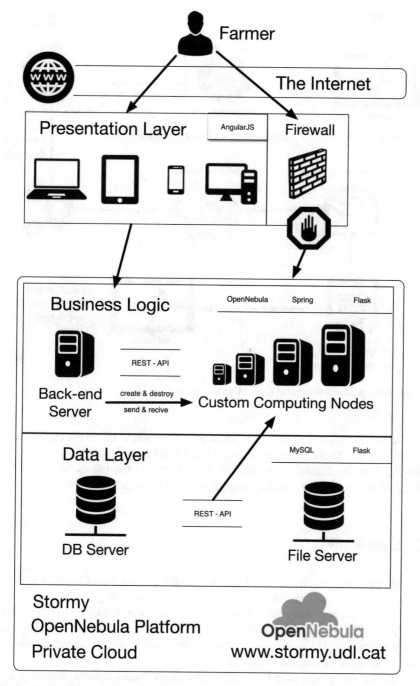

Fig. 4 Cloud architecture divided into their three main layers. Highlighting the communication and the technology used to deploy the platform

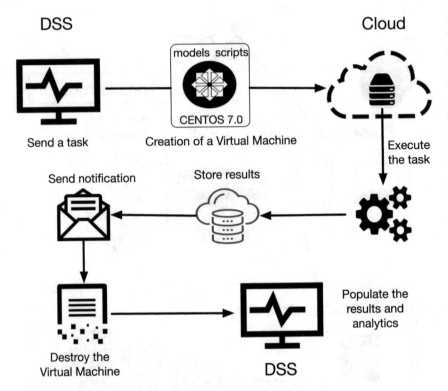

Fig. 5 Life-cycle of a computing node. Steps to execute the tasks

The computing nodes host the virtual machines where the mathematical models are solved. These nodes are elastic virtual machines, dynamically created and destroyed under demand. For instance, Fig. 5 depicts the life-cycle of a virtual machine inside the platform that manages the execution of a farmer task.

5 Case Study

A realistic case study is presented to illustrate the capabilities of eSow; the cloud-based DSS presented in this chapter. The potential advantages of using specific individual models in real environments have already been presented in other works [2, 7, 32]. The focus of this work is to close the gap between model development and practical use. In this section, we present the main outputs from where the user can extract information and knowledge.

The case study presented here uses realistic data from a Spanish company with 900 sows. Some of the data displayed has been truncated to preserve the privacy of the information. For illustrative purposes, the reader can test the application at

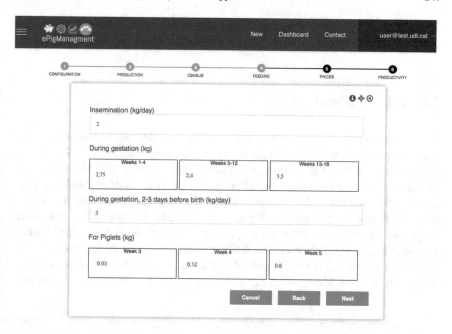

Fig. 6 Step-by-step form. Parameters used in the case study

https://stormy02.udl.cat/sowmodule (ePigManagement website) with a test user (details are given on the main page). Input data is not required by the test user, although the test user can check the data that is required to an active user.

5.1 Data Integration

Data integration is one of the most important features of the eSow DSS. After registration, the user is required to introduce their own specific farm parameters through a simple step by step form. The steps are organized by areas of information: Production, Census, Feeding, Prices and Production (see Fig. 6). Some of the information refers to farm's policy on management and production system (such as the number of weeks of lactation) while others refer to the farm-specific performance (historical census of sow productivity or mortality and fertility rates among others). All the information required can be easily imported from almost all commercial farm management software used in Spain. For a registered user, this step may be automatized with a background script that collects and sends the required information to our application.

Note that the input boxes provide automatically default values to speed up the input procedure. The first time default values are those defined by the system, while for subsequent times default values will be the values introduced by the user in the last calibration. The input data step is equipped with basic error-checking filters to avoid involuntary errors.

5.2 Data Explorer

After input data is provided, the calibration step is performed transparently to the user. The most informative results are the ideal structure of the farm and its associated performance indexes. These results, compared with the current state of the farm, help to understand better the latter. Figure 7 compares the structure of the herd by parities in the ideal situation with the current state. The structure of the herd in the ideal situation shows the distribution of sows over different parities with a descendent shape, i.e. less (older) sows for higher parities. However, the real situation observed as a solid line in Fig. 7 shows an unbalanced situation in the second parity.

Based on the ideal structure, several performance indicators of the farm can be computed, as those presented in Fig. 8 and representing a benchmark. These indi-

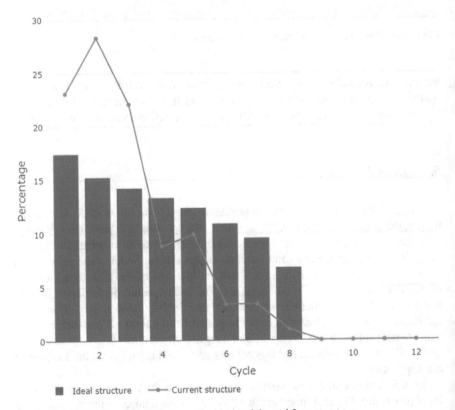

Fig. 7 Case Study. Comparison between the ideal and the real farm structure

Census	
Average lifetime of sows (in years)	2.22
% Sow Replacement per year	45.05

Reproductive Rate	
Number of parities (per sow and year)	2.47
Average number of days between parities	147.85

Productivity	
Average number of piglets weaned per sow	56.22
Average number of piglets weaned per sow and year	25.19
Average litter size at weaning	10.26
Average number of litters produced per sow	5.48

Income	
Average net returns per year and sow (EUR)	178.85

Fig. 8 Performance indicators under the ideal structure

cators are of common usage to farmers and farm managers. In this way, the final user can compare the current farm's performance and the ideal situation. Although reaching and remain in the ideal situation may be a utopia, the comparison helps the user to position the state of the farm with respect to the desirable situation provided their own specific parameters.

5.3 Optimal Recommendations

The most relevant feature of the eSow DSS is the ability to offer a list of sows candidates to be replaced, ranked according to an economic index representing the relative potential value. This economic value is related to the relative values that appear when solving infinite-horizon average-revenue Markov decision processes and they are also named in our context retention-pay-off (RPO) values [19]. Then,

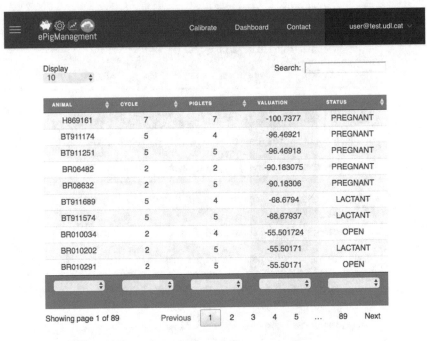

Fig. 9 Sow ranked according to the retention pay-off

the eSow DSS solve a replacement problem [7, 22] and based on the resulting optimal replacement policy deliver a list of sows with her current parity and reproduction state, historical productivity and corresponding RPO (see Fig. 9). Sows with negative valuation mean a reproductive potential lower than the expected potential offered by a new gilt. The RPO summarizes the future potential production of a sow with respect the farm average (i.e. it is a relative value) in economic terms.

Figure 9 shows how the ranking is presented to the farmer on a table. Once optimization is run, each sow present on the farm is evaluated, and a RPO (shown as VALUATION in Fig. 9) is computed. The rank list is accompanied by extra information such as the current parity (i.e. CYCLE in Fig. 9) of the sow, the number of piglets produced in the last parity and the current reproductive state of the sow which are: PREGNANT, OPEN or LACTANT. By using the ranked list, the farmer can sort it by any other column and gets access to almost all the crucial information needed to evaluate whether or not to replace a sow. Hence, not only the economic value is important, but also the current state and history are required to make a smart decision. With the values obtained for this case study, the farmer has a tool to better support sow replacement decisions, corroborating their intuition or being complemented by additional information not considered by eSow like health status or critical diseases affecting a sow.

5.4 Scenario Analysis

From a managerial point of view, the farm manager may have interest in exploring future scenarios or performing *what-if* analysis based on the results of the optimal replacement policy. The eSow DSS offers the possibility of simulating scenarios and comparing them with the current business-as-usual situation. For instance, we can consider the sensitivity of changing the maximum number of parities a sow can stay on farm. This is a strategic management parameter that farmers usually fix. Hence, the eSow DSS allows the farmers to simulate different scenarios based on the maximum number of parities and comparing resulting herd structure (see Fig. 10) or expected average revenue to check the convenience or non-convenience to make this decision and the effects that would cause to the production flow, e.g. lactation facilities occupation.

Figure 10 depicts the comparison between the base scenario and the situation where the maximum parities are reduced from 8 to 7. A lower life span implies an increment of primiparity sows, less productive than older ones. Thus, the prolificity curve play an important role to balance economically this situation [32]. As in Sect. 5.2, the graphic is backed by the technical and economic indicators, and so that the comparison between the two situations can be enriched.

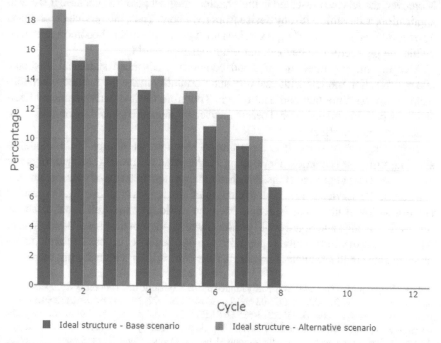

Fig. 10 Case Study. Simulating scenarios based on parameters

6 Conclusions

The transfer of models devised to support decisions in practice from the research community to final users within an economic sector is not an easy endeavour. The novelty of this research is the development of eSow, a usable, flexible and scalable DSS to support the decision making process in a pig production context. The eSow has been developed as a cloud-based service aimed to better support management decisions of farmers on sow farms.

Four main causes of failure in the adoption of DSS tools by end-users have been identified in literature and avoided in eSow. In this sense, the application presented in this chapter only shows those results that are really informative for the decision-making process. The learning curve for using the model is flat, as the user has to provide basic information already collected while the complex parts of the model, i.e. calibration and optimization, are performed transparently. Regarding the output part, the results are also concise, reporting principal measures or KPIs familiar to farmers. The interface with main commercial sow herd management systems in Spain and eSow is enabled utilizing a standard vector of information, allowing most farmers to benefit from eSow capabilities. Finally, the barrier concerning the system maintenance is overcome by developing a cloud-based service. This feature also avoids problems in terms of computation power needs and system-versions nuances. Moreover, the service presented in this chapter has great potential because it is a web application, accessible with any device from anywhere. Thus, the service presented is a powerful seed for a much larger service with a great potential to become a reference within the pig sector and beyond, in the agribusiness world.

Merging the potential use of cloud computing, advanced statistical methods, and optimization models with usability and portability make the farmer's decision-making process smoother and data-driven. Therefore tactical and operational decisions can be explored on a model before they are implemented, avoiding or mitigating undesired counter-effects.

Regarding future work, it is important to highlight the ability of the model to simulate different scenarios. Developing further, this feature will allow farmers to explore better strategic decisions. Another enhancement in terms of usability and the comfort of the service is the implementation of an alert service. Finally, the storage and evaluation of historical farm data open the world for a non-relational database. Integrating Mongo and Spark could be a key factor in view of the arrival of new sensor data enriching the information capabilities to make better decisions. The exploitation of these data could stimulate much further the competition within the pig sector.

Acknowledgements This work was partly funded by the Ministerio de Economía y Competitividad under contracts TIN2011-28689-C02-02 and TIN2014-53234-C2-2-R and by the European Union FEDER (CAPAP-H5 network TIN2014-53522-REDT). Some of the authors are members of the research group 2017-SGR363 or 2017-SGR1193 acknowledged by the Generalitat de Catalunya. Anders R. Kristensen was financially supported by the Danish Council for Strategic Research (Copenhagen, Denmark; the PigIT project, grant no. 11-116191).

References

1. Lytos A, Lagkas T, Sarigiannidis P, Zervakis M, Livanos G (2020) Towards smart farming: systems, frameworks and exploitation of multiple sources. Comput Netw 172:107147
2. Cornou C, Kristensen AR (2013) Use of information from monitoring and decision support systems in pig production: collection, applications and expected benefits. Livest Sci 157(2–3):552–567
3. Plà LM (2007) Review of mathematical models for sow herd management. Livest Sci 106:107–119
4. Esteve N-R, Plà-Aragonès Lluís M, Antonio A-A (2019) Production planning of supply chains in the pig industry. Comput Electron Agric 161:72–78
5. Dubeau F, Julien PO, Pomar C (2011) Formulating diets for growing pigs: economic and environmental considerations. Ann Oper Res 190(1):239–269
6. Gray J, Banhazi TM, Kist AA (2017) Wireless data management system for environmental monitoring in livestock buildings. Inf Process Agric 4(1):1–17
7. Kristensen AR, Algot Søllested T (2004) A sow replacement model using Bayesian updating in a three-level hierarchic Markov process: II. Optimization model. Optim Model Livest Prod Sci 87(1):25–36
8. Rodríguez SV, Jensen TB, Plà LM, Kristensen AR (2011) Optimal replacement policies and economic value of clinical observations in sow herds. Livest Sci 138(207–219):2011
9. Fountas S, Carli G, Sørensen CG, Tsiropoulos Z, Cavalaris C, Vatsanidou A, Liakos B, Canavari M, Wiebensohn J, Tisserye B (2015) Farm management information systems: current situation and future perspectives. Comput Electron Agric 115:40–50
10. Plà-Aragonés Lluis M (ed) (2015) Handbook of operations research in agriculture and the agri-food industry, vol 224. International series in operations research and management science. Springer, New York
11. Manos B, Paparrizos K, Matsatsinis N, Papathanasiou J (2010) Decision support systems in agriculture, food and the environment: trends, applications and advances. IGI Global
12. Fielke S, Taylor B, Jakku E (2020) Digitalisation of agricultural knowledge and advice networks: a state-of-the-art review. Agric Syst 180:102763
13. Zhai Z, Martínez JF, Beltran V, Martínez NL (2020) Decision support systems for agriculture 4.0: survey and challenges. Comput Electron Agric 170:105256
14. Kamp JALM (1999) Knowledge based systems: from research to practical application: pitfalls and critical success factors. Comput Electron Agric 22(2–3):243–250
15. Jones JW, Antle JM, Basso B, Boote KJ, Conant RT, Ian Foster H, Godfray CJ, Herrero M, Howitt RE, Janssen S, Keating BA, Munoz-Carpena R, Porter CH, Rosenzweig C, Wheeler TR (2017) Toward a new generation of agricultural system data, models, and knowledge products: state of agricultural systems science. Agric Syst 155:269–288
16. Mell PM, Grance T (2011) The NIST definition of cloud computing. Technical report, National Institute of Standards and Technology, Gaithersburg, MD
17. Sara Rodríguez SV, Plà LLM, Faulin J (2014) New opportunities in operations research to improve pork supply chain efficiency. Ann Oper Res 219(1):5–23
18. Dijkhuizen AA, Morris RS, Morrow M (1986) Economic optimization of culling strategies in swine breeding herds, using the "porkchop computer program". Prev Vet Med 4(4):341–353
19. Huirne RBM, Dijkhuizen AA, Renkema JA (1991) Economic optimization of sow replacement decisions on the personal computer by method of stochastic dynamic programming. Livest Prod Sci 28(4):331–347
20. Huirne RBM, Hardaker JB (1998) A multi-attribute utility model to optimise sow replacement decisions. Eur Rev Agric Econ 25(4):488–505
21. Kirchner K, Tölle KH, Krieter J (2004) Decision tree technique applied to pig farming datasets. Livest Prod Sci 90(2–3):191–200
22. Kristensen AR, Søllested TA (2004) A sow replacement model using Bayesian updating in a three-level hierarchic Markov process: I. Biological model. Livest Prod Sci 87(1):13–24

23. Plà LM, Pomar C, Pomar J (2004) A sow herd decision support system based on an embedded Markov model. Comput Electron Agric 45(1–3):51–69
24. Rodríguez SV, Albornoz VM, Plà LM (2009) A two-stage stochastic programming model for scheduling replacements in sow farms. TOP 17(1):171–189
25. Cloudfarms (2020) https://cloudfarms.com/. Accessed May 2020
26. Agrivi (2020) https://www.agrivi.com/. Accessed May 2020
27. Agroptima (2020) https://www.agroptima.com/. Accessed May 2020
28. Farmlogics (2020) https://farmlogics.com/. Accessed May 2020
29. Wolfert S, Ge L, Verdouw C, Bogaardt M-J (2017) Big data in smart farming - a review. Agric Syst 153:69–80
30. O'Grady MJ, O'Hare GMP (2017) Modelling the smart farm. Inf Process Agric 4(3):179–187
31. Hindsborg J, Kristensen AR (2019) From data to decision - implementation of a sow replacement model. Comput Electron Agric 165:104970
32. Fernández Y, Bono C, Babot D, Plà LM (2015) Impact of prolificity on sow replacement policies I Impacto de la prolificidad en la política de reemplazamiento de cerdas reproductoras. ITEA Informacion Tecnica Economica Agraria 111(2):127–141
33. OpenNebula (2020) http://opennebula.org/. Accessed May 2020
34. Stormy (2020) http://stormy.udl.cat. Accessed May 2020

Jordi Mateo Ph.D. in Computer Science. He is a professor at the University of Lleida and a member of the Distributed Computing Group. His research interest and expertise involve differents topics related to data science fields such as Operations Research (stochastic optimization); High-Performance Computing (parallelization of algorithms); Cloud computing; Big data; Internet of things (IoT), Decision Support Systems, eHealth and Agriculture 4.0.

Dídac Florensa received the B.S. and M.S. in Computer Science degree from the Escola Politècnica Superior of the Universitat de Lleida (UdL), in 2017 and 2019 respectively. Currently, he is a Ph.D. student of the Distributed Computing group and he is researching about different artificial intelligence algorithms to detect and analyse the mortality and incidence of cancer. Furthermore, his research is related to Big Data, eHealth, Machine Learning and Data Analysis.

Adela Pagès-Bernaus received the Ph.D. degree in Statistics and Operations Research from the Polytechnical University of Catalunya, Barcelona, Spain in 2006. Her main areas of research interest are the development of optimization models and solution algorithms to support decision making in real-life applications. The fields of application range from Agriculture, Sustainable Transport, Logistics or Energy.

Lluis M. Plà-Aragonès is an associate professor in the Department of Mathematics at University of Lleida (UdL), researcher at Agrotècnio and a Senior Researcher in the Area de Producción Animal at the extincted UdL-IRTA Center. His research interests include operational research methods applied in agriculture and forest management, with special reference to simulation, dynamic programming, planning, Markov decision processes and production planning.Co-ordinator of the EURO-Working group: Operational Re-search in Agriculture and Forest management and the CYTED network BigData and DSS in Agriculture. He is a member of INFORMS and EURO.

Francesc Solsona received the B.S., M.S. and Ph.D. degrees in computer science from the Universitat Autònoma de Barcelona, Spain, in 1991, 1994 and 2002 respectively. Currently, he is a full professor in the Department of Computer Science at the University of Lleida (Spain). He is the cofounder of the Hesoft Group company. His research interest include distributed processing, cluster computing, coscheduling, administration and monitoring tools for distributed systems, cloud computing, linear programming, big data, data analysis, simulation, optimization, social networks analysis and bioinformatics.

Anders Ringgaard Kristensen is a Professor of Animal Husbandry, Pigs, at the Department of Veterinary and Animal Sciences at University of Copenhagen. He obtained his master's degree (animal science) in 1982 and later PhD (1985) and Doctor of Agricultural Science degree (1993) from The Royal Veterinary and Agricultural University (now University of Copenhagen). In 2003 he, furthermore, obtained the MITS degree (Master of Information Technology, Software development) from The IT University of Copenhagen. His research has focused entirely on model based production monitoring and decision support in animal production.

Automatic Plant Leaf Disease Detection and Auto-Medicine Using IoT Technology

Channamallikarjuna Mattihalli, Fikreselam Gared, and Lijaddis Getnet

Abstract Leaf diseases in plants cause substantial production and economic losses, as well as a decrease in both the quality and quantity of the crops. It is easier to identify leaf diseases early in leaf health and can promote disease control through proper management strategies. This chapter provides a method for the early detection of leaf diseases in plants based on some important features taken from their images. Beaglebone Black (BBB) is interfaced with a digital camera used to take images of plant leaves to detect diseases in leaves. The image of the leaves is captured in the proposed framework and contrasted with images in the leaves database which are pre-stored. After image comparisons, if the plant leaves are found to be infected, this device automatically supplies medicine by a sprinkler to the area of the plant leaves. Leaf diseases are identified by initial or final phase tests. The responsible administrator may allow the medicine by sending a message back to GSM (Global System for Mobile Communication) if necessary. The diseases are at the final stage, the system does not wait for the admin message, and it allows the medicine or water to flow to the farm automatically.

Keywords Relay · Beaglebone black · Auto-medication · GSM · Sensors

List of Acronyms with Explanation

GSM   Global System for Mobile Communications
BBB   Beaglebone Black

C. Mattihalli (✉) · F. Gared · L. Getnet
Faculty of Electrical & Computer Engineering, Bahir Dar Institute of Technology, Bahir Dar
University, Bahir Dar, Ethiopia
e-mail: ckmattihalli@gmail.com

F. Gared
e-mail: fikreseafomi@gmail.com

L. Getnet
e-mail: lijaddisg@gmail.com

© Springer Nature Switzerland AG 2021
P. Krause and F. Xhafa (eds.), *IoT-based Intelligent Modelling for Environmental and Ecological Engineering*, Lecture Notes on Data Engineering and Communications Technologies 67, https://doi.org/10.1007/978-3-030-71172-6_11

IoT   Internet of things
HSV   Hue Saturation Value
IO   Input Output
UART   Universal Asynchronous Receiver/Transmitter
IDE   Integrated Development Environment
RGB   Red, Green, Blue
CCTV   Closed-circuit television
ARM   Acorn RISC Machine
HIS   Hue, Saturation, Intensity
CBIR   Content-Based Image Retrieval
AUIPC   Area Under the Infection Progress Curve

1   Introduction

We know that Ethiopia is one of the fastest growing countries in Africa, and the signature piece of our rate of growth depends on agriculture alone. Horticulture certainly brings important revenue to Ethiopia. In rural areas, plant leaf diseases cause severe development and financial malaise. The recognizable knowledge of plant diseases is the way to keep the misfortunes in yield and quantity of the rural element. Plant sickness inquiries indicate the discovery of externally noticeable examples found on the farm is taking off. Wellbeing observation and detection of infection on plants is especially important for horticulture and can be treated. Agriculturists need continuous supervision by experts who may be expensive and tediously restrictive. Physically, plant diseases are extremely hard to track. The search for a simple, more affordable, and more accurate approach to detect diseases as a result of side effects on the plant leaf is a pressing practical necessity. Not withstanding this physically supplying the polluted plant zone with drugs or pesticides is a worrying undertaking.

The fundamental objective of this chapter is to screen the well-being of plants and to identify the diseases of their leaves at an early period. Bearing this in mind the end goal of decreasing the spread of diseases by empowering the plants with auto-medication through the processor Beaglebone Black. The camera that is interfaced with the Beaglebone Black processor continually captures the images of the leaf and compares them with the database of the leaf. When some sicknesses are detected then Auto-solution is used according to the circumstances.

What is more, due to progress in climatic conditions, soil dampness and temperature sensors are used to maintain a strategic distance from disease spread. If the humidity and temperature levels exceed the predefined ride, the Beaglebone Black processor authorizes the plants to take auto-medication.

As part of the request, Global System for Mobile Communications (GSM) innovation is used to communicate something specific about the well-being of plants and the data identified with dirt humidity and temperature. Because of the condition, this allows the client to monitor and supply of drugs or water.

2 Scope of the Project

We distinguish the diseases normally occurring in the leaf plants by the attributes of the leaves such as shading and its tendency. By looking at the leaf influenced by the sound leaf infection it is conceivable that the plant's disease is recognized.

We introduce the point-by-point outline of an effective framework using a Beaglebone Black processor that helps to recognize ailments using innovation in image handling. In this proposed framework, the calculation of Bhattacharyya Distance is used to the obtained image which helps to recognize the plant's disease.

Leaf segmentation is critical in plant disease detection that affects the reliability of extraction features. We may use a hybrid clustering approach or superpixel clustering in which adjacent pixels are clustered into homogeneous regions with some characteristic in terms of brightness, texture, and color. Which can reduce image complexity from more pixels [1]. Several papers indicate that the expectation–maximization (EM) algorithm could be a reasonable approach for the segmentation of color images.

3 Literature Survey

Ananthi and Varthini [2] proposed plant diseases that are only recognized by using the K-map technique, the use of K-implies grouping by Neural Networks (NNs) has been designed for bunching and ordering the diseases that affect the plant. The perception of the illness is the reason for the proposed approach for the most part. The test comes about to demonstrate that the suggested solution is a viable method. It is a little theoretical effort that will essentially help an effective discovery of leaf diseases. Not considering the advantages of the crossover technique devote the future to determine the extent of the distinguished condition.

Al-Hiary et al. [3] proposed an article, which focuses on the CAP-LR preprocessing company. Leaf photos often shift to vague images due to the closeness of commotion and low or high complexity in both the edge region and the picture field. In this paper, an approach is displayed that changes differentiation all the while and improves limits. Since numerous shocks harm the nature of nature pictures, an enhanced upgrade strategy is needed to enhance the complexity of the leaf images.

Al Bashish et al. [4] discussed a clarification of the use of surface investigations in identifying plant diseases. The perception of sickness is mainly the reason for the approach proposed. The trial comes about showing the proposed approach with minimal computational effort, which helps to identify the leaf diseases. The examination subject is specified in this paper, of course, to identify the diseases from the symptoms that appear on the plant leaves. The increase in this undertaking is intended to improve the identification rate of the grouping phase.

Valliammal and Geethalakshmi [5] Presents Crop sicknesses are caused by pathogens fungi, viruses, and microbes yet there is a question of multi-sided visual example quality. Infection is caused by a pathogen that causes illness by any specialist. Attributes or diseases on the plant's leaves or stems are seen in the greater part of the cases. There are noticeable illness of plants, leaves, stems, and then finding the illnesses, and prevention of disease occurrence. There was a tremendous interest in the interpretation of specific and current picture designs.

Beyyalal and Beyyalal [6] discussed to explore the use of PC vision and image handling (image processing) procedures in farming applications. In natural science, a huge number of pictures are sometimes produced in a solitary test. These images may be required to help analyze plant diseases.

Padmavathi [7] proposed the actual job database to be installed in the PC The leaf affected by the infection to be attempted is contrasted and the PC database gives results and displays. The framework created consists of four phases that incorporate Hue, Saturation, Intensity (HIS) change, investigation of the histogram, and alteration of force. Highlight extraction is the third stage which mainly manages three highlights; shading, size, and spot status. The Content-Based Image Retrieval (CBIR) structure and basic strategy for discovering leaf diseases are not included in important papers.

Wu et al. [8] proposed the system with image and data processing techniques, they employ Probabilistic Neural Network (PNN) in this paper to introduce a general-purpose automatic leaf recognition for plant classification. 12 Leaf characteristics are extracted and orthogonalized into 5 main variables consisting of the PNN input vector. The PNN is equipped by 1800 leaves to identify 32 plant types with a precision exceeding 90%.

Mattihalli et al. [9, 10] proposed the system using the Raspberry Pi processor using Python language they programmed the processor. They used Bhattacharyya Distance to find healthy and infected leaves. They also used temperature and Humidity sensor to water the land, since variation in humidity and temperature leads to some diseases in the plants.

Vinita and Janwe [11] based on their proposal, the plants of medicinal value were studied. Through this, they demonstrated how computers could recognize medicinal plants. Based on the techniques of color space, histogram, and edge detection, we can find out the plant diseases. Medicinal plants are the foundation of a medicinal method called *Ayurveda* and are useful for treating many chronic diseases. The most significant part of plant disease research is to define the CBIR-based disease (content-based image retrieval) that is primarily concerned with the accurate identification of the diseased plant.

Elad and Pertot [12] in this paper they studied the effects of climate change on plant pathogens, and in some pathosystems the diseases they cause were examined. It is projected that future climate changes will influence pathogen growth and survival levels and alter host susceptibility, leading to improvements in the effect of diseases on crops. The effects of these climatic changes will vary according to pathosystem and geographic region. These changes can affect not only optimal infection conditions but also host specificity and plant infection mechanisms. They describe research on the effects on the biology of pathogens and their ability to infect plants and survive in natural and agricultural environments of changes in temperature, CO_2 and ozone concentrations, precipitation, and deficiency.

Swan et al. [13] proposed the influence of surface soil moisture and stubble management practices on the progress of *Fusarium pseudograminearum* infection with wheat, the cause of crown rot, was evaluated in a field trial at *Moree* in northern New South Wales. Wheat was sown onto warm surface soil during the 1994 dry season. Increases the infection incidence followed by rainfall events that elevated the surface soil's water content above the equivalent of water potential of -1.5 MPa. The area under the infection progress curve (AUIPC) was consistently larger when stubble was kept on the surface compared to a disk plow incorporation, and this difference was significant in 2 out of 3 years. A comparison of AUIPCs indicated greater epidemiological differences between treatments for stubble management than comparisons of infection incidence at single points during the season did.

Azath and Mattihalli [14] proposed a system to interface a mobile and PC using the GSM module. We can read our mails even we are far away from our PC.

4 Plant Diseases Analysis and Its Symptoms

Following are some basic manifestations of parasitic, bacterial and viral plant leaf disorders.

4.1 Bacterial Disease Symptoms

Minor light green spots that soon come into focus as water-splashed represent the ailment. The injuries escalate and eventually appear as dry dead spots as seen in Fig. 1, an e.g. bacterial leaf spot has darker or white water-drenched spots on the leaves, often with yellow radiance, by and wide indistinguishable in estimation. The spots get a spotted look under dry conditions.

Fig. 1 Bacterial leaf spot

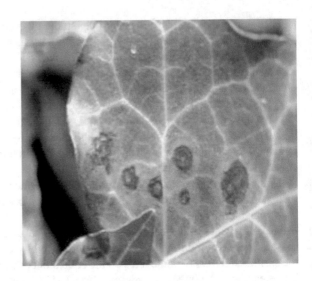

4.2   Viral Disease Symptoms

The hardest to be studied is among all plant leaf infections caused by pathogens. Infections have no hint that can be detected promptly and mistook routinely easily for inadequacies in the supplement and damage to herbicides. Aphids, leafhoppers, whiteflies, and creepy cucumber crawlies bugs are natural carriers of this ailment, for example, mosaic virus, look for yellow or green streaks, or spots on leaves, as shown in Fig. 2. Leaves can be wrinkled, twisted, and can impede growth.

Fig. 2 Mosaic virus

4.3 Fungal Disease Symptoms

Fungal disease symptoms caused by species are discussed below in all plant leaf diseases and appear in figures, an illustration of the late scourge caused by the parasite infesters of *Phytophthora* appears in Fig. 3. At first, it occurs on lower, more developed leaves such as water-splashed, dark green spots. The spots advance toward being invisible at the stage where infectious infection grows up, and after that white parasitic growth frames on the leaf's downside. The organism *Alternaria Solani* has shown in Fig. 4 triggers the early curse. This first occurs on the lower side of the leaf, on more seasoned leaves, like tiny dark-colored spots with clustered circles forming an example of a dead center. At the stage where ailment grows-up; the leaf ends up yellow as the disease is introduced outwardly. This happens in delicate mold yellow to white spots on the surfaces above of more developed leaves. These places were secured to the undersides with white to grayish as shown in Fig. 5.

Fig. 3 Late blight

Fig. 4 Early blight

Fig. 5 Downy mildew

5 Proposed Methodology

The proposed methodology includes the detection of plant leaf diseases at an early stage and enabling and disabling of medicines/pesticides automatically.

The proposed system square map appears underneath in Fig. 6. In the Beaglebone Black processor, the database of the leaf images (i.e., healthy and unhealthy) is prepositioned. This proposed system consists of a computer named Beaglebone Black which is a 1.2 GHz ARM processor Quad-core 64-bit. It also contains co-processor

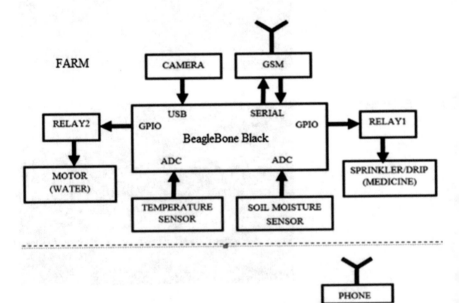

Fig. 6 Block diagram of the proposed system

dual-core multimedia that support multimedia applications. The camera is interfaced and set in an agrarian farm with Beaglebone Black processor gadget. The captured images compared with the picture database that is pre-stored in the gadget using photo planning strategies. What's more, soil humidity and temperature sensors are demanded to refrain from transmitting disease due to climatic development. Two switches are set to disable the water stream and medicine stream on the valves. Data on the stage of the disease and the type of the disease are implied to the customer using GSM.

The complete proposed system is given in the form of a flow chart in Fig. 7.

The picture handling program involves picture acquisition by an advanced camera, picture pre-preparation integrates image enhancement and picture separation, morphological systems, and arrangement. Finally, the closeness of diseases on the plant leaf will be recognized as shown image processing steps in Fig. 8, which demonstrates the fundamental advances used for the location of plant leaf diseases.

Right off the bat, pictures of various leaves are procured using a camera with the dedication needed for better quality. In the second step, this picture is pre-handled to enhance the picture details that smother unwanted twists, boost some important picture highlights for further preparation and inspection. It integrates the transformation of shading space, improvement of the image, and division of the frame. The Red, Green, Blue (RGB) images of the leaves are turned into pictures of shading space. The reason behind the shading space is to standardly encourage the detail of the hues. RGB images have changed to the Hue Saturation Value (HSV) shading picture of the space. Because RGB is for shading edge and its descriptor for shading. HSV demonstration is the optimal shading observation apparatus. The shade is a shading property that an onlooker depicts untouched shading as evident. Immersion called relative feature or white light measure applied to shade and value implies an excess of color. The tone section used to facilitate investigation after the shading space change phase. Additionally, HSV images are converted into dim scale images.

Here the green pixel settled steady is 90. The dark scale image can now have two qualities: 0 (black) and 255 (white). From the dim scale picture, thresholding can be used to replace each pixel in a picture with a dark pixel (0) if the image power is not as steady (if less than 90) or white (255) if the image intensity is more prominent than stable (if more notable than 90). Clamor and surface contour double positions provided by thresholding. Morphological image management is enhanced by reflecting the image structure for expelling these flaws. Widening and decomposition are the two important morphological practices. The expansion adds pixels to the boundaries of objects in an image, while disintegration evacuates pixels on boundaries. Usually, in situations such as expulsion from the clamor, enlargement follows the disintegration. Shaping is a device linked to computerized images with a common end goal of extricating their limit for further handling.

Bhattacharyya is handled separately in the last advance. Bhattacharyya removal is yet another simple histogram computing technique. It requires the database picture histogram and the current image. Contrasts are processed between the database and the current image. In the wake of finding whether the present picture histogram has the least similarities or comparison with any of the histogram of the database image that

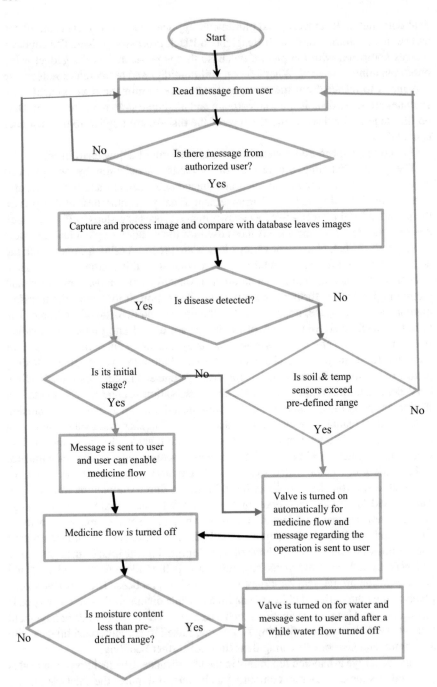

Fig. 7 Flow chart of the proposed system

Fig. 8 Image processing steps

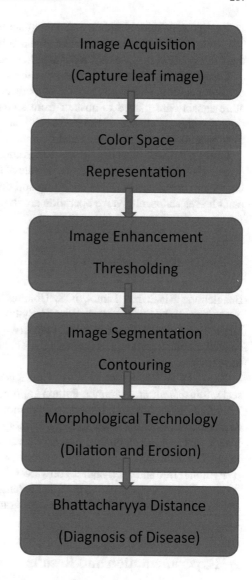

is pre-placed in the Beaglebone Black processor. The Beaglebone Black processor and flow image of the leaf are eventually known whether or not it is affected by the disease.

On the off chance that leaves are affected, it determines whether it is at the early stage or the last stage, as well as manifestations of its type. It is insinuated to the client via GSM in case infections are at the introductory level. If necessary, the client can stimulate the pharmaceutical stream to the leaves of the plant by sprinkler, by sending a message to the GSM as "product on." The gadget must supply 3.3v to the relay1 that turns the drug supply valve on.

It empowers auto-medicine and data about the solution stream is inferred to the client via GSM if illnesses are detected at a definitive point. The client can monitor the valves only if illnesses are recognized in the initial stage.

Consequently, valves are turned on at the definitive process of ailments without customer control. Due to advances in climatic conditions, soil dampness and temperature sensors are placed to abstain from spreading the ailments at the beginning time. On the off chance of the sensor estimates exceeding the predefined run, the pharmaceutical stream is consequently empowered.

Given pharmaceutical flow, water flow is additionally regulated and water wastage is restricted. Relay 2 is shut off in the middle of the solution stream which means that the water stream ceases. Data regarding soil dampness content, temperature range, plant leaves status, and valve operation are implied via GSM to the admin.

6 Software Description

Beaglebone Black has Linux, iOS, UbuntuCloud9 IDE device compatibility on Node.js w / Bone Script library, WinCE. Python programming is utilized for Beaglebone Black creation. For processing, Python would read the signal value via Universal Asynchronous Receiver/Transmitter (UART) and then collect the received signal to the database.

The language provides structures designed to support both small and large-scale, transparent programmer. Python supports multiple programming paradigms, including object-oriented, imperative, functional, or procedural styles. It features a dynamic form structure and automated memory management and has a wide standard library that is comprehensive. Figure 9 demonstrates the Beaglebone Black Setup I/O Python Library.

Python is free and open-source software, with a community-based model of development. Python was created to be rather extensible. Python can also be incorporated into existing applications that require a programmable interface.

7 Experimentation and Results

The fully assembled working model is as shown in Fig. 10, after choosing the components used for the project. We have performed an operational prototype based on the concept configuration. The medicine flow is enabled automatically by turning on the valve when the BBB device detects the diseases with the help of a camera.

The image processing operation is shown in Fig. 11 as it includes the original, HSV, gray, threshold, dilates, and eroding image.

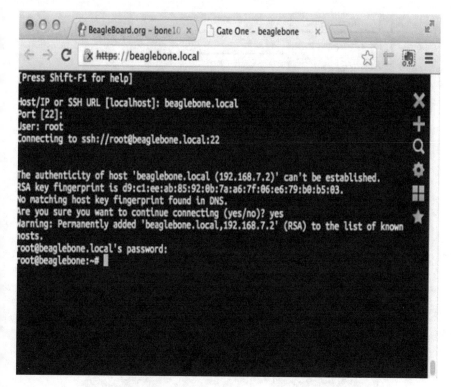

Fig. 9 Setting up I/O python library

7.1 Advantages of the Proposed System

Initially, plant sicknesses may be established. Disease dissemination can be kept a strategic distance from this. Data on the plants and the system will be suggested via GSM interfacing with the client. Depending on the requirement, this can be configured to all kinds of leaves. This can be altered to all kinds of leaves that depend on the set precondition.

8 Conclusion

The proposed IoT platform assists the early identification of plant infections. If any disease is detected, auto-medicine is activated by turning on the valve, and medication is splashed to the rural ranch through a sprinkler.

Installed framework for programmed medicine in a horticultural field provides a potential answer to aid site-specific water system administration that enables manufacturers to improve their productivity by recognizing the diseases at the start.

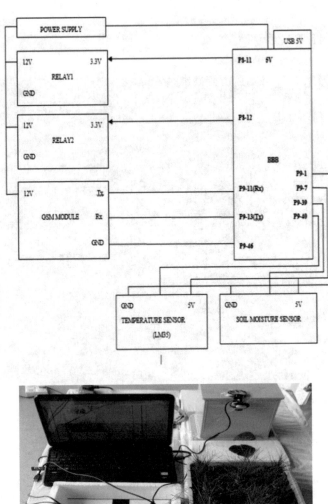

Fig. 10 Circuit diagram and assembled model

Fig. 11 Image processing

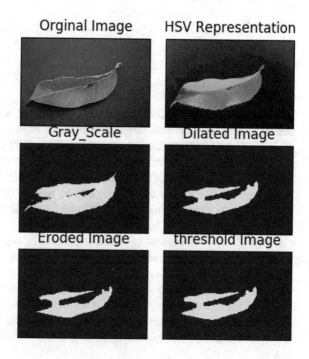

Orginal Image   HSV Representation

Gray_Scale   Dilated Image

Eroded Image   threshold Image

Similarly, soil humidity and temperature sensors are used to limit ailment spread due to progression in climatic conditions. In the case that the humidity or temperature levels exceed the predefined go, the BBB empowers the plants to take auto-medicine or water. It can be customized to any kind of leaves that depend on the client's needs.

The scheme is illustrated using gadget BBB. Information about sickness discovery and valve operation is sent to the client via GSM. This system is robust, simple to use, and requires minimal effort.

Discuss briefly if the proposed system would be the suitable choice for a small to medium size Farming enterprise? What is the area to be covered by the proposed system? What would be the cost of implanting such a system? Please expand on this part.

Beaglebone Black	100$
Submersible Water Pump 2no.	50$
12v Relay 2no.	10$
GSM Module	50$

(continued)

(continued)

Web Camera	15$
Pipe	15$
Total	240$

The estimated cost of this workaround 250USD. If we are going to increase the better camera which covers larger area then we have to go for better water pump which can give more pressure that can cover the area covered by the cameras.

The existing method for the detection of plant disease is naked eye observation by experts by which plant disease identification and detection is achieved. To do so, there is a need for a large team of experts as well as continuous plant monitoring, which will cost very high as we do with large farms. Around the same time, farmers in some countries do not have sufficient facilities or even the knowledge that they can contact experts. Because of which expert consulting costs high as well as time-consuming too. The suggested technique is proving beneficial in tracking large crop fields under these conditions. By simply seeing the symptoms on the plant leaves, automatic detection of the diseases makes it both easier and cheaper.

Visual detection of plant disease is a more laborious and at the same timeless effective process, which can only be performed in specific areas. Whereas it will take less effort, less time and become more effective if an automatic detection technique is used.

The area covered by this system depends on how many comers we are interfacing with the BBB device and how much area each camera can cover. It can be best implemented for small-scale agricultural lands.

In this work, we are providing a database which contains healthy and unhealthy leaf. When the system is going to take a new image, it will process i.e. image processing, and compares the features which are already there in the database. Predictive analytics is a category of data analytics that aims to make predictions of future results based on historical data and analytical techniques such as statistical modeling and machine learning. The science of predictive analytics with a large degree of accuracy will produce potential insights. With the assistance of advanced methods and models for predictive analysis. Initially, we tried with around 40 images in the database the accuracy will be less. Then we increase the database size to around 400 images we got better accuracy compares to the previous database. It shows that if we have a large database, we will get better accuracy.

References

1. Zhang S, You Z, Wu X (2017) Plant disease leaf image segmentation based on superpixel clustering and EM algorithm. Springer
2. Ananthi S, Varthini VS (2012) Detection and classification of plant leaf diseases. Int Res Eng Appl Sci 2(2):763–773. ISSN: 2249-3905

3. Al-Hiary H, Bani-Ahmad S, Reyalat M, Braik M, AlRahamneh Z (2011) Fast and accurate detection and classification of plant diseases. Int J Comput Appl (0975–8887) 17(1):31–38
4. Al Bashish D, Braik M, Bani-Ahmad S (2010) A framework for detection and classification of plant leaf and stem diseases. In: International conference on signal and image processing, pp 113–118
5. Valliammal N, Geethalakshmi SN (2011) Hybrid method for enhancement of plant leaf recognition. World Comput Sci Inf Technol J (WCSIT) 1(9):370–375
6. Beyyala A, Beyyala SP (2012) Application for the finding of illnesses in crops utilizing picture handling. Int J Life Sci Biotechnol Pharma Res 1(2):172–175
7. Padmavathi K (2012) Investigation and observing for leaves ailment discovery and assessment utilizing picture preparing. Int Res J Eng Sci Technol Innov (IRJESTI) 1(3):66–70
8. Wu SG, Bao FS, Xu EY, Wang YX, Chang YF (2007) A leaf recognition algorithm for plant classification using probabilistic neural network. In: IEEE seventh international symposium on signal processing and information technology
9. Mattihalli C, Mehretie E, Endalamaw F, Necho A (2018) Real time automation of agriculture land, by automatically detecting plant leaf diseases and auto medicine, pp 325–330. https://doi.org/10.1109/WAINA.2018.00106
10. Mattihalli C, Mehretie E, Endalamaw F, Necho A (2018) Plant leaf disease detection and auto-medicine. Internet Things 1–2:67–73. https://doi.org/10.1016/j.iot.2018.08.007
11. Tajane VM, Janwe NJ (2013) CBIR in image based medicinal plant diseases retrieval. Int J Eng Res Technol (IJERT) 2(11)
12. Elad Y, Pertot I (2014) Climate change impact on plant pathogens and plant diseases. J Crop Improv 28(1):99–139. https://doi.org/10.1080/15427528.2014.865412
13. Swan LD, Backhouse D, Burguess LW (2000) Surface soil moisture and stubble management practice effects on the progress of infection of wheat by Fusarium pseudograminearum. Aust J Exp Agr 40:693–698
14. Azath M, Mattihalli C (2012) Mobile based E-mail reading system at the international conference on advances in computing-ICADC 2012, July in Bangalore, India. Springer

Dynamic Shift from Cloud Computing to Industry 4.0: Eco-Friendly Choice or Climate Change Threat

Manmeet Singh, Shreshth Tuli, Rachel Jane Butcher, Rupinder Kaur, and Sukhpal Singh Gill

Abstract Cloud computing utilizes thousands of Cloud Data Centres (CDC) and fulfils the demand of end-users dynamically using new technologies and paradigms such as Industry 4.0 and Internet of Things (IoT). With the emergence of Industry 4.0, the quality of cloud service has increased; however, CDC consumes a large amount of energy and produces a huge quantity of carbon footprint, which is one of the major drivers of climate change. This chapter discusses the impacts of cloud developments on climate and quantifies the carbon footprint of cloud computing in a warming world. Further, the dynamic transition from cloud computing to Industry 4.0 is discussed from an eco-friendly/climate change threat perspective. Finally, open research challenges and opportunities for prospective researchers are explored.

M. Singh
Centre for Climate Change Research, Indian Institute of Tropical Meteorology (IITM), Pune, India

Interdisciplinary Programme (IDP) in Climate Studies, Indian Institute of Technology (IIT), Bombay, India

M. Singh
e-mail: manmeet.cat@tropmet.res.in

S. Tuli
Department of Computer Science and Engineering, Indian Institute of Technology (IIT), Delhi, India
e-mail: shreshthtuli@gmail.com

R. J. Butcher · S. S. Gill (✉)
School of Electronic Engineering and Computer Science, Queen Mary University of London, London, UK
e-mail: s.s.gill@qmul.ac.uk

R. J. Butcher
e-mail: r.j.butcher@se19.qmul.ac.uk

R. Kaur
Department of Chemistry, Guru Nanak Dev University, Amritsar, Punjab, India
e-mail: rupinderchem@gmail.com

© Springer Nature Switzerland AG 2021
P. Krause and F. Xhafa (eds.), *IoT-based Intelligent Modelling for Environmental and Ecological Engineering*, Lecture Notes on Data Engineering and Communications Technologies 67, https://doi.org/10.1007/978-3-030-71172-6_12

Keywords Cloud computing · Industry 4.0 · Climate change · Environment ·
Cloud data center · Carbon footprint

List of Acronyms

CDC	Cloud Data Centres
AI	Artificial Intelligence
IoT	Internet of Things
6G	6th Generation
VMs	Virtual Machines
SLA	Service Level Agreements
QoS	Quality of Service
NAS	Network Attached Storage
IPCC	Intergovernmental Panel on Climate Change

Glossary of Terms

Terms	Description
Cloud Computing	It is on-demand cloud service (using resources such as processor, storage, memory and network) to the users via Internet
Industry 4.0	Current trend of automation and data transfer in manufacturing technologies
Climate Change	Change in the mean or basic state of the climate
Eco-friendly	Anything that does not harm the environment
Quality of Service (QoS)	It is a measurement in terms of performance parameters to evaluate the service quality
Service Level Agreements (SLA)	SLA is an official document, which is signed between cloud user and cloud provider based on QoS requirements
Agriculture 4.0	Atomization of Agriculture related aspects such as precision agriculture and big data analytics
Healthcare 4.0	Management of vast amount of healthcare data efficiently
Carbon Footprints	It is the amount of carbon released in the environment by the computing system
Digitization of Economies	The transition of financial systems towards digital platforms
COVID-19 Pandemic	It is a coronavirus disease 2019, which is caused by SARS-CoV-2 (severe acute respiratory syndrome coronavirus 2)

1 Introduction

The world is waking up to the serious disaster caused by the ongoing global warming by the awareness efforts of young activists like Greta Thunberg. Climate change is a fact that threatens entire living ecosystems. Global web users are growing daily, and Internet demand is building exponentially with every passing day [14]. Google is effectively and efficiently using emerging technologies and digital transformations such as Artificial Intelligence (AI), the Internet of Things (IoT) and the 6th Generation (6G) that require large cloud data centers to fulfil this user generated demand [11]. The expansion of cloud data centres furthers the rise in carbon emissions by enhancing energy consumption.

The large-scale consumption of energy and release of carbon have a negative effect on the Earth's climate, directly or indirectly, leading to melting of glaciers, rising sea levels and the ice-free Arctic. Rising temperatures are accompanied by changes in weather and climate [14]. Existing technologies are increasingly becoming fast and automated by the advancements in cutting-edge scientific research, however, these developments have come at the cost of an increase in the consumption of energy and resulting negative impacts on the ecosystems.

On 26 July 2019, in the heatwave event that broke records across Western Europe [11], Paris saw a record high temperature of 42.6 °C (108.6 °F). While extreme weather events such as heatwaves happen naturally, according to the UK Met Office "research shows that climate change is more likely and perhaps as frequent as every other year". As early as 14 August 1912, the possibility of future warming by coal-burning through the introduction of carbon dioxide in the air was published by a New Zealand daily [10]. Previously estimates of the amount of carbon in the atmosphere are on the rise as cloud data centres are expanded to meet consumer demand. In order to minimize the carbon emissions by the increasing use of cloud data centres, the state-of-the-art cloud data centre management techniques were investigated by [14]. Moreover, an innovative solution to the global challenge of climate change has been developed to enable holistic resource management for reliable, energy-efficient and cost-effective cloud computing [16]. This study proposes the use of renewable sources of energy which could help to reduce the environmental impact of cloud data centres.

In brief, this chapter has two main objectives. The first is the development of various pathways based upon the demand of cloud computing in the future to quantify its role in a warming climate, and the second is the use of artificial intelligence-based techniques to take intelligent decisions for the cloud to minimize its negative impact on the climate.

The rest of the chapter is organized as follows. In Sect. 2, we discuss cloud computing. After that, we discussed Industry 4.0 in Sect. 3 and climate change in Sect. 4. Section 5 quantifies the carbon footprints of cloud computing in a warming world. Section 6 presents the dynamic shift from cloud computing to Industry 4.0. Section 7 explores future trends of cloud computing research, which can be an eco-friendly choice or climate change threat. Section 8 presents impact of carbon footprints on climate change and Sect. 9 discusses open research challenges and opportunities. Finally, we summarize the findings and conclude the chapter in Sect. 10.

2 Cloud Computing

Cloud computing is a dynamic provisioning model that supplies clients on an on-demand basis from a shared pool of computing resources [30]. Using virtualization technology, several Virtual Machines (VMs) are created on each physical server to allow for efficient resource allocation and reductions in the amount of hardware in use [46]. The prominent cloud providers such as Facebook, Amazon and Google use a huge amount of Cloud Data Centres (CDC) and offer a very reliable and cost-effective cloud service to the end-users [14].

Recent years have seen the dramatic migration of applications and services to the cloud [18]. This paradigm allows organizations to outsource computer requirements and resources, promising features such as high scalability, agility and reliability [29]. However, as the popularity of cloud computing increases, energy consumption by its data centres is becoming more prominent. Cloud infrastructures consume a great deal of energy, accounting for 3% of global electrical energy consumption as of 2017 [14]. This high energy usage of has been attributed to energy inefficiencies and wastage. One analysis of energy wastage found data centres to use only 6–12% of electricity powering their servers to perform computations, with much of the remaining energy used to keep servers idle, ready to deal with potential activity surges [32]. There is a spectrum of disadvantages to this; from increasing environmentally detrimental carbon emissions, to driving up operating costs for cloud providers and the subsequent reduction in profit margin. Curbing the energy budgets of data centres is therefore both sustainably and economically incentivized but is not without challenges.

Despite brown energy becoming less favorable, it is evident that the solution will not be as simple as making the swap to green renewables. The production of energy from wind turbines and solar panels is weather limited. It remains a challenge to balance intermittent green energy production with continuous functioning requirements of the data centre, without compromising on Service Level Agreements (SLA) or Quality of Service (QoS) [46]. Other potential solutions explore green architectures involving client-orientated green cloud middleware to assist in energy-efficient decision making. For example, Hulkury and Doomun propose a user interface application which gathers system specifications such as SLAs and QoS, and returns the estimated carbon footprint for each business operation as performed on the client's machine, a private cloud or the public cloud with the aim of recommending the 'greenest' option for the job [30]. Kuppusamy also proposes an adapted middleware for managing the selection of Cloud provider with the lowest carbon emissions, based on parameters such as data centre cooling efficiency [24]. Future challenges and opportunities are explored further later in the chapter.

3 Industry 4.0

As the efficiency of hardware and software systems have increased in the last few decades, modern computational frameworks are able to perform tasks with much lower result delivery times, cost and energy consumption [5]. On the other hand, the service requirements of different industrial sectors like manufacturing, supply networks, agriculture and healthcare have become more demanding over the years [17, 21]. Moreover, modern cloud-based infrastructures have contributed to a significant portion of global energy consumption and have played a critical impact on the carbon footprint [9, 14]. For the growing user demands, by leveraging novel technologies, the industry and academia have started the fourth industrial revolution.

The fourth industrial revolution, named as Industry 4.0 marks the complete shift in the technological backbone of industrial applications. This revolution is marked by the upcoming paradigms such as the IoT, AI and Blockchain [17]. The new paradigm on IoT allows end-to-end integration of sensors and actuators close to the user with geographically distributed cloud servers via a large number of intermediary smart edge/fog nodes [53]. This allows the development of a hierarchical and more robust infrastructure for modern industrial applications [13]. The new field of AI allows users to utilize and exploit large amounts of sensor information using big data analytics to properly and efficiently manage resources, prevent service failures and provide high-quality results [35]. Furthermore, technologies like Blockchain allow strong data integrity to prevent fraudulent data manipulation and provide strong support to various Machine learning algorithms which depend on this data [7]. New paradigms require the development of novel security measures to prevent attacks like evasion, data poisoning and model stealing [28]. All these challenges need to be solved in diverse areas with a primary focus on sustainability.

3.1 Healthcare 4.0

Healthcare is one of the most challenging domains with respect to providing efficient services using innovative computation models and advanced algorithms [49]. This difficulty primarily arises from the critical need for ultra-precise results in extremely small amounts of time with low latency, response time, and high accuracy [47, 49].

Many recent works aim at providing effective healthcare services leveraging next-generation technologies at affordable costs [6, 40, 47]. This requires optimizing the existing models used for leveraging diverse computational frameworks in a seamless fashion. This is done to enhance the abilities and provide higher quality disease detection results, improved automated prescription generation and reduce service costs.

To effectively leverage computational technologies, prior works use advanced scheduling and offloading approaches like genetic-algorithms, machine learning and reinforcement learning [47, 49]. Many prior works utilize data sharing capabilities of Network Attached Storage (NAS) to further reduce the allocation times [51]. Recent works also leverage microservices and container-based deployments to further enhance the efficacy of intelligent scheduling policies. Furthermore, with the advent of computationally powerful edge devices researchers have been able to run sophisticated federated learning policies for holistic improvement of QoS parameters like energy and response time which are crucial for time-critical healthcare applications [51]. Other works aim to minimize SLA violations. Violating SLA could lead to catastrophic impacts on critical applications like e-healthcare. Hence, novel approaches are required which are able to not only manage large-scale healthcare solutions but also have extremely low overheads.

The driving idea of the future-proof model is to utilize and build upon techniques of AI and machine learning for efficient task scheduling and placement to achieve the above-mentioned objectives [17]. However, there still exist several pragmatic challenges remaining to be solved. Most of the AI-based approaches are suitable for cloud setups which have GPUs for faster training of the underlying neural networks and results generation [39]. This not only increases the cost and power consumption of such deployments but also limits such frameworks to be only deployed in resource-intensive data centres [25]. This is not suitable for low resource-based Edge deployments which are the upcoming paradigm, being readily leveraged by modern industries. Thus, AI not only allows the development of precise and accurate computational services crucial for the healthcare sector, but it also poses challenges like scalability, affordability, security, high-availability, energy and budget limitations.

3.2   Agriculture 4.0

The traditional approach of the agriculture industry is undergoing a fundamental transformation. Advance research-driven worldwide global trends are affecting the agriculture industry. Various researchers, academicians and industries are developing new agriculture systems [15, 42], enabling the automation of agriculture industry and improving agriculture as a service using new digital transformations such as AI, 5G, IoT and edge computing. Cloud computing uses a large number of CDCs to provide the required computational capacity. Due to the use of a huge number of data centres, a huge quantity of carbon-di-oxide is generated, which further impacts the climate. A large amount of energy is required to run the CDCs continuously to maintain the QoS. To solve this problem, renewable sources (wind energy, solar energy etc.) should be utilized to run the CDCs, which can reduce the impact of cloud computing on climate saving a large amount of brown energy.

4 Climate Change: Projected Warming in the 21st Century

The increase in greenhouse gas emissions across the globe is expected to inevitably lead to 1.5-degree warming and there is a very small window available to avoid reaching the stipulated warming which is also estimated to be a tipping point for various climatic systems across the globe [34]. The projected increase in greenhouse gases can eventually lead to an increase in global surface temperatures on Earth, changing the regional patterns and amount of rainfall, reduction in snow and ice cover and leading to damages in permafrost, raising the mean sea levels, increasing extreme storms and rainfall events, increased acidification of the oceans, change ecosystems and eventually threatening human life. These impacts are not just limited to future projections but have rather started showing up in various forms in nature. These changes in future heavily depend upon the role of anthropogenic activities and are quantified by different pathways by Intergovernmental Panel on Climate Change (IPCC) based upon projected emissions under different scenarios. A recent study [22] has shown that the changes in sea-level rise could be thrice as compared to previous estimates endangering major coastal cities with flooding worldwide. There is no doubt that climate change is already a global problem created by humanity and it is leading to catastrophic impacts all over the world. The first solution is a reduction in emissions but quantifying the role of different sectors remains a challenging task. At this stage when global warming has become an international calamity, increasing sources such as, for example by cloud computing, is the last thing humanity can afford.

5 Quantifying Carbon Footprints of Cloud Computing in a Warming World

According to the IPCC Assessment Report 5 [36, 44], the rise in global average surface temperatures have been primarily driven by anthropogenic activities. Until 2010, the rise has been 0.6° relative to 1950 [36]. The warming has been on the rise and it has been recently projected that a 1.5° rise which may be unavoidable would trigger tipping points [33]. In the past 3 decades, there has been a lot of research focusing on various sources of greenhouse gases-led global warming, such as the burning of fossil fuels, the release of methane etc. However, studies focusing on the role of emerging technologies in driving the warming have been limited. Further, we can quantify the role of cloud computing infrastructure and its maintenance activities as drivers of global warming acting as carbon sources. There has been an increase in the establishment of new cloud-based servers globally, and these servers require a continuous supply of electricity. Thus, quantifying the role of cloud data centres for their impact on the global climate is an important question that needs to be asked for decision-makers. This activity would employ data from various cloud servers and an

AI-based inverse modelling approach to quantify their role in warming the climate. This would require expertise in climate, data science and cloud computing.

6   Dynamic Shift from Cloud Computing to Industry 4.0

Cloud computing has been readily adopted by most industrial and academic sectors in the last few decades. Computation on the cloud allowed users to leverage geographically distributed services and give application developers or service provider's hassle-free computational framework. The original cloud enterprise model consisted of four main layers: personal device layer, an interface layer, management layer and physical layer. The device layer, which included personal laptops, desktops and mobile phones, allowed users to interact with a frontend interface of the cloud service. Many advancements have taken place allowing diverse devices to be able to interact with the cloud platform [4]. The interface was supported by various technologies of key-exchange, encryption and signature-based certification to maintain security [27]. Many works have also focused on providing different types and more-user friendly interfaces to allow easy integration and quick deployment of cloud services in the industrial pipeline [4]. Above the interface layer is the management layer which manages the tasks across different computational nodes by either a form of load balancer or scheduler [1]. The management layer was also developed to be robust and provide high-availability using recovery management for fail-prone cloud servers [31]. Many works have also allowed the integration of large-scale database management systems with the cloud through the management layer [47, 48]. Finally, there is the physical layer which provides the underlying computational infrastructure which computing nodes being either physical servers of virtual machines deployed on such servers. However, the Industry 4.0 revolution revamps this design for the next-generation industrial demands.

Industry 4.0 employs state-of-the-art technologies for providing optimum quality with a scalable computational platform which is strongly integrated with the new IoT paradigm. The user-level devices are now sensors or actuators with limited resource capabilities in embedded devices. This not only provides low latency response but also brings computation closer to the edge of the network allowing high scalability using hierarchical architecture [48]. The interface now consists of multiple microservices instead of monolithic cloud service, catering to different user needs efficiently [12]. Data and interaction security are maintained with smart contacts, smart management and sophisticated data recovery mechanisms [26]. Further, the management layer now consists of AI and ML-powered scheduling algorithms [52] with high integrity Blockchain mechanisms [48] for fast, accurate and tamper-proof application deployment. Hybrid infrastructure in the physical layer consisting of both edge and cloud nodes allow efficient allocation of low latency tasks to edge devices and high accuracy tasks to cloud servers [52]. Virtualization has now shifted to lightweight containers for supporting resource-constrained fog nodes and quick task migration

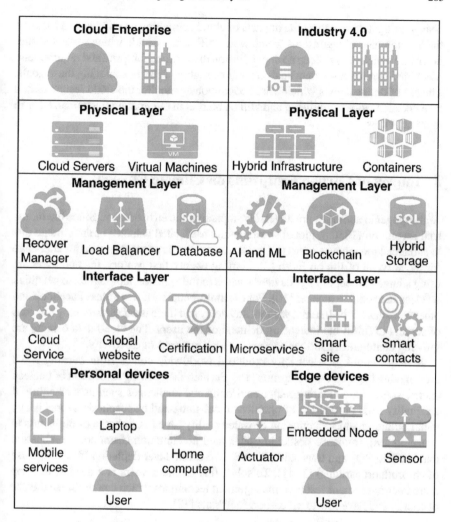

Fig. 1 Summary of a dynamic shift from the cloud through Industry 4.0

to prevent application failures [45]. A summary of the paradigm shift from cloud enterprise through Industry 4.0 is shown in Fig. 1.

7　Eco-friendly Choice or Climate Change Threat

Whether cloud computing is an eco-friendly choice or a climate change threat is yet to be determined. This would require quantification of the contribution of cloud computing towards carbon emissions. In future, the new methodologies for climate change mitigation to negate the impacts of cloud computing on climate will focus on the development of algorithms which would efficiently decide which servers to shut down and which ones to keep up depending on their carbon footprint and requirement. Machine learning based data analysis will be highly useful in achieving these goals. The use of online agents which would continuously monitor the cloud resources and take decisions looks promising and will be helpful in eliminating the human error in these activities.

8 Impact of Carbon Footprints on Climate Change

Anthropogenic activities are adversely impacting the environment. Since the industrial revolution (1850s), global warming and associated changes in the climate due to enhanced greenhouse gases (CO_2, CH_4 and others) has become a major challenge for the survival of life on Earth [2]. Starting twenty-first century, the end users are using a large number of edge devices such as mobile phones, laptops etc. to get quick services in a reliable manner [16]. For example, people want to access Facebook and other websites on the Internet without any delay and there is a need to deploy millions of servers to fulfill the increasing demand of the users. The cloud data centers are growing continuously further consuming high amounts of electricity.

A major chunk of electricity produced worldwide comes from burning fossil fuels having large carbon footprints [14]. Further, the growing utilization of unclean energy sources are creating major environmental challenges such as worsening of air quality (fog, production of photochemical smog and tropospheric ozone, heavy metal pollution, etc.), fluctuation of water quality (due to discharge of pollutants to water bodies), immense pesticide usage, acid precipitation (from coal combustion that leads to SO_2 and thus sulfuric acid) and ozone layer depletion (due to the use of chlorofluorocarbons) [3, 41]. To solve, this problem, there is a need to develop more energy-efficient resource management techniques which can optimize the use of energy and have low to nil carbon footprints [18].

9 Future Research Challenges and Opportunities

In this section, we discuss future challenges and opportunities for prospective researchers. Figure 2 shows the promising future directions and opportunities.

Fig. 2 Future research directions and opportunities

9.1 Role of Cloud Computing in Global Warming and Its Quantification

The role of cloud computing as a source in emitting greenhouse gases via the consumption of dirty electricity has to be quantified for devising future projections on climate change. In this aspect, this study aims to calculate numbers for various pathways that would be developed considering the demand for cloud computing in the future. The outcomes will lead to a better understanding of the percentage of electricity consumed by the emerging technologies associated with the cloud originating from fossil fuel burning. The development of such pathways will be useful for policymakers and also for incorporating cloud computing as a source in the IPCC assessments in future. The requirement of Facebook and Google of today to continuously keep their servers up with an increasing population adopting the Internet is expected to make cloud computing as having a major carbon footprint in the future.

9.2 Increasing Digitization of Economies

The increasing digitization of economies across the globe is leading to the development of cloud data centres which are important for driving the digital world. However, the development and maintenance of these cloud data centres have a carbon footprint which is increasing day by day and contributing to the anthropogenic sources responsible for global warming. In future, there is a need to quantify the role of these data centres in their capacity to cause global warming, helping stakeholders in Europe, USA and across the globe to develop policies for the future.

9.3 Development of Novel Frameworks for Large-Scale Hybrid Infrastructures

As the demands in service industry increase, there is a need for novel frameworks that can exploit the low latency of edge nodes and also high computation power of cloud servers. Such frameworks must maintain a careful balance between accuracy requirements and response times for different healthcare tasks. For such a framework, real-time analysis of critical patients can be done at the most powerful edge resources or the closest cloud servers. Other healthcare services with long deadlines can be run on distant cloud data centres, and those with low compute and low response time requirements can be run on cheaper and resource-constrained edge nodes. This itself requires robust scheduling and task placement algorithms to be built for seamless integration of different paradigms and meeting the SLAs.

9.4 Pandemic Induced Reduction in Emissions Due to Cloud Computing

It is known that the climate has been changing towards a hotter planet hitherto since the industrial revolution. Increasing temperatures have also been associated with degrading air quality across major industrial centres around the globe [23]. The air quality has been declining due to rampant and unsustainable development activities undertaken by humans. However, recent COVID-19 [20] enforced global lockdowns have led to a drastic improvement in the quality of air around the world. These improvements in air quality were last observed during the global economic recession of 2008–2009 [50]. However, it was seen that some countries took on a path of steep development leading to the coinage of the term "revenge pollution" [37]. Suppression of pollutants in the air has hence been associated with economic slowdowns or pandemic related lockdowns which are expected to be followed by periods of enhanced aerosols in the atmosphere. The quantification of the role cloud computing plays in this suppression and enhancement of pollutants would be important for devising policies in the future.

9.5 Role of Artificial Intelligence in Tackling Climate Change

One of the biggest problems facing mankind is climate change machine learning can help in solving the multi-faceted challenge of global warming and climate change [38]. Machine Learning can deliver important contributions in addressing climate change challenges across domains (for example, see [8]). It can allow automated monitoring by remote sensing satellites helping the task of scientific exploration and maximizing productivity systems. Artificial intelligence can speed costly physical simulations by hybrid modeling. It must be remembered that AI is only part of the solution allowing other methods used in climate change science across fields. Applying machine learning to combat climate change related problems has the potential to support humanity and advance machine learning. Meaningful progress on climate challenges involves dialog with disciplines within and beyond computer science, and may lead to interdisciplinary methodological developments, such as enhanced data-driven approaches that are limited to physics. Unlike traditional AI-benchmarks, wherever climate data exists, it may not be structured for a single purpose. Datasets can contain heterogeneous source information and must be combined with domain expertise. The existing data does not reflect global usage cases. As an example, the data for US weather forecasting or energy demand is plentiful, however for Africa it is scarce. Transfer learning techniques and domain adaptation would undoubtedly prove necessary in low-data environments. For certain tasks training with carefully generated simulation data might be feasible. However, the better alternative is still the more actual observational datasets.

9.6 Earth System Modelling and Advances Statistical Techniques Using Cloud Computing for Climate Change Research

Earth System models are global environmental models with the integrated potential to directly reflect biogeochemical mechanisms that communicate with the real climate, thereby altering its reaction to forcing like the one associated with anthropogenically-driven greenhouse gas emissions. Representing the climate system requires feedback between the physical atmosphere and the water and land chemical-biological cycles that eat up the carbon dioxide produced, thereby minimizing warming. Such models offer important insights into climate instability and transition, and the impact of human activity and potential mitigating measures on future climate change. The use of ESMs and the advanced statistical techniques such as the one by [43] can help in estimating the climate response to natural and anthropogenic forcing. Various advanced statistical methods such as causality, complex networks, recurrence plots, deep learning, deep reinforcement learning and others offer hope in providing interesting insights into the mechanisms behind climate change and science. The explosion of climate data in the last two to three decades and the advent of cloud computing offers with the correct analysis, storage and reproducibility becoming big problems in climate science, cloud computing solutions such as Google Earth Engine offer new paradigms for the climate science community.

9.7 Role of Quantum Computing in Climate Control

Quantum computing is expected to become very effective in handling different problems related to climate change such as energy production, carbon capture and storage of carbon, which are the main reasons of production of CO_2 emissions [19]. Further advancements in the quantum computing can aid to simulate large complex molecules, which further helps to discover new catalysts for carbon capture economically. Moreover, Quantum computing can aid in extremely high-resolution simulations of weather and climate models of the future. Cloud-computing will play a crucial role in driving the advances driven by quantum computing in future.

10 Summary and Conclusions

In this chapter, we discussed the impact of cloud computing transformations on climate and presented the role of cloud computing in global warming. This chapter quantifies the role of cloud computing on climate change in the future depending upon various representative concentration pathways that would be developed. There is a need to develop a system which would be targeted at reducing the carbon footprint of cloud computing infrastructure. This chapter can potentially help future researchers in two ways: First, it would help the policymakers involved in the business of cloud computing to develop climate-aware policies for the future. Second, the development of a system to assist in cloud computing activities will also aid in reducing the carbon footprint of the cloud in the future. Moreover, we proposed some promising research directions based on the analysis that can be pursued in the future.

Acknowledgements We would like to thank the Editors (Prof. Paul Krause and Prof. Fatos Xhafa) and anonymous reviewers for their valuable comments and suggestions to help and improve our research paper.

Glossary of Terms

Terms Description
Cloud Computing It is on-demand cloud service (using resources such as processor, storage, memory and network) to the users via Internet

Industry 4.0 Current trend of automation and data transfer in manufacturing technologies

Climate Change Change in the mean or basic state of the climate

Eco-friendly Anything that does not harm the environment

Quality of Service (QoS) It is a measurement in terms of performance parameters to evaluate the service quality

Service Level Agreements (SLA) SLA is an official document, which is signed between cloud user and cloud provider based on QoS requirements

Agriculture 4.0 Atomization of Agriculture related aspects such as precision agriculture and big data analytics

Healthcare 4.0 Management of vast amount of healthcare data efficiently

Carbon Footprints It is the amount of carbon released in the environment by the computing system

Digitization of Economies The transition of financial systems towards digital platforms

COVID-19 Pandemic It is a coronavirus disease 2019, which is caused by SARS-CoV-2 (severe acute respiratory syndrome coronavirus 2)

References

1. Al Nuaimi K, Mohamed N, Al Nuaimi M, Al-Jaroodi J (2012) A survey of load balancing in cloud computing: challenges and algorithms. In: Proceedings of the second symposium on network cloud computing and applications. pp 137–142
2. Banipal TS, Kaur R, Banipal PK (2017) Interactions of diazepam with sodium dodecylsulfate and hexadecyl trimethyl ammonium bromide: conductometric, UV–visible spectroscopy, fluorescence and NMR studies. J Mol Liq 236:331–337
3. Banipal TS, Kaur R, Banipal PK (2018) Effect of sodium chloride on the interactions of ciprofloxacin hydrochloride with sodium dodecyl sulfate and hexadecyl trimethylammonium bromide: conductometric and spectroscopic approach. J Mol Liq 255:113–121
4. Bittencourt L, Immich R, Sakellariou R, Fonseca N, Madeira E, Curado M, Villas L, DaSilva L, Lee C, Rana O (2018) The internet of things, fog and cloud continuum: integration and challenges. Internet Things 3:134–155
5. Buzzi S, Chih-Lin I, Klein TE, Poor HV, Yang C, Zappone A (2016) A survey of energy-efficient techniques for 5G networks and challenges ahead. IEEE J Sel Areas Commun 34(4):697–709
6. Catarinucci L, Cappelli M, Colella R, Tarricone L (2008) A novel low-cost multisensor-tag for RFID applications in healthcare. Microwave Opt Technol Lett 50(11):2877–2880
7. Crosby M, Pattanayak P, Verma S, Kalyanaraman V (2016) Blockchain technology: beyond bitcoin. Appl Innovation 2(6–10):71
8. Dasgupta P, Metya A, Naidu CV, Singh M, Roxy MK (2020) Exploring the long-term changes in the Madden Julian Oscillation using machine learning. Sci Rep 10(1):1–13
9. Dayarathna M, Wen Y, Fan R (2015) Data center energy consumption modeling: a survey. IEEE Commun Surv Tutorials 18(1):732–794
10. Did a 1912 Newspaper Article Predict Global Warming? [Accessed 2 Feb 2020] Available at: https://www.snopes.com/fact-check/1912-article-global-warming/
11. Europe heatwave: Paris latest to break record with 42.6C, [Accessed 2 Feb 2020] Available at: https://www.bbc.co.uk/news/world-europe-49108847

12. Familiar B (2015) IoT and microservices. In: Microservices, IoT, and Azure (pp 133–163). Apress, Berkeley, CA

13. Gill SS, Buyya R (2018) SECURE: Self-protection approach in cloud resource management. IEEE Cloud Comput 5(1):60–72

14. Gill SS, Buyya R (2019) A taxonomy and future directions for sustainable cloud computing: 360 degree view. ACM Comput Surv (CSUR) 51(5):1–33

15. Gill SS, Chana I, Buyya R (2017) IoT based agriculture as a cloud and big data service: the beginning of digital India. J Organ End User Comput (JOEUC) 29(4):1–23

16. Gill SS, Garraghan P, Stankovski V, Casale G, Thulasiram RK, Ghosh SK, Ramamohanarao K, Buyya R (2019) Holistic resource management for sustainable and reliable cloud computing: an innovative solution to global challenge. J Syst Softw 155:104–129

17. Gill SS, Tuli S, Minxian Xu, Singh I, Singh KV, Lindsay D, Tuli S et al (2019) Transformative effects of IoT, blockchain and artificial intelligence on cloud computing: evolution, vision, trends and open challenges. Internet Things 8:1–26

18. Gill SS, Buyya R (2019) Sustainable cloud computing realization for different applications: a manifesto. In: digital business, Springer, Cham, pp 95–117

19. Gill SS, Kumar A, Singh H, Singh M, Kaur K, Usman M, Buyya R (2020) Quantum computing: a taxonomy, systematic review and future directions. arXiv preprint arXiv:2010.15559

20. Jamshidi S, Baniasad M, Niyogi D (2020) Global to USA county scale analysis of weather, urban density, mobility, homestay, and mask use on COVID-19. Int J Environ Res Public Health 17(21):7847

21. Krug L, Shackleton M, Saffre F (2014) Understanding the environmental costs of fixed line networking. In: Proceedings of the 5th international conference on Future energy systems. 87–95

22. Kulp SA, Strauss BH (2019) New elevation data triple estimates of global vulnerability to sea-level rise and coastal flooding. Nat Commun 10(1):1–12

23. Kumar A, Sharma K, Singh H, Naugriya SG, Gill SS, Buyya R (2020) A drone-based networked system and methods for combating coronavirus disease (COVID-19) pandemic. Futur Gener Comput Syst 115:1–19

24. Kuppusamy P (2014) A Cloud-oriented green computing architecture for E-Learning applications. Int J Recent Innovation Trends Comput Commun 2:3775–3783

25. Lane ND, Bhattacharya S, Georgiev P, Forlivesi C, Jiao L, Qendro L, Kawsar F (2016) Deepx: a software accelerator for low-power deep learning inference on mobile devices. In: Proceedings of the 15th ACM/IEEE International Conference on Information Processing in Sensor Networks (IPSN), 1–12

26. Lee J, Azamfar M, Singh J (2019) A blockchain enabled cyber-physical system architecture for industry 4.0 manufacturing systems. Manufact Lett 20:34–39

27. Lee CC, Chung PS, Hwang MS (2013) A survey on attribute-based encryption schemes of access control in cloud environments. IJ Netw Secur 15(4):231–240

28. Liu Q, Li P, Zhao W, Cai W, Yu S, Leung VC (2018) A survey on security threats and defensive techniques of machine learning: a data driven view. IEEE Access 6:12103–12117

29. Liu J, Wang S, Zhou A, Xu J, Yang F (2020) SLA-driven container consolidation with usage prediction for green cloud computing. Frontiers Comput Sci 14(1):42–52

30. Hulkury MN, Doomun MR (2012) "Integrated green cloud computing architecture," In: 2012 International Conference on Advanced Computer Science Applications and Technologies (ACSAT), Kuala Lumpur, pp 269–274

31. Mahallat I (2015) Fault-tolerance techniques in cloud storage: a survey. Int J Database Theory Appl 8(4):183–190

32. Masdari M, Zangakani M (2019) Green cloud computing using proactive virtual machine placement: challenges and issues. J Grid Computing 1–33

33. Masson-Delmotte V, Zhai P, Pörtner HO, Roberts D, Skea J, Shukla PR, Pirani A, Moufouma-Okia W, Péan C, Pidcock R, Connors S (2018) Global warming of 1.5 C. An IPCC Special Report on the impacts of global warming of, 1

34. Millar RJ, Fuglestvedt JS, Friedlingstein P, Rogelj J, Grubb MJ, Matthews HD, Skeie RB, Forster PM, Frame DJ, Allen MR (2017) Emission budgets and pathways consistent with limiting warming to 1.5 C. Nature Geoscience 10(10):741–747

35. O'Leary DE (2013) Artificial intelligence and big data. IEEE Intell Syst 28(2):96–99

36. Pachauri RK, Allen MR, Barros VR, Broome J, Cramer W, Christ R, Church JA, Clarke L, Dahe Q, Dasgupta P, Dubash NK (2014) Climate change 2014: synthesis report. Contribution of Working Groups I, II and III to the fifth assessment report of the Intergovernmental Panel on Climate Change. p 151

37. Rebecca Wright, "There's an unlikely beneficiary of coronavirus: the planet", Available at: https://edition.cnn.com/2020/03/16/asia/china-pollution-coronavirus-hnk-intl/index.html Accessed 20 Mar 2020

38. Rolnick D, Donti PL, Kaack LH, Kochanski K, Lacoste A, Sankaran et al (2019) Tackling climate change with machine learning. arXiv preprint arXiv:1906.05433

39. Roopaei M, Rad P, Jamshidi M (2017) Deep learning control for complex and large scale cloud systems. Intel Autom Soft Comput 23(3):389–391

40. Rosenthal B (2006) "Method and system for providing low cost, readily accessible healthcare." U.S. Patent Application 11/105,220, filed October 19, 2006

41. Shaikh A, Uddin M, Elmagzoub MA, Alghamdi A (2020) PEMC: Power efficiency measurement calculator to compute power efficiency and CO2 emissions in Cloud Data Centers. IEEE Access

42. Singh S, Chana I, Buyya R (2020) Agri-info: cloud based autonomic system for delivering agriculture as a service. Internet Things 9:100131

43. Singh M, Krishnan R, Goswami B, Choudhury AD, Swapna P, Vellore R et al (2020) Fingerprint of volcanic forcing on the ENSO–Indian monsoon coupling. Sci Adv 6(38):eaba8164

44. Stocker TF, Qin D, Plattner GK, Tignor M, Allen SK, Boschung J, Nauels A, Xia Y, Bex V, Midgley PM (2013) Climate change 2013: the physical science basis. In: Contribution of working group I to the fifth assessment report of the intergovernmental panel on climate change, 1535

45. Tang Z, Zhou X, Zhang F, Jia W, Zhao W (2018) Migration modeling and learning algorithms for containers in fog computing. IEEE Trans Serv Comput 12(5):712–725

46. Thi MT, Pierson JM, Da Costa G, Stolf P, Nicod JM, Rostirolla G, Haddad M (2019) Negotiation game for joint IT and energy management in green datacenters. Futur Gener Comput Syst. https://doi.org/10.1016/j.future.2019.11.018

47. Tuli S, Basumatary N, Gill SS, Kahani M, Arya RC, Wander GS, Buyya R (2020) HealthFog: an ensemble deep learning based smart healthcare system for automatic diagnosis of heart diseases in integrated IoT and fog computing environments. Futur Gener Comput Syst 104:187–200

48. Tuli S, Mahmud R, Tuli S, Buyya R (2019) Fogbus: A blockchain-based lightweight framework for edge and fog computing. J Syst Softw 154:22–36

49. Tuli S, Tuli S, Wander G, Wander P, Gill SS, Dustdar S, Sakellariou R, Rana O (2020) Next generation technologies for smart healthcare: challenges, vision, model, trends and future directions. Internet Technol Lett 3(2)e145:1–6

50. Tuli S, Tuli S, Tuli R, Gill SS (2020) "Predicting the growth and trend of COVID-19 pandemic using machine learning and cloud computing." Internet Things 100222

51. Tuli S, Gill SS, Casale G, Jennings NR (2020) iThermoFog: IoT Fog based automatic thermal profile creation for cloud data centers using artificial intelligence techniques. Internet Technol Lett 3(5):e198

52. Yang R, Yu FR, Si P, Yang Z, Zhang Y (2019) Integrated blockchain and edge computing systems: A survey, some research issues and challenges. IEEE Commun Surv Tutorials 21(2):1508–1532

53. Zanella A, Bui N, Castellani A, Vangelista L, Zorzi M (2014) Internet of things for smart cities. IEEE Internet Things J 1(1):22–32

Manmeet Singh is a Scientist at the Centre for Climate Change Research at the Indian Institute of Tropical Meteorology (IITM), Pune. He is presently pursuing his Ph.D. in Climate Studies from the Indian Institute of Technology Bombay. His main research interest is in the computational modelling of earth system and its processes. He is presently involved in studying the Monsoons using coupled general circulation models and non-linear methods. He also has experience in the development of Direct Numerical Simulation model and using it for the basic understanding of jets. He is passionate about the use of Deep Learning and Deep Reinforcement Learning as tools to solve the problems in Climate Science.

Shreshth Tuli is an undergraduate student at the Department of Computer Science and Engineering at Indian Institute of Technology—Delhi, India. He worked as a visiting research fellow at the CLOUDS Laboratory, School of Computing and Information Systems, the University of Melbourne, Australia. His research interests include Internet of Things (IoT), Fog Computing, Blockchain, and Deep Learning.

Rachel Jane Butcher is an M.Sc (Computing and Information Systems) student at Queen Mary University of London, UK. She has finished her B.Sc in Biological Sciences form Queen's University Belfast, UK. Rachel has done work as a Technology Intern at the Bright Network (UK) and Data Governance Graduate intern at the Queen's University Belfast (UK).

Rupinder Kaur is a senior research fellow in the Department of Chemistry, Guru Nanak Dev University, Amritsar, India. She received her M.Sc. degree in Chemistry from the Punjabi University, Patiala, India. Her area of research is to study the micellization behavior of surfactants in the presence of various additives.

Sukhpal Singh Gill is a Lecturer (Assistant Professor) in Cloud Computing at School of Electronic Engineering and Computer Science, Queen Mary University of London, UK. Prior to this, Dr. Gill has held positions as a Research Associate at the School of Computing and Communications, Lancaster University, UK and also as a Postdoctoral Research Fellow at CLOUDS Laboratory, The University of Melbourne, Australia. His research interests include Cloud Computing, Fog Computing, Software Engineering, Internet of Things and Healthcare. For further information, please visit https://www.ssgill.me.

A Research Roadmap for IoT Monitoring and Computational Modelling for Next Generation Agriculture

Paul Krause and Fatos Xhafa

Abstract In this final chapter, we outline a vision for a technological revolution in agriculture that would work to regain a sense of balance between food production and natural ecosystems. We promote an ecological engineering approach to crop production that draws on experience from the organic and conservation agriculture movements. However, we expand on this by promoting in addition:

1. An Internet of Things enabled biomonitoring system that enables key (above and below ground) environmental indicators to be automatically monitored across an agricultural unit.
2. A combination of network and thermodynamic ecosystem modelling approaches to enable a deep understanding of the response of ecosystem service functioning to changes in biodiversity, and the abiotic context, in any given agroecosystem.

We also call for an explicit recognition of the "ethnosphere" (the sphere of human social and cultural experience) as a fifth geosphere which emerged from the biosphere, and the other three geospheres, and whose continued existence is therefore contingent on the health and stability of the other four geospheres.

List of Acronym

CA	Conservation Agriculture
EU	European Union
IoT	Internet of Things
Pg	Petagram (10^{12} kg)

P. Krause (✉)
University of Surrey, Guildford, UK
e-mail: p.krause@surrey.ac.uk

F. Xhafa
Universitat Politècnica de Catalunya, Barcelona, Spain
e-mail: fatos@cs.upc.edu

SDG   Sustainable Development Goal
SIC   Soil Inorganic Carbon
SOC   Soil Organic Carbon

1 Introduction

In this chapter, we try to knit together above and below-ground systems, modelling, bio-monitoring and artificial intelligence to build a vision for a framework for the ecological engineering of agricultural systems and promotion of a more sustainable agriculture. Our long-term goal is to provide hard evidence of the benefits of a radical change to agricultural practice for policy makers and industry leaders, and to develop a prototype monitoring system to support the widespread implementation of ecological engineering in agriculture.

The Green Revolution, initiated in the 1960s, is credited with saving over a billion people from starvation. As such, it was a very necessary response to the massive increase in the global population during the twentieth century. However, the core technologies of the Green Revolution, high-yield varieties, chemical fertilizers and other agrochemicals, controlled irrigation, mechanisation, are not without associated adverse side effects. These include but are not limited to: significant reduction in captured soil carbon; increased emission of greenhouse gases (Carbon dioxide, Methane and Nitrous oxide); residual fertilizer run off leading to eutrophication of watercourses and extensive coastal areas; loss of social cohesion in agricultural communities; severe socio-economic issues in developing countries due to the high costs of seeds and agrochemicals.

Work is under way to address some of these issues with intensive agriculture. Precision agriculture, for example, can significantly reduce wastage and excessive run off from agrochemicals. But the fundamental problem still remains that agricultural systems predicated on extensive anthropogenic inputs lead to progressive degradation of the structure and biodiversity of soil, suggesting that some level of the above side effects is unavoidable without a major change in paradigm. This, together with the increasing number of successful examples of conservation and organic farming is suggesting that a move to *ecological intensification*, rather than intensification of anthropogenic inputs, may lead to an agricultural system that is closer to the closed cycles of an extensive natural ecosystem, but with yields that are close to or even better than intensive agriculture.

The challenge is to learn how to "engineer" the above and below ground biomes so that the ecosystem services provided by biodiversity can be integrated into crop production systems.

In this chapter, we will discuss how to bring together a range of research strands to establish a set of technologies that will facilitate an "ecological engineering" approach to sustainable agriculture. At a high level, this will require the development of two key technologies:

1.   An Internet of Things enabled biomonitoring system that enables key (above and below ground) environmental indicators to be automatically monitored across an agricultural unit.
2.   A combination of network and thermodynamic ecosystem modelling approaches to enable a deep understanding of the response of ecosystem service functioning to changes in biodiversity, and the abiotic context, in any given agroecosystem.

Achieving this will require a highly interdisciplinary team to bring together techniques from computational modelling, internet engineering, controlled experimental studies of ecosystems, longitudinal studies of agroecosystems, theoretical and empirical ecology, next generation sequencing, as well as agricultural practitioners. We believe we have identified the right team to achieve our goals.

We need to fundamentally change the way in which we produce food. Whilst the media focus has been primarily on the impact of transport and industry on climate change and other environmental impacts, globally agriculture is arguably a much stronger cause of environmental issues. However, we can change a negative to a positive. Whilst it seems impossible to conceive of a transport system that has no environmental impact, we can conceive of an agricultural system that is not only carbon neutral but one that is environmentally (and socially) net positive.

Over the last 40 years, we have seen emerging communities of practice in both organic and conservation agriculture that are collectively able to reduce the consumption of anthropogenic inputs and residual pollution, whilst maintaining the profitability of their agricultural units through a combination of higher value products and the maintenance of similar yields to those units that use intensive cultivation practices. Indeed, some practitioners of conservation agriculture in the USA midwest see the same yields as those using high levels of artificial fertilizers, pesticides and weedkillers *except* in drought years where their yields remain high; in contrast to those with intensive use of anthropogenic inputs who typically need to call on their crop insurance (Fig. 1).

What is clear is that the practices of organic and conservation agriculture lead to increases in soil biodiversity and captured carbon compared to intensive agriculture. Furthermore, the research community is gaining an increased understanding of the role biodiversity plays in the provision of the services that support a sustained ecosystem in a given abiotic context. Our goal is to develop a suite of computational modelling tools, and monitoring tools, to facilitate the engineering of a specific agroecology into a sustainable system for a given abiotic context that optimises yield, soil organic carbon content, and the provision of other ecosystem services. This will enable us to unify the experiences of organic and conservation agriculture into a generalised framework for sustainable agriculture that will offer significant measurable benefits over existing practices, as will be discussed in the next section.

Our overall goal is to augment the current state of art in ecological modelling with advanced techniques from Machine Learning and Internet of Things research programmes to facilitate the creation of practical tools for ecological engineering that can be applied by agricultural professionals.

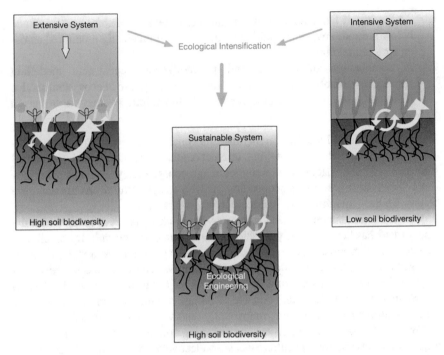

Fig. 1 Soil and ecological intensification. Redrawn after [2]

In order to achieve this, we will need to:

- Develop network and thermodynamic models of agroecosystems that enable us to predict the effect of abiotic context and specific management activities on ecosystem service provision;
- Develop machine learning techniques that enable us to identify appropriate ecological models from key measurable environmental indicators;
- Develop a robust internet-based monitoring system that enables semantically enriched data from IoT devices in the field to be synthesised into real-time monitors for managers of agricultural units;
- Develop a globally diverse network of agricultural units to collect data for training the models, and for evaluating the resulting models;
- Initiate controlled experiments in both laboratory conditions and in operational agricultural units to test interventions that the models predict could lead to enhanced ecosystem service provision.

As well as bringing together world-class researchers to work towards these scientific objectives, we will also need to engage with experienced team members to prepare and execute an effective dissemination, exploitation and communication strategy.

2   The Breakthrough Vision

Following extensive work by EU projects such as EcoFINDERS, and global initiatives such as the Global Soil Diversity Initiative and the Global Soil Partnership, amongst others, procedures for diagnosing the biological state of soils are now available. Extensive and intensive data collection activities over the last two decades, together with the development of sampling protocols now enable robust diagnoses to be performed based on, firstly, analysis of the abundance of microbial communities and specific bioindicators, and secondly the interpretation of the corresponding analyses through comparison with reference values [13]. These diagnostic procedures are leading to recommendations that can restore and improve soil fertility and associated ecosystem services. Essentially, there are two classes of ecological engineering approach that arise from this work.

The first approach is to propose the introduction of microbial strains selected for their beneficial effects. This has seen major successes, such as the inoculation of fodder crops and grain legumes with different species of *Rhizobia*. For example, the inoculation of soyabean has extended its cultivation to soils devoid of its corresponding symbiotic bacteria, *Bradyrhizbium japonicum*. As a result, in less than 50 years the cultivation of this crop now covers more than 100 million hectares [14]. There is also potential to use strains of Rhizobium not only to fix nitrogen through symbiosis with leguminous crops, but also to convert N_2O to N_2 and so help in reducing emissions of this currently major contribution of agriculture to greenhouse gas emissions.

The second approach is a more systemic approach, that is aligned with the broad ambitions of the organic and conservation agriculture movements. This is to redefine the agricultural management processes in order to favour microbial populations and facilitate soil processes that are beneficial to plant growth and health and other ecosystem services, whilst at the same time reducing adverse environmental effects. This is much less well developed than the former approach, but the potential benefits to transitioning agricultural practice to a net beneficiary in the move to living within planetary boundaries make it a compelling approach to take. It is the approach that we support in this chapter.

Large scale empirical studies are showing that ecological networks play a strong role in operationalising the link between biodiversity and ecosystem function. The pan-European study of [5], for example, showed that basal respiration, molecular microbial biomass and fungal richness were strong indicators associated with the ability of an ecosystem to cycle and store soil organic carbon (SOC). Their network analysis showed that the connectivity of fungal taxonomic units was greatest in forest sites, a finding reinforced by the work of [11] which demonstrated significant carbon exchange between forest trees via ectomycorrhizal networks.

The key challenge remains, however, to identify a suite of indicators that link biodiversity to soil functioning. We believe this can only be achieved by developing a strong understanding of the causal relationships that are responsible for this link. This will require us to make a stronger link between mathematical ecological models and the empirical work. Network models are providing a common paradigm for both approaches to gaining understanding of ecosystem functioning. Indeed, evaluating the impact of changes in agricultural practice from a network perspective rather than (as currently) at the species or assemblage level is vital in order to avoid overlooking species interactions and compensatory effects [16]. It is important to consider species' *roles* in networks, rather than focusing on species identities. Cirtwill et al. [4] advocate the Eltonian niche as an overarching framework, and the need to be precise about which aspect of a species' niche that is being addressed in a specific study.

Ecological engineering can also have major benefits to the above ground ecosystem. Systemic control of pests (including invasive weeds) and diseases can also lead to significant reduction in the need for anthropogenic inputs. The experience of many in the conservation agriculture community is that they are able to progressively reduce the need for pesticides and weedkillers as the increase in above ground biodiversity rebuilds ecosystem services that once again maintain a natural stability in the agroecosystem. The important question now is to understand the mechanisms that are at work here. Petit et al. [20] have shown that over a four-year period the adoption of conservation agriculture (CA) practices significantly advances the abundance of carabid beetles and hence the amount of weed seed predation in arable fields. It is imperative that we gain increased understanding of the mechanisms by which ecosystem engineering (as exemplified by CA practices) can rebuild services that replace or minimise the needs for pesticides and weedkillers. It is not only their negative impact on the natural environment that is of concern, but also the social and human health issues (e.g. [8]).

The challenge to date has been the low number of highly resolved above and below ground ecological networks. Recent work has shown how the use of machine learning with advanced molecular biology techniques provides a new methodology for the construction of large-scale replicated networks [3]. However, system level responses remain largely unexplored. This is where mathematical modelling approaches can help us. Network models can show us how ecosystems may be sensitive to changes in the flow connections (of energy, carbon or nutrients) within a network. Work in the ecological modelling community has shown us the equivalence between this view and the thermodynamic/information theoretic perspective in which an ecosystem is assumed to develop in a direction that maximises thermodynamic efficiency, e.g. [18]. The challenge now is to integrate the mathematical modelling work with the ability to generated highly resolved agroecological networks in order to gain a deeper understanding of the system level response to inform the engineering of more thermodynamically efficient systems.

The scientific advances we believe that now need targeting are to:

- Extend the mining of ecological networks from trophic networks to a broader range of attributes of Eltonian niches;
- Integrate highly resolved networks into mathematical models that enable us to explore their behaviour at a system level;
- Use state of the art machine learning approaches to enable us to identify a subset of indicators that enable low-cost routine monitoring of agroecosystems.

The breakthrough we are targeting is to use the above scientific advances to develop an end-to-end tool set to support routine agricultural management using an *ecological engineering* approach.

This will require the development of a suite of agroecological models that can be selected from and customised to specific abiotic contexts. This can then be used for routine monitoring of agroecosystems, and for the identification of interventions that will optimise the delivery of ecosystem services from the respective agroecosystems.

The challenge is to mature these advances in ecological modelling to enable them to be used to identify candidate interventions that could be tested and subsequently used to maintain yields at the levels of intensive agriculture, whilst pushing the need for external inputs other than natural light and water down towards zero.

3 Building the Community

In order to achieve the targeted breakthrough, an interdisciplinary community needs to be established with experience in:

- Empirical ecology, for small, controlled studies of ecological systems and evaluations of functioning agroecosystems;
- Theoretical ecology, to support the analysis and simulation modelling of ecological networks;
- Machine learning and data mining, to support the extension of the learning of highly resolved ecological networks from molecular biology data, and the identification of measurable indicators of ecosystem health;
- High performance computing to support the development of large-scale simulation models;
- Semantic data enrichment to aid interpretation of streamed data;
- Internet of things and edge computing infrastructure;
- The regulatory and policy landscape for the agriculture industry;
- The use of biotechnology in developing countries to facilitate economic advance whilst restraining impact within planetary boundaries.

As mentioned in the introduction, there are two possible approaches that can be taken to ecological engineering for agroecosystems. The first is to introduce microbial strains that are selected for beneficial effects. This approach has been demonstrated to increase yields, and even allow certain crops to grow in areas where they could not be previously cultivated. However, this approach does not necessarily lead to any additional benefits, such as increased carbon capture, reduced emissions of greenhouse gases and natural controls of pests and diseases. The alternative approach is to redefine the agricultural management processes in order to favour microbial populations and facilitate terrestrial processes that are beneficial to plant growth and health and other ecosystem services, whilst at the same time reducing adverse environmental effects. This is the approach we are supporting. Whilst we are seeing a steady growth in numbers of practitioners of this approach (organic and conservation agriculture), it is still not mainstream, and the systematic reviews of their benefits are so far equivocal due to limitations in the comparability of the respective experimental environments. These challenging scientific and technological objectives need to be addressed in order to establish a firm foundation which can subsequently be used to transition the agricultural industry into a net-positive contributor to the environment whilst maintaining, even enhancing, its profitability.

4 Expected Impacts

Given the real-world experience and academic research in conservation agriculture so far, we believe it should be possible to produce empirical and theoretical evidence that enable agricultural units to reduce usage of fertilisers, pesticides and weed-killers to less than 10% of their usage in conventional agriculture in a wide range of abiotic contexts, whilst maintaining yields at existing levels through the application of ecological engineering. We would also expect to see corresponding reductions in emissions of CO_2 and N_2O, with demonstrated potential to increase soil organic carbon to 32% above conventional arable fields. These figures are taken from claims made from within the conservation agriculture movement, and discussed extensively in, for example, [17]. The SOC figures are plausible given studies such as [6]. Given the strength of discussion within the community around the possible reductions in anthropogenic inputs, we feel it is appropriate to set these claims as a societal goal.

Prediction of reductions in methane will be harder to generate as this will be impacted to a large extent by a willingness of the population at large to reduce their consumption of meat. However, low intensity grazing appears to be an important component of systemic management of agroecosystems and another challenge is to assess the potential to redirect the food market to lower levels of consumption, but of higher quality, meat products.

The primary societal impact will be to provide a strategy for moving the agriculture industry back to within planetary boundaries, and become a major contributor to achievement of Sustainable Development Goals (SDGs) 1, 2, 3, 6, 8, 10, 11, 12, 13, 14 and 15 [22].[1] Indeed one could argue that successfully transitioning agriculture to a sustainable and equable system is a necessary (but not sufficient) precondition to achieving *all* the SDGs.

Our core ambition is to stimulate a move to collective action amongst ecologists from a broad range of backgrounds and computer scientists to accelerate the transition of our agricultural industry into a net positive contributor to the sustainability of

---

[1] The most directly relevant are SDGs 2, and 12–15, which are respectively: End hunger, achieve food security and improved nutrition and promote sustainable agriculture; Ensure sustainable consumption and production patterns; Take urgent action to combat climate change and its impacts; Conserve and sustainably use the oceans, seas and marine resources for sustainable development; Protect, restore and promote sustainable use of terrestrial ecosystems, sustainably manage forests, combat desertification, and halt and reverse land degradation and halt biodiversity loss.

both the biosphere and a vibrant ethnosphere.[2] We will now end with a discussion of why we feel it is so important to identify this "fifth geosphere" and explicate its relationship to the other four.

5 The Fifth Geosphere

In the early phases of Earth's history, three "geospheres" emerged or perhaps one might say, condensed out as the universe cooled: the atmosphere; the lithosphere; and, the hydrosphere. They are not distinct but interact, with the interfaces between these geospheres being major sources of innovation as the earth evolved.

The *atmosphere* contains the air that we breath and extends from about 1 m below the earth's surface to about 10,000 km above it. The *lithosphere* contains the solid land at the surface of the earth's crust and extends down to the molten core. As we have seen, the atmosphere overlaps with the lithosphere, as does the *hydrosphere* which contains all the solid, liquid and gaseous water of the planet. It extends from the Earth's surface downwards, sometimes in quite a distinct layer and sometimes tightly integrated with the lithosphere. It extends upwards about 12 km into the atmosphere and overall ranges from 10 to 20 km in thickness.

We don't yet fully understand how, but somehow the energy imbalances at these interfaces catalysed the generation of complex molecules that eventually led to the formation of living matter. This progressively developed into the fourth geosphere; the *biosphere* that contains the amazing diversity of life forms that we humans are part of. Although the detailed mechanisms are as yet poorly understood, at a high level the creation of the biosphere could be seen as a phase transition as the complex earth system cooled further from its primordial phase.

One can breathe the atmosphere, touch the lithosphere and swim in the hydrosphere. Even though they have strong and sometimes complex overlap at their interfaces, they are easy to conceptualise as distinct regions. The biosphere is harder to separate out conceptually as it is strongly intertwined with all three. This is perhaps not surprising as it is, in a sense, a product of all three. Nevertheless, it does have processes (metabolism, respiration, for example) and entities of its own (microbes, plants, animals) although their survival is critically dependent on the specific relationships with the other three geospheres.

The above world view is discussed in more detail in books such as [15, 21]. The important point is that the biosphere is a product of and embedded within the three other geospheres; a change in any one of the latter three geospheres will impact on the current state of the biosphere. This is an obvious point, but even now the significance of it is not universally recognised. Of course, processes within the biosphere can also

---

[2]Here we take the biosphere as having emerged from interactions between the lithosphere, atmosphere and hydrosphere, with the ethnosphere (the sphere of human social and cultural experience) as having emerged from the four preceding geospheres.

impact on the states of the other three geospheres with the impact of plant life on the composition of the atmosphere perhaps being the most significant.

There is more, though. A separate strand of research has been gaining a better understanding of how human society has emerged, e.g. [9]. With all its complexities, and pathologies, we believe it stands recognition as a *fifth* geosphere; the ethnosphere.[3] Again, it has a complex of processes (legal, social, economic, industrial, political) and a suite of entities (more than just humans, but also books, works of art, music, religions, institutions). Even harder to isolate, it is a product of and embedded within the four other geospheres; a change in any one of the latter four geospheres will impact on the current state of the ethnosphere. As with the biosphere, processes within the ethnosphere can and do impact on the stability of the other four geospheres.

The embeddedness of the ethnosphere within the four other geospheres is a critical feature. The microbiome within our gastro-intestinal tract is as important to feed and nourish as that within the soil that grows our food, and both in turn can sustain our physical and mental health. Equally, the natural world around us nourishes our physical, mental and spiritual health, as well as providing inspiration for our art, music and literature.

This is why we suggest that the climate crisis and the biodiversity crisis need to be solved together. It is simply not possible to solve one without addressing the other as they are both symptomatic of a disruption to the earth's systems.

To set the scene, let us first take a look at the global picture of carbon pools and fluxes between them. Figure 2 is taken from [12].

Notice that to first order, there is a closed cycle of carbon within the atmospheric, biotic, pedologic pool clique. The three main disruptors to this are deforestation (reducing the biotic pool), soil erosion (reducing the pedologic pool) and fossil fuel combustion (Fig. 2).

Reducing fossil fuel combustion is not the primary focus of this chapter. However, increasing the biotic and pedologic carbon pools is. The real value will be in terms of demonstrating and furthering best practice in working with ecological systems to optimise the natural capital contained within open spaces and show how their management can be integrated into an ecological engineering approach to management of agricultural, amenity and urban landscapes to demonstrate a carbon negative approach to land management and restore the stability of both the biosphere and its embedded ethnosphere.

---

[3]The choice of name has been motivated by the discipline of ethnobiology, by which is meant engaging in "the scientific study of dynamic relationships among peoples, biota, and environments". There is a common presumption that any name that begins with "ethno" refers to a study of "other" peoples, populations, and societies, but this is incorrect; "ethno" simply refers to culture/people. Literally the word means 'nation'. The etymology of the word refers to nation, not in patriotic sense, but in the sense of tradition, practices, and overall culture of the people. So, in a way it refers to people together with their 'lives'.

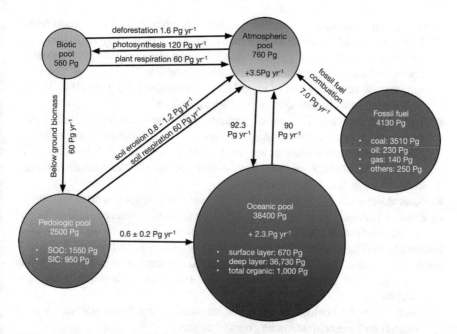

Fig. 2 Principal Carbon pools and fluxes between them. Redrawn after [12]. The data on carbon pools in the major reservoirs are from [1, 7, 19]. The data from fluxes are from [10]

6 Conclusion

The collective work in the earlier chapters of this book demonstrates a tremendous technological potential for transforming agriculture. Just as Gordon E. Moore set his famous "law" as a challenge to stimulate innovation in the semiconductor industry, so we have tried to propose a collective vision that could trigger a vitally important transformation to the agricultural sector. Given the tremendous advances that have been shown in the chapters of this book, we believe that the time is ripe to build a collective vision that will bring us all together and empower a new agricultural revolution that will help restore a sense of balance with the world and among ourselves.

Glossary of Terms

Agroecology The study of ecological processes applied to agricultural production systems
Anthropogenic An entity or effect that has been generated by humans
Biodiversity The variety and variability of life on Earth

Conservation agriculture "A farming system that promotes minimum soil distur-
bance (i.e. No-till farming), mainte-nance of a permanent soil cover, and diversifi-
cation of plant species. It enhances Biodiversity and natural biological processes
above and below the ground surface, which contribute to increased water and
nutrient use efficiency and to improved and sustained crop production." (Food
and Agriculture Organization of the United Nations)

Ecological engineering The use of ecology and engineering to predict, design,
construct or restore, and manage ecosystems that integrate "human society
with its natural environment for the benefit of both" (W.J. Mitsch and S.E.
Jorgensen (Editors), *Ecological Engineering: An Introduction to Ecotechnology.*
John Wiley & Sons, New York)

Ecological network A representation of the biotic interactions in an ecosystem

Ecosystem Services The many and varied benefits to humans provided by the
natural environment and from healthy ecosystems

Ethnosphere The sphere of human social and cultural experience

Green Revolution The set of research technology transfer initiatives occurring
between 1950 and the late 1960s, that increased agricultural production
worldwide

Eltonian niche An ecological niche that emphasizes the functional attributes of
animals and their corresponding trophic position

Eutrophication Over enrichment of a body of water with minerals and nutrients
which induce excessive growth of algae

Pedologic Of the soil

Precision agriculture A farming management concept based on observing,
measuring and responding to inter and intra-field variability in crops

Systemic Relating to a system rather than an individual component

References

1. Batjes NH (1996) Total C and N in soils of the world. Eur J Soil Sci 47:151–163
2. Bender SF, Wagg C, van der Heijden MG (2016) An underground revolution: biodiversity and soil ecological engineering for agricultural sustainability. Trends Ecol Evol 31(6):440–452
3. Bohan DA et al (2017) Next-generation global biomonitoring: large-scale, automated recon-struction of ecological networks. Trends Ecol Evol 32(7):477–487
4. Cirtwill AR et al (2018) A review of species role concepts in food webs. Food Webs 16:e00093
5. Creamer RE et al (2015) Ecological network analysis reveals the inter-connection between soil biodiversity and ecosystem function as affected by land use across Europe. Appl Soil Ecol 97:112–124
6. Edmondson JL, Davies ZG, Gaston KJ, Leake JR (2014) Urban cultivation in allot-ments maintains soil qualities adversely affected by conventional agriculture. J Appl Ecol 51(4):880–889
7. Falkowski P et al (2000) The global carbon cycle: a test of our knowledge of earth as a system. Science 290(5490):291–296
8. Greenpeace   (2015)   es.greenpeace.org/espana/Global/espana/2015/Report/transgenicos/Plaguicidas_Y_Nuestra_Salud_ingles.pdf

9. Henrich J (2017) The secret of our success: how culture is driving human evolution, domesticating our species, and making us smarter. Princeton University Press
10. IPCC (2001) Climate change 2001: the scientific basis. Internationals government panel on climate change. Cambridge UK: Cambridge University Press
11. Klein T, Siegwolf RTW, Körner C (2016) Belowground carbon trade among tall trees in a temperate forest. Science 352(6283):342–344
12. Lal R (2008) Carbon sequestration. Phil Trans R Soc B 363:815–830
13. Lemanceau P et al (2015) Understanding and managing soil biodiversity: a major challenge in agroecology. Agron Sustain Dev 35:67–81
14. Lindström K, Murwiri M, Willems A, Altier N (2010) The biodiversity of beneficial microbe-host mutualism: the case of rhizobia. Res Microbiol 161(6):453–463
15. Luisi PL (2016) The emergence of life: from chemical origins to synthetic biology. Cambridge University Press, Cambridge UK
16. Ma A et al (2019) Ecological networks reveal resilience of agro-ecosystems to changes in farming management. Nat Ecol Evol 3:260–264
17. Montgomery DR (2018) Growing a revolution: bringing our soil back to life. WW Norton & Company
18. Nielsen SN, Ulanowicz RE (2000) On the consistency between thermodynamical and network approaches to ecosystems. Ecol Model 132:23–31
19. Pacala S, Socolow R (2004) Stabilization wedges: solving the climate problem for the next 50 years with current technologies. Science 305:968–972
20. Petit S, Trichard A, Biju-Duval L, McLaughlin OB, Bohan D (2017) Interactions between conservation agricultural practice and landscape composition promote weed seed predation by invertebrates. Agr Ecosyst Environ 240:45–53
21. Smith E, Morowitz HJ (2016) The origin and nature of life on earth: the emergence of the Fourth Geosphere. Cambridge University Press, Cambridge UK
22. UNGA (2015) Transforming our world: the 2030 Agenda for Sustainable Development. United Nations General Assembly

Paul Krause is Professor in Complex Systems at the Department of Computer Science, University of Surrey. He has over forty years' research experience in the study of complex systems in a wide variety of domains, in both indus-trial and academic research laboratories. Currently his research work focuses on distributed systems for the Digital Ecosystem and Future Internet domains. He has over 120 publications and is author of a textbook on reasoning under uncertainty.

He has been working in and leading strong interdisciplinary teams since 2006 in the EU funded DBE and OPAALS projects, more recently in the RCUK funded ERIE (for Evolution and Resilience of Industrial Ecosystems) and MILES projects. These last two were funded under the Complexity in the Real World, and Bridging the Gaps, programmes respectively. He also has forty years' experience as a volunteer in practical nature conservation projects.

He is a Fellow of the Institute of Mathematics and its Applications, and a Chartered Mathematician.

Fatos Xhafa is Professor of Computer Science at the Technical University of Catalonia (UPC), Barcelona, Spain. He received his PhD in Computer Science in 1998 from the Department of Computer Science of BarcelonaTech. He was a Visiting Professor at University of Surrey (2019/2020), Visiting Professor at Birkbeck College, University of London, UK (2009/2010) and a Research Associate at Drexel University, Philadelphia, USA (2004/2005). Prof. Xhafa has widely published in peer reviewed international journals, conferences/workshops, book chapters and edited books and proceedings in the field (Google h-index 52 / i10-index 270, Scopus h-index 41, ISI-WoS h-index 33, as of December 2020). He is awarded teaching and research merits by Spanish Ministry of Science and Education, by IEEE conference and best paper awards. Prof. Xhafa has an extensive editorial and reviewing service for international journals and books from major publishers as a member of Editorial Boards and Guest Editors of Special Issues. He is Editor in Chief of the Elsevier Book Series "Intelligent Data-Centric Systems" and of the Springer Book Series "Lecture Notes in Data Engineering and Communication Technologies". He is a member of IEEE Communications Society, IEEE Systems, Man & Cybernetics Society and Emerging Technical Subcommittee of Internet of Things. His research interests include parallel and distributed algorithms, massive data processing and collective intelligence, IoT and networking, P2P and Cloud-to-thing continuum computing, optimization, security and trustworthy computing, machine learning and data mining, among others. He can be reached at fatos@cs.upc.edu and more information can be found at WEB: http://www.cs.upc.edu/fatos/ Scopus Orcid: http://orcid.org/0000-0001-6569-5497

Printed in the United States
by Baker & Taylor Publisher Services